Physical Processes in Laser–Materials Interactions

NATO ADVANCED STUDY INSTITUTES SERIES

A series of edited volumes comprising multifaceted studies of contemporary scientific issues by some of the best scientific minds in the world, assembled in cooperation with NATO Scientific Affairs Division.

Series B: Physics

Recent Volumes in this Series

This series is published by an international board of publishers in conjunction with NATO Scientific Affairs Division

A Life Sciences	Plenum Publishing Corporation
B Physics	London and New York
C Mathematical and Physical Sciences	D. Reidel Publishing Company Dordrecht, The Netherlands and Hingham, Massachusetts, USA
D Behavioral and Social Sciences	Martinus Nijhoff Publishers The Hague, The Netherlands
E Applied Sciences	

Physical Processes in Laser–Materials Interactions

Edited by
M. Bertolotti
Institute of Physics
University of Rome
Rome, Italy

PLENUM PRESS • NEW YORK AND LONDON
Published in cooperation with NATO Scientific Affairs Division

Library of Congress Cataloging in Publication Data

Main entry under title:

Physical processes in laser-materials interactions.

   (NATO advanced study institutes series. Series B, Physics; v. 84)
   "Proceedings of the 1980 NATO Advanced Study Institute on Physical Processes in
Laser-Materials Interaction, which was the ninth course of the Europhysics School of
Quantum Electronics, held July 13–25, 1980, in Pianore, Italy." — Verso t.p.
   Includes bibliographical references and index.
   1. Laser materials — Effect of radiation on — Congresses. 2. Laser beams — Con-
gresses. I. Bertolotti, Mario. II. NATO Advanced Study Institute on Physical Processes
in Laser-Materials Interaction (1980: Pianore, Italy) III. Title. IV. Series.
QC374.P46 1982   621.36′6   82-13147
ISBN 978-1-4684-4324-0   ISBN 978-1-4684-4322-6 (eBook)
DOI 10.1007/978-1-4684-4322-6

Proceedings of the 1980 NATO Advanced Study Institute on Physical Processes in
Laser-Materials Interaction, which was the Ninth Course of the Europhysics School
of Quantum Electronics, held July 13–25, 1980, in Pianore, Italy

© 1983 Plenum Press, New York
Softcover reprint of the hardcover 1st edition 1983
A Division of Plenum Publishing Corporation
233 Spring Street, New York, N.Y. 10013

FOREWORD

It is a pleasure to write a few words as an introduction to the
proceedings of the 1980 NATO ASI on "Physical Processes in Laser-
Material Interaction."

This ASI is the ninth course of a series devoted to lasers and
their applications, held under the responsibility of the Quantum
Electronics Division of the European Physical Society, and for this
reason known as the "Europhysics School of Quantum Electronics."

Since 1971 the School has been operating with the joint direc-
tion of myself as representative of the academic research, and
Dr. D. Roess (formerly with Siemens AEG, Munich, and now with Sick,
Optik und Electronik, GmbH, Munich) for the industrial applications.

Indeed the aim of the School is to alternate fundamental and
applied frontier topics in the area of quantum electronics and modern
optics, in order to introduce young research people from universities
and industrial R & D laboratories to the new aspects of research
opened by the laser.

This is clearly illustrated by the sequence of titles of the
past courses:

    (1) - 1971 Physical and technical measurements with lasers
    (2) - 1972 Nonlinear optics and short pulses
    (3) - 1973 Laser frontiers: high powers and short wavelengths
    (4) - 1974 Cooperative effects in multi-component systems
    (5) - 1975 Molecular spectroscopy and photochemistry with lasers
    (6) - 1976 Coherent optical engineering
    (7) - 1977 Coherence in spectroscopy and modern physics
    (8) - 1979 Lasers in biology and medicine

NATO has become the main sponsor under the program of the
Advanced Study Institute, starting with the third course, and then
continuously since the fifth. Therefore the School may be considered
as part of the NATO enterprises in the area of physics and applied

sciences. Due to the potentialities of the laser in many areas of applications, it became necessary to cover interdisciplinary areas. Such was the case of the eight course, where the participants were coming from physics, biology, and medicine.

Such is also the case of the ninth course, devoted to the application of lasers in metallurgy (a field monopolized by mechanical engineers) as well as in semiconductor annealing (a field thus far practiced mainly in solid-state physics laboratories).

Besides use of the laser as a tool, another area of conceptual unification between the two fields is the common heading of "nonequilibrium thermodynamics."

As during past years, the School Directors, once having chosen a given subject, have chosen one (or a few) outstanding person(s) active in the field and have asked him (or them) to act as "scientific responsible(s)," selecting the lecturers and editing the proceedings of the volume. The scientific responsible for this course was Prof. M. Bertolotti, leader of a research group engaged in laser annealing of semiconductors. He will detail the motivation of his choices and show the correlations among the different contributions.

Prof. F.T. Arecchi, Director

Istituto Nazionale di Ottica
Firenze, Italia

INTRODUCTION

This ninth course of the "Europhysics School of Quantum Electronics" is devoted to the study of physical processes in laser-materials interactions. The field, which is very broad indeed, has been restricted to the effects of laser light in metals and semi-conductors having in mind applications such as metal working, drilling, cutting, alloying, hardening, and semiconductor annealing. The main body of the school is therefore centered on the coupling of laser radiation to these materials, heating, and phase transformations. Practical implications of these techniques have also been considered in some detail.

Laser treatment of materials opens also a new field of the thermodynamics of fast processes.

To complete the scenario some information on other materials, e.g., refractory materials, and some general background information of laser effects in materials have also been included.

The matter has been divided into regular lessons, seminars, and contributions from participants to the course.

I wish to express my sincere thanks for the assistance and help given by the co-directors of the course, D. Smith, and S. Solimeno.

Miss A. De Cresce is also due for special thanks for her help during the editing of this volume, and Mrs. F. Medici, and C. Sanipoli for drawings.

June 1981   M. Bertolotti

CONTENTS

CONTENTS

CONTRIBUTIONS

CHARACTERISTICS OF LASER BEAMS FOR MACHINING

I.J. Spalding

UKAEA Culham Laboratory
Abingdon, Oxon, OX14 3DB

INTRODUCTION: CHOICE OF LASER

Important requirements for lasers which are used for industrial machining include:

i.   Sufficient power available at the work (cw or pulsed).
ii. Controlled focal intensity profile (hence well-characterised spatial-mode).
iii. Reproducibility (of power, mode, polarisation, pointing-stability etc).
iv. Reliability (long interval between services).
v.   Capital and running costs which are economic for the application.

For a few of the applications discussed at this school the choice of wavelength (λ) will also be important: either because of the way in which the workpiece absorbs the light (e.g. preferential absorption in a thin film or a transmitting substrate), or because a particularly thin kerf-width (i.e. small focal spot) is required. However, the majority of machining applications currently utilize either argon-ion (λ = 488.0, 514.5nm etc); ruby $Cr^{3+}:Al_2O_3$ (at λ = 694.3mm); Nd^{3+}:YAG (yttrium aluminium garnet) and Nd^{3+}:glass (at $\lambda \sim$ 1064nm); or CO_2 molecular gas ($\lambda \sim$ 10.6 μm) lasers. All of these systems are normally operated at fixed, and well-defined, wavelengths. For wavelength tunable applications optically-pumped dye lasers may also be used. The width of the gain-spectrum ($\Delta\nu$) of the laser can also be an important consideration in pulsed laser design. (It follows from the uncertainty principle that the output pulse duration $\tau \gtrsim (\Delta\nu)^{-1}$, i.e. that short pulses require broad laser band-widths).

The physics of these particular lasers will not be discussed

1

in depth here – our emphasis in the first lecture will be on the general characteristics of such systems as unique heat-sources for a variety of machining applications. Standard texts, such as Maitland and Dunn (1969), Charschan (1972), Duley (1976), or Svelto (1976), should be consulted for detailed consideration of excitation mechanisms etc. We shall also note here that particle-beams do afford alternative, localized, and high-intensity, heat-sources; however, they differ in normally requiring a vacuum for efficient transmission, in their energy (and particle) deposition characteristics in the target, and in their deflection by local fields (e.g. by $\underline{E}$ at re-entrant surfaces and $\underline{B}$ from ferritic workpieces).

Although the various laser systems differ tremendously in detail (e.g. in their phase-solid, liquid, gas or plasma; and their excitation – electrical, optical, gas dynamic, chemical etc.) they have one thing in common: an optical resonator. This will be discussed in detail in the second lecture, as a necessary pre-liminary to discussing in the third lecture the focusing and manipulation of the laser beam onto the workpiece.

Lasers for Machining Applications

Table 1 summarizes the characteristics of some typical 'production-engineering' lasers. Most of the gas lasers are electrically pumped, normally by a discharge along the length of the containing tube. For the argon-ion system the current density must be sufficiently high that A^+ is created by multiple electron collisions within a plasma having a temperature of many electron volts, so that high input powers and high thermal conductivity cooling tubes (of BeO, ceramic etc) are required. For the He–Cd$^+$ system, pumping is achieved via Penning collisions between excited metastable He atoms having energies exceeding 10 eV and the (lower ionization potential) Cd atoms, so that only modest cooling is required. In contrast the CO_2 molecule lases from its electronic ground-state, and is normally excited by collision with vibrationally-excited metastable N_2 molecules produced in a (0.8:1:7) CO_2/N_2/He glow-discharge, at much lower temperatures. Because of the resulting high overall efficiency, η, of the CO_2 laser (10µm laser power out/total pump power in = 5 to 15%) various industrial versions suitable for heavy metal-working applications will be detailed below.

Each of the other lasers listed in Table 1 is optically-pumped: the ruby and Nd solid-state lasers by flash lamps, and the liquid (dye) lasers by flash lamp or another laser. Because of the high number-density of lasant ions or molecules, the available optical energy stored per unit volume tends to be higher than for gas lasers – thus facilitating the construction of very compact units. Although η tends to be much lower for these systems

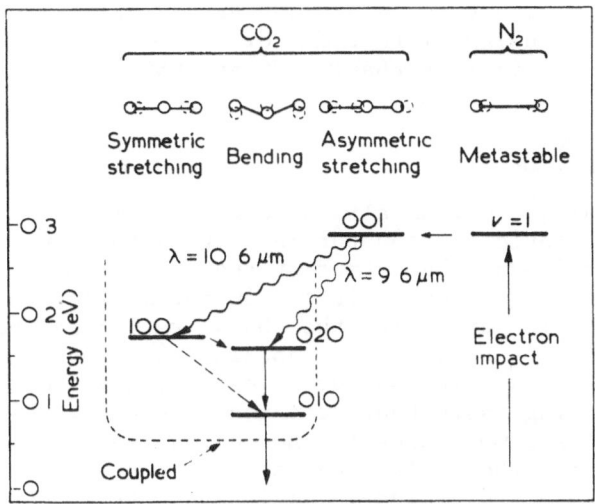

Fig 1. CO_2 Laser: Vibrational Energy Levels

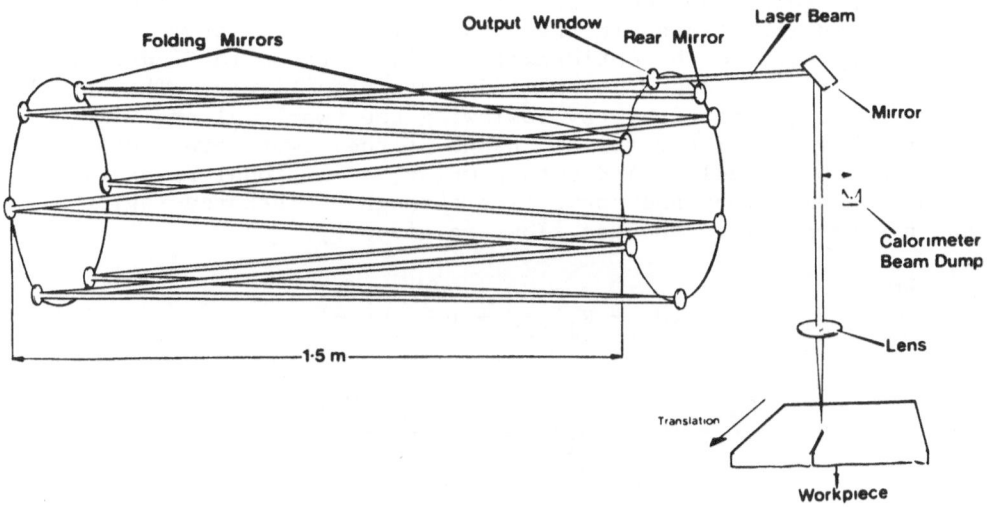

Fig 2.   Slow Axial-Flow CO_2 Laser (courtesy Ferranti Ltd)

(approaching 2-3.5% for state-of-the-art YAG lasers) the electrical and flash-lamp replacement-costs can nevertheless be perfectly acceptable for a wide variety of industrial applications. (For example, with flash-lamp life-times of $10^6 \sim 10^8$, the cost per NdYAG welding-shot is less than £10^{-4} at 1980 prices, cf. Weedon 1980).

CO_2 Laser Technology

With the slow-flow, axial-discharge systems discussed above, the axis of the optical resonator is aligned for convenience along the discharge axis and the output of the CO_2 laser, per unit length of glow-discharge, is then typically 50-80W/m after the gas composition (and in particular the CO_2:N_2 ratio) has been optimised for the particular tube diameter, absolute pressure and electrical conditions. The resulting output is essentially limited by the problem of removing waste-heat from the discharge, which is cooled by diffusion of gas molecules to the cooled inner wall of the tube. This unwanted heat thermally populates the CO_2 01'0 level (lying only $670cm^{-1}$ above the ground level), thus bottle-necking the lower (100, 020) lasing levels shown in Figure 1. Nevertheless, very convenient subkilowatt systems can be made by 'folding' the discharge-tube in ways similar to that shown in Figure 2; here a compact 400 watt laser system is achieved by a very ingenious technique for keeping a multiplicity of reflecting mirrors in good optical alignment.

If the gas is now convected more quickly along the length of the tube, as illustrated in the 'fast-axial-flow' system of Figure 3, or transverse to the optic axis, as illustrated in Figure 4, it can be cooled (and hence de-excited) more effectively. This can be demonstrated qualitatively by following the two-level analysis of De Maria (1973). The rates of change of the lower level population (N_1) and upper level (N_2) due to pumping, collisional relaxation, stimulated emission, and convection through a discharge-region of length x parallel to the gas flow are given by:

$$\frac{\partial N_1}{\partial t} = R_1 - \frac{N_1}{\tau_1} + (N_2 - N_1) \frac{\sigma I}{h\nu} + \left(\frac{N_{10} - N_1}{\tau_F}\right) \tag{1.1}$$

$$\frac{\partial N_2}{\partial t} = R_2 - \frac{N_2}{\tau_2} - (N_2 - N_1) \frac{\sigma I}{h\nu} + \frac{(N_{20} - N_2)}{\tau_F} \tag{1.2}$$

with R_1, R_2 appropriate volume pumping rates; τ_1, τ_2 collisional relaxation times; N_{10}, N_{20} level population, when I (the laser beam intensity) = 0; σ stimulated emission cross-section; $h\nu$ photon energy; $\tau_F = x/V_F$, where V_F = gas flow velocity. When the system is in equilibrium $\dot{N}_1 = \dot{N}_2 = 0$, and the gain

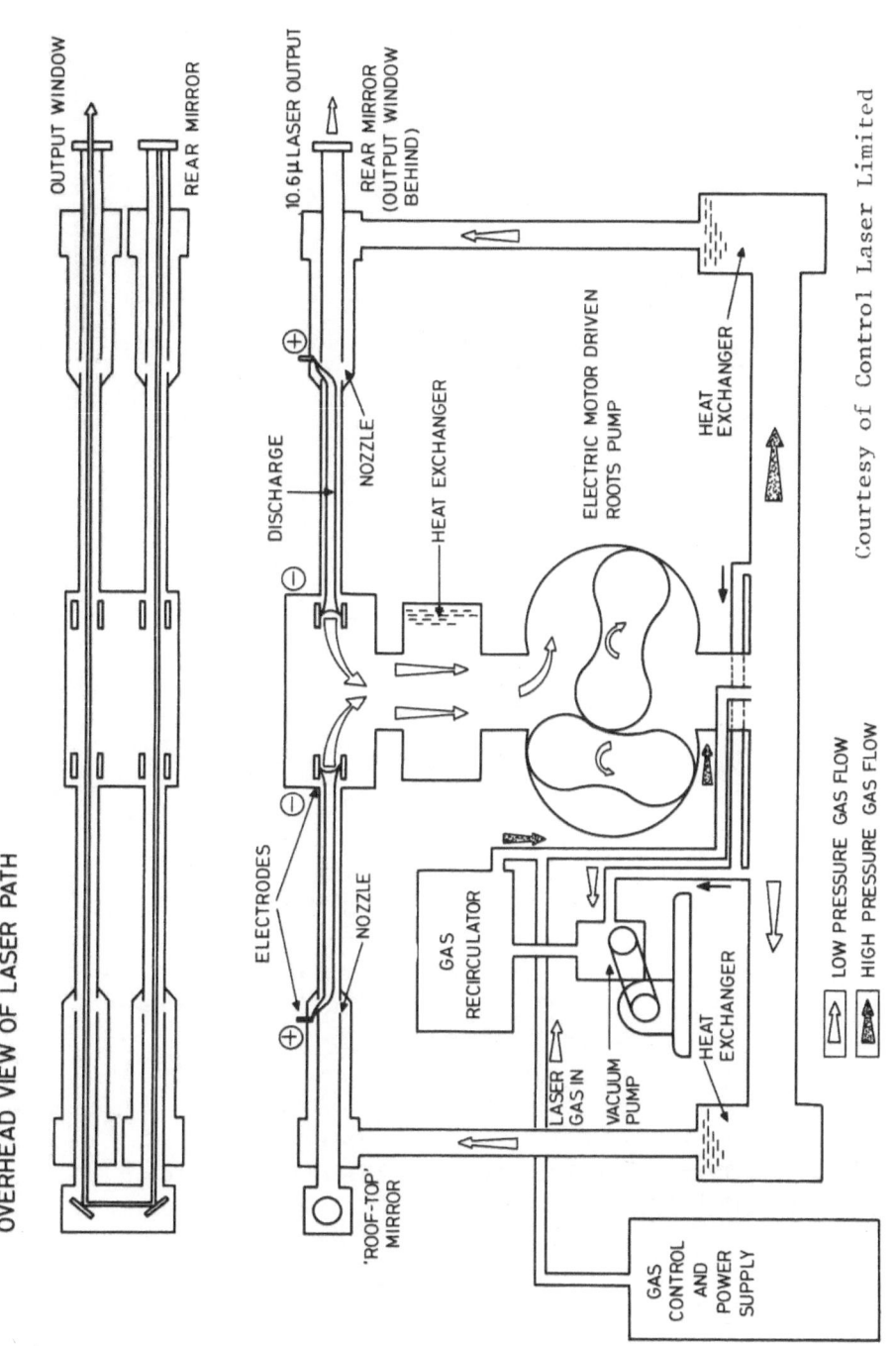

Fig. 3. Fast axial-flow, after Willis 1977

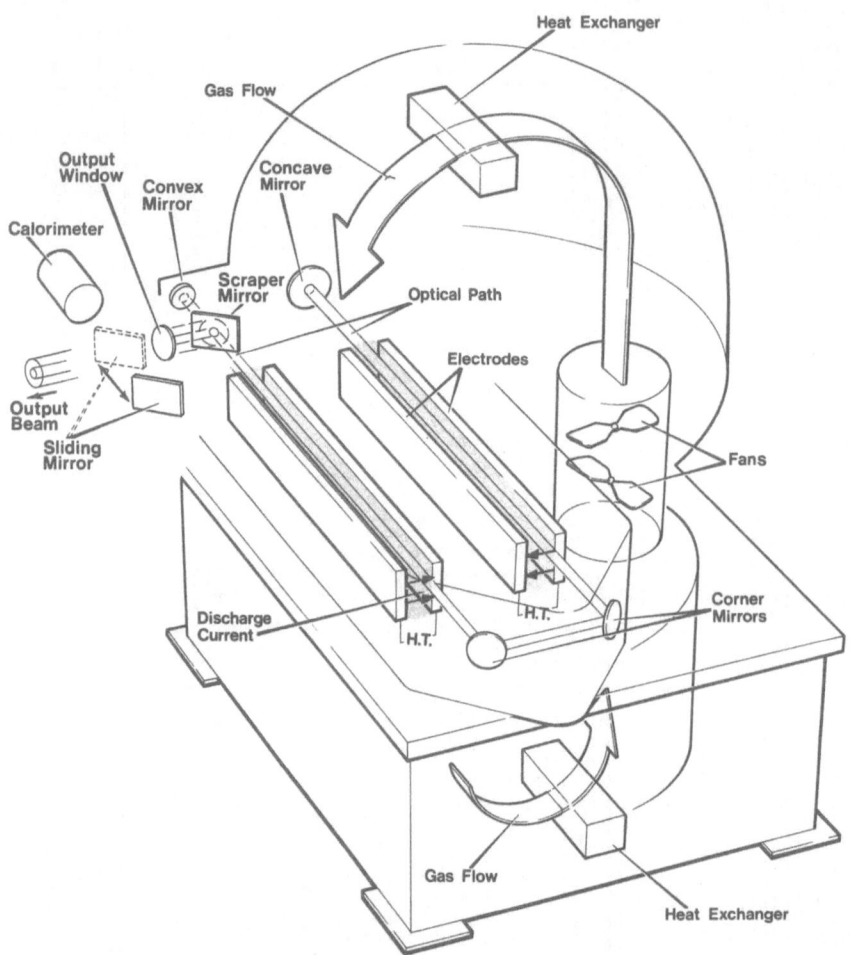

Fig 4.   Fast Transverse-Flow 5kW CO_2 Laser (schematic) -
         Culham Laboratory

coefficient $\alpha = \sigma(N_2 - N_1)$ and saturation intensity $I_s = I\alpha$ $[\alpha(o)-\alpha]^{-1}$ are given by:

$$\alpha = \frac{\sigma(R_2\, \tau_2 - R_1\, \tau_1)}{1 + (I\sigma/h\nu)\, [\tau_2\, \tau_F/(\tau_2 + \tau_F) + \tau_1\, \tau_F/(\tau_1 + \tau_F)]} \qquad (1.3)$$

$$I_s = \frac{h\nu}{\sigma}\, [\tau_2\, \tau_F/(\tau_2 + \tau_F) + \tau_1\, \tau_F/(\tau_1 + \tau_F)]^{-1} \qquad (1.4)$$

Now the maximum extractable power density $P = \alpha I_s$, so that;

$$\frac{P(\text{as } V_F \to \infty)}{P(\text{as } V_F \to o)} = (\frac{\tau_2 + \tau_1}{2\tau_F})\; (\frac{1 + \sigma I(\tau_2 + \tau_1)/h\nu}{1 + 2\sigma T(\tau_F/h\nu)}) \qquad (1.5)$$

Thus the ratio of the maximum extractable power density increases linearly with flow velocity (i.e. with τ_F^{-1}) at high flow rates. Quantitative calculations require numerical techniques, such as those discussed by Armandillo & Kaye (1980). In practice the maximum laser output from fast axial-flow systems tends to be determined by the very high (subsonic) flow velocities required, whilst the long thin geometry of the discharge tube proves suitable for the 'stable', Gaussian-mode, resonators to be discussed in § 2.1. In contrast the maximum output of a transverse flow device is rarely limited by mass flow (typical pressures and flow velocities are of order 30-50 Torr and 50 ms^{-1}, respectively) but is determined by the maximum electrical power density at which uniform volumetric excitation can be sustained; its short, fat geometry is appropriate for the unstable optical resonators discussed in § 2.4. A detailed discussion of the origin of electrical instabilities, and means for their suppression, will not be undertaken here. However, it is important to note that auxiliary electron-beam (or radio frequency) ionization techniques offer the potential advantage of an 'additional knob' to control the local electron density (independently of the main electrical discharge parameter E/N), but at the cost of additional technical complexity and/or factory safety precautions. The usable life time of the lasing gas (i.e. the plasma chemistry of the particular discharge) will also be an important factor when comparing the merits of alternative deivces. The various applications lectures at this school should provide an ideal vehicle for assessing the current state-of-the-art. (My own seminar will summarize experience with Culham's 5 kW low-order-mode, CL5, laser and with other CO_2 systems in the power range 75-15,000 watts).

Table 1.1 Typical Machining Lasers (CW)

λ (μm)	Medium	Type	$\phi^{(a)}$ (mm)	$\theta^{(b)}$ (mRad)	Mode	Max. Power (watts)
.325	He Cd⁺	Gas	1.3	0.5	TEMoo	0.01
.33–0.8	Krypton	Gas	1–2	0.6–1.2	TEMoo	5
.39–1.0	Dye	Liquid	.5–.7	1.5–2	Multi	1
.45–.53	Argon Ion	Gas	0.6–2.0	0.6–1.6	TEMoo	40
1.06	Nd-YAG	Solid	1–6	2	TEMoo	20
1.06	Nd-YAG	Solid	1–8	10–20	Multi	800
2.6–3	HF (f)	Gas	35	15	Multi	10,000
5–7	CO (i)	Gas	5–7	1.5–4	TEMoo	30
					Multi	50
9–11	CO_2 Waveguide	Gas	1.4	10		10
10.6	Sealed tube		1.3–7	2–10	TEMoo	25
10.6	Slow Axial Flow		5–10	1–3	TEMoo	1000
10.6	Fast Axial Flow		10–30	1–3	(c)	4000
10.6	Transverse Flow		25–70	0.5–3	(d)	10,000
10.6	Transverse Flow (e)				(d)	15,000

Notes:

(a) Near field beam diameter (1/e² intensity points).

(b) Beam divergence.

(c) Approximately TEMoo mode.

(d) Unstable cavity low order mode.

(e) Utilizing e-beam sustainer.

(f) Chemical laser.

(g) Multimode.

(h) Frequency doubled.

(i) Normally cryogenically cooled.

Table 1.1 Typical Machining Lasers (Pulsed)

λ (μm)	Medium	Type	φ(a) (mm)	θ(b) (mRad)	Energy (j)	Max. Rep. Rate (Hz)	Pulse Length (μsecs)	Max. Mean Power (watts)
0.19-0.35	Rare gas Halide (Excimer)	Gas	10 x 20	2 x 5	10^{-2}-1	10^2	0.02	50
0.337	Nitrogen	Gas	8 x 25	2 x 5	10^{-5}-10^{-2}	10^2	0.005-0.01	1
.45-.53	Argon	Gas	1.3-1.6	0.6-0.7	10^{-8}-10^{-5}	10^8	2.10^{-4}-0.015	10
1.06	Nd-Glass	Solid	5-25	0.2-0.5	10^{-2}-10^2(c)	10^{-1}	5.10^{-4}-10^4	
				2-7	1-10^3(g)	10	10^{-2}-10^4	50
1.06	Nd-YAG	Solid	2-20	0.1-5	10^{-3}-10(c)	10^2	10^{-2}-10^4	500
	"	"	5-20	1-10	10^{-1}-500(g)	10^2	10^{-2}-10^4	10
.53	(Harmonic)	"	5-10	.3-.6	10^{-3}-1(h)	20	10^{-2}	1
0.694	Ruby	Solid	1.5-16	0.4-1	10^{-2}-10	1	$.02$-10^3	20
0.2-5.0	Dye	Liquid	2-20	1-7	1-500	1	$.02$-10^3	
10.6	CO₂ TEA	Gas	5	1.5	10^{-2}	300	2.10^{-3}-2	
"	TEA	"	3-30	1-10	10^{-5}-10	10^3	0.1-1	200
"	Low pressure	"	5-20	1-3	0.1-1	$2.5.10^3$	50-cw	700
"	High energy	"	50-350	0.3-1	10-10^4	1	0.1-3	150

GEOMETRICAL AND WAVE THEORIES OF OPTICAL RESONATORS

The most common type of laser resonator consists of two spherical mirrors, separated by a distance L (Figure 1). In general, the mirrors will have unequal radii of curvature R_1 and R_2, one of which may be partially transmitting. Light rays which lie close to the optic axis i.e. the line joining the centres of curvature of the two mirrors, and which make only small angles with respect to it, are termed paraxial rays. Paraxial rays reflected back and forth between the two mirrors may experience a periodic focusing action, and the resonator is then termed stable (Boyd and Kogelnik 1962). Alternatively, the rays may become more and more dispersed on each transit, and the resonator is then termed unstable. We shall find that the intensity of the standing wave set up in the stable resonator tends to be greatest near its optic (z) axis, and diffraction losses at the edges of the mirrors are therefore normally less important than in unstable resonators, where the ray bounces out of the resonator after a finite number of transits – even if both mirrors are fully reflecting. Although both classes of resonator are now of practical importance, it is very important to make a firm distinction between them, as their characteristics are in general very different.

The optical stability of a resonator can be conveniently investigated using the ray transfer matrix techniques appropriate to geometrical optics (Brower 1964). These techniques are used by Svelto, following Bertolotti (1964), to show that, if the mirror curvature R_1 be counted positive when the mirror is concave inwards toward the resonator, and negative when convex, the condition for stability can be written:

$$0 \leq (1 - \frac{L}{R_1})(1 - \frac{L}{R_2}) \leq 1 \tag{2.1}$$

or, more conveniently,

$$0 \leq g_1 g_2 \leq 1 \tag{2.2}$$

where the parameter $g_i = 1 - L/R_i$. Figure 2 illustrates the position of various resonator configurations on a stability diagram plotted in the plane (g_1, g_2); the shaded region encloses all stable systems. The transition from stability to instability, which occurs near the boundaries defined by (2.2) is gradual for resonators having transverse dimensions sufficiently small that the wave nature of the laser beam is of primary importance, but is more abrupt for resonator having such large diameters $(2a_1, 2a_2)$ that the corresponding Fresnel Number

$$N = \frac{a_1 a_2}{\lambda L} \tag{2.3}$$

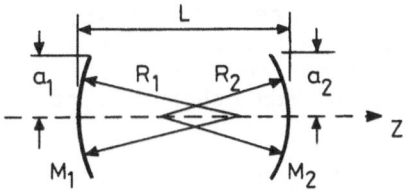

Fig 1. Typical laser resonator, comprising two spherical mirrors M_1, M_2, of curvature R_1, R_2 respectively.

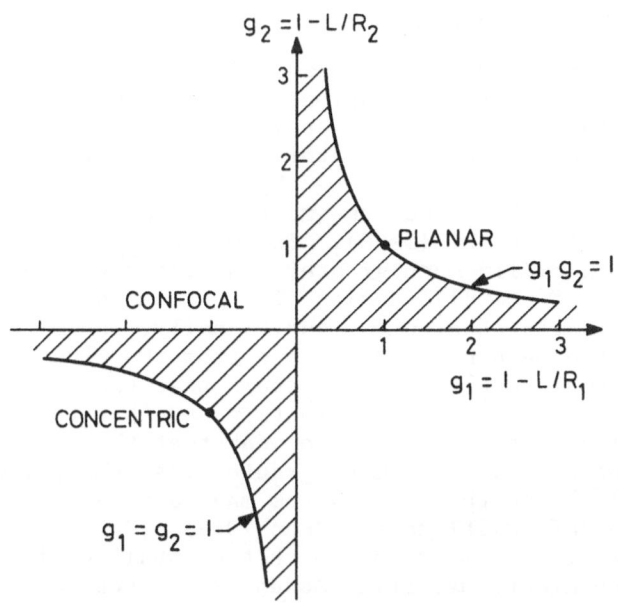

Fig 2. Stability diagram for resonators in the g_1, g_2 plane - applies to the general optical resonator of Fig 1. Shaded region is stable; an unstable resonator corresponds to points elsewhere.

considerably exceeds unity (Kogelnik and Li 1966). It should be
particularly noted that relatively minor deviations from nominal
resonator specifications are sufficient to displace symmetric confocal ($R_1 = R_2 = L$) and plane-plane cavities ($R_1, R_2 \to \infty$) away from
their positions on the boundary of the g_1, g_2 stability diagram.
The resonator could then inadvertently lie in one of the unshaded
regions representing unstable configurations; it would then exhibit
the high diffraction losses and other characteristics of such
systems.

Although resonators formed from other mirror shapes, such as
cones and paraboloids, have been analysed in the literature
(Bergstein 1968), we shall arbitrarily restrict the discussion in
this lecture to resonators formed from spherical components,
because spherical mirrors and related components of high optical
quality are readily and economically available. However,
numerically controlled diamond-machining techniques may eventually
offer a much greater flexibility to the laser designer in his
choice of optical configuration.

Stable Resonators of Large Cross Section

Consider first a stable resonator formed with spherical
mirrors having transverse dimensions so large that only a negligible
proportion of the total laser power is intercepted at their edges.
The radiation field in the resonator can be calculated using the
wave approximations first discussed by Fox and Li, Boyd and Gordon
(1961). We shall follow the treatment adopted in the excellent
review by Kogelnik and Li (1966). The mirror separation and other
dimensions are normally much greater than the wavelength so that
a scalar theory of the diffraction of light can be used, instead
of solving Maxwell's equations. Additionally, the mirror curvatures
are such that the wave normals deviate from the direction of the
optic axis by small angles only. Assume that the radiation field
is linearly polarized in the y direction and that the losses on
each traversal of the cavity are so weak that the optical decay
constant of the resonator is much greater than the period of
oscillation (λ/c), so that the system may be regarded as being in
a steady state (cf. Maitland and Dunn 1969). The electric field
in the xz plane is nearly zero, so that we shall be investigating
TEM (transverse electromagnetic) modes of the resonator.

The amplitude (u), of a field component (E or B) of the
coherent light then satisfies the scalar, time-independent,
(Helmoholtz) wave equation (Born and Wolf 1966).

$$\nabla^2 u + k^2 u = 0 \qquad\qquad (2.4)$$

where $k = 2\pi/\lambda$ is the propagation constant in the medium, which is
assumed to be uniform and not amplifying. For light travelling in

the z direction we write

$$\frac{u}{u_o} = \psi(x,y,z) \exp(-jkz) \tag{2.5}$$

where ψ is a slowly varying complex function which represents the following differences between a real laser beam and the (ideal) properties of a plane wave:

(a)   non-uniform intensity distribution;
(b)   expansion of the beam with distance of propagation;
(c)   curvature of the phase front.

From (2.4) and (2.5), and neglecting the second derivative $\frac{\partial^2 \psi}{\partial z^2}$ (because nearly all laser beams have high directivity), one finds

$$\frac{\partial^2 \psi}{\partial x^2} + \frac{\partial^2 \psi}{\partial y^2} - 2j\,k\,\frac{\partial \psi}{\partial z} = 0 \tag{2.6}$$

This differential equation is reminiscent of the time-dependent Schrödinger equation, and can be solved by methods familiar from quantum theory (Schiff 1955). A trial solution for (2.6) is

$$\psi = \exp[-j(P + \frac{k}{2Q}\,r^2)] \tag{2.7}$$

where $r^2 = x^2 + y^2$ $\qquad\qquad$ (2.8)

Inserting (2.7) into (2.6) and comparing terms of equal power in r, we find that (2.7) is an acceptable solution providing that

$$\frac{\partial Q}{\partial z} = 1 \tag{2.9}$$

and $\frac{\partial P}{\partial z} = j/Q$ $\qquad\qquad$ (2.10)

Thus the parameter P(z) in (2.7) represents a complex phase shift associated with the propagation of the beam, and Q(z) represents a complex beam parameter describing both the intensity variation with r and the curvature of the phase front. However, our physical understanding can be made clearer by introducing two real beam parameters R and w, which are related to the complex parameter Q by the expression

$$\frac{1}{Q} = \frac{1}{R} - j\,\frac{\lambda}{\pi w^2} \tag{2.11}$$

Comparing (2.11) with (2.7) one sees that R(z) is the radius of curvature of the wavefront that intersects the axis at z, and w(z) is a measure of the decrease of the field amplitude with r. This decrease is, of course, Gaussian in form; w is the distance at which the amplitude is $(1/e)$, or the intensity $(1/e^2)$, times that on the axis (cf. Figure 3).

Inspection of (2.11) reveals that the complex beam parameter $Q(z)$ is purely imaginary when the phase front is plane $(R \to \infty)$, and it will therefore be convenient to measure z from this position where

$$Q_o = j \pi w_o^2 / \lambda \qquad (2.12)$$

Integration of (2.9) then gives

$$Q(z) = Q_o + z = z + j \frac{\pi w_o^2}{\lambda} \qquad (2.13)$$

Combining (2.13) with (2.11) and equating real and imaginary parts, one obtains

$$w^2(z) = w_o^2 [1 + \{\frac{\lambda z}{\pi w_o^2}\}^2] \qquad (2.14)$$

and $R(z) = z[1 + \{\frac{\pi w_o^2}{\lambda z}\}^2]$ \qquad (2.15)

To calculate the complex phase shift, one uses (2.13) with (2.10):

$$\frac{\partial P}{\partial z} = -\frac{j}{Q} = \frac{-j}{z + j(\pi w_o^2/\lambda)} \qquad (2.16)$$

Integration then yields

$$jP(z) = \ln[1 - j(\pi z/\pi w_o^2)]$$

$$= \ln\sqrt{1 + (\lambda z/\pi w_o^2)^2} - j \text{ arc tan } (\lambda z/\pi w_o^2) \qquad (2.17)$$

Thus one solution of the scalar equation (2.4) is the expression

$$u(r,z) = u_{ro} \{\frac{w_o}{w}\} \exp\{-j(kz - \Phi) - r^2(\frac{1}{w^2} + \frac{jk}{2R})\} \qquad (2.18)$$

where $\Phi = \tan^{-1} (\lambda z/\pi w_o^2)$ \qquad (2.19)

Before we discuss other permissible solutions of (2.4), let us consider some physical implications of equations (2.18) and (2.19), which represent the solution of greatest practical importance.

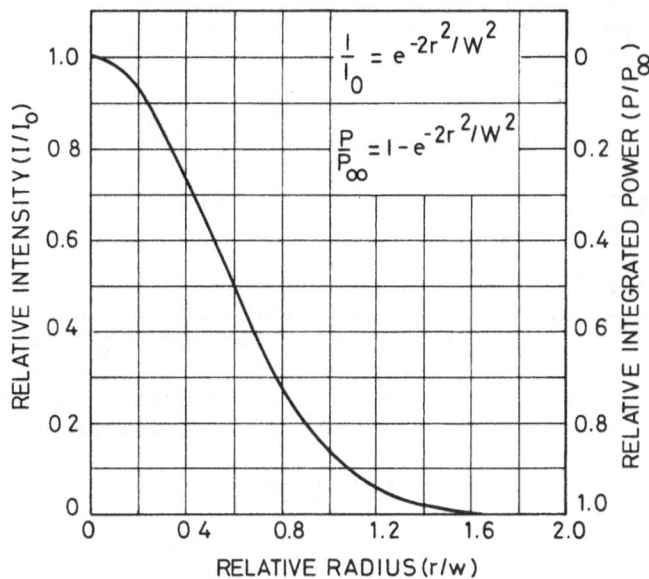

Fig 3. Relative intensity and integrated power for cross-section
        of a Gaussian beam.

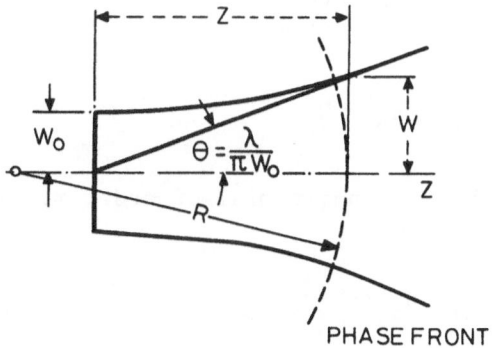

Fig 4. Contour of a Gaussian beam.

From (2.14) we note that although the width of the intensity
profile changes along the z axis it always retains its Gaussian
form. The laser beam contracts to a minimum diameter $2w_o$ at the
beam waist, at the position z = 0 where its phase front is plane.
(However, the position of the beam waist with respect to the
mirrors, and also the magnitude of w_o remain arbitrary until we
consider the boundary conditions). The beam contour w(z) is the
hyperbola shown with solid lines in Figure 4, with asymptotes
inclined to the z axis at an angle

$$\tan \theta = \frac{w(z)}{z} = \frac{\lambda}{\pi w_o} \sim \theta \tag{2.20}$$

Thus, in the far field (z $\rightarrow \infty$) the phase front, shown as a dotted
line in Figure 4, approximates to a spherical wave diverging from
a point source located at the origin. The term $\{\frac{w_o}{w}\}$ in (2.18)
describes the way in which the axial intensity decreases with
expansion of the beam, whilst the Φ term describes the phase shift
difference between the Gaussian beam and an ideal plane wave.

Higher Order Solutions of the Wave Equation

     More general solutions of (2.4) are possible, and their
derivation in Cartesian and cylindrical coordinate systems will be
summarized below. For resonators constructed from rectangular
mirrors one naturally employs a Cartesian (x, y, z) geometry, and
generalizes the trial solution (2.7) to the form

$$\psi = g\{\frac{x}{w}\}.h\{\frac{y}{w}\}.\exp\{- j[P + \frac{k}{2Q} (x^2 + y^2)]\} \tag{2.21}$$

where g is a real function of x and z, and h is a real function of
y and z. Inserting (2.21) into (2.6) one finally arrives (Kogelnik
1965) at differential equations for g and h of the form

$$\frac{d^2 H_m}{dx^2} - 2x \frac{dH_m}{dx} + 2m H_m = 0$$

where $H_m(x)$ is the Hermite polynomial of order m. Thus (2.6) can
be satisfied if

$$g.h = H_m \{\sqrt{2}\frac{x}{w(z)}\} H_n \{\sqrt{2}\frac{y}{w(z)}\} \tag{2.22}$$

The same scaling parameter w(z) therefore applies to each of the
orthogonal solutions characterized by the integers m and n.

     The first few Hermite polynomials are:

$$H_0(x) = 1$$

$$H_1(x) = x$$

$$H_2(x) = 4x^2 - 2$$

$$H_3(x) = 8x^3 - 12x$$

$$H_4(x) = 16x^4 - 48x + 12 \qquad (2.23)$$

The intensity pattern normal to the z axis is thus described by a product of Hermite and Gaussian functions; the product has m zeros in the x direction and n in the y direction, cf. Figure 5. Manipulation of the equations shows that the phase-front curvature R(z) is identical for all orders of m and n, but that the phase shift Φ takes the more general form

$$\Phi(m,n;z) = (m + n + 1) \tan^{-1} (\lambda z/\pi w_o^2) \qquad (2.24)$$

(2.23) implies that the phase velocity increases with m or n, and thus that the various transverse modes of oscillation in a resonator will have different frequencies. In Cartesian co-ordinates, the generalized solution is thus

$$u(x,y,z) = u(o)[\frac{w_o}{w}][g.h] .\exp\{- j(kz - \phi) - r^2[\frac{1}{w^2} + \frac{jk}{2R}]\} \qquad (2.25)$$

where w(z), g(x,z).h(y,z), ϕ(m,n;z), R(z) are evaluated in equations (2.14), (2.22), (2.24) and (2.15) respectively.

A cylindrical (r,θ,z) geometry is more appropriate to mirrors of circular cross section, and one then adopts a general trial solution of the form

$$\psi = g[\frac{r}{w}] . \exp\{- j[P + \frac{kr^2}{2Q} + \ell\theta]\} \qquad (2.26)$$

Inserting (2.26) into (2.6) one obtains, after some calculation (Kogelnik 1965)

$$g(r,z) = \{\sqrt{2}\frac{r}{w}\}^\ell . L_p^{\ell} \{2 \frac{r^2}{w^2}\} \qquad (2.27)$$

where L_p^{ℓ} is a generalized Laguerre polynomial, and p and ℓ are integers characterizing the radial and angular modes respectively. Some Laguerre polynomials of low order are

$$L_o^{\ell}(x) = 1$$

$$L_1^{\ell}(x) = \ell + 1 - x$$

$$L_2^{\ell}(x) = \tfrac{1}{2}(\ell + 1)(\ell + 2) - (\ell + 2)x + \tfrac{1}{2}x^2 \qquad (2.28)$$

As we might expect from the Cartesian calculations, the beam parameters $w(z)$ and $R(z)$ are the same for all cylindrical modes. The phase shift is, however, dependent on the mode numbers p and ℓ, and is given by

$$\phi(p,\ell;z) = (2p + \ell + 1) \tan^{-1}(\lambda z/\pi w_o^2) \qquad (2.29)$$

Thus,

$$u(r,z) = u_o \{\tfrac{w_o}{w}\} g(r,z) \exp\{-j[kz - \phi(p,\ell;z)] - r^2\{\tfrac{1}{w^2} + \tfrac{jk}{zR}\}\} \qquad (2.30)$$

Evaluation of w_o from The Boundary Conditions

Ignoring diffraction losses at the edges of the (large) mirrors, the modes of oscillation within the resonator are those self-consistent field configurations for which the complex beam parameter Q remains unchanged after successive round trips of the beam within the resonator. The variation of Q with axial position can be described formally using the ABCD law discussed by Kogelnik (1965) with the mathematical formulation of ray transfer matrix techniques. We shall here employ identical physical arguments, but avoid the matrix formalism, to calculate the transformation of paraxial rays as they are reflected back and forth within the cavity.

Let the laser beam be described by the complex parameter Q_1 as it leaves mirror 1. When is reaches mirror 2 we find from (2.13) that its value is

$$Q_2 = Q_1 + L \qquad (2.31)$$

[$Q(r,w)$ therefore transforms in a similar manner to the single, real, parameter r which characterises a spherical wave diverging from a stationary point source. For example, after propagation over a distance z, the radius of curvature of a spherical phase front becomes $r_2 = r_1 + z$. More importantly, the transformation of a spherical wave r_1 by a thin lens of positive focal length F, i.e. one which converges a plane wavefront to a concave wavefront when viewed from the $z = +\infty$ side of the lens is described by the relation

$$(1/r_2) = (1/r_1) - (1/F)$$

In the paraxial approximation, the wavefront curvature of the stable cavity modes are also spherical, and so one treats the transformation of Q by a mirror of curvature R_2 in similar fashion.]

Since $F = R_2/2$, it follows that after reflection at the mirror 2,

$$\frac{1}{Q_3} = \frac{1}{Q_2} - \frac{2}{R_2} \qquad\qquad (2.32)$$

On reaching mirror 1 again,

$$Q_4 = Q_3 + L \qquad\qquad (2.33)$$

Finally, after reflection at mirror 1,

$$\frac{1}{Q_5} = \frac{1}{Q_4} - \frac{2}{R_1} \qquad\qquad (2.34)$$

The beam has now completed a round trip, and it is therefore required of a self-consistent mode that

$$Q_5 = Q_1 \qquad\qquad (2.35)$$

Re-arranging equations (2.31) - (2.35), one obtains

$$Q_5 = \frac{R_1 R_2 (Q_1 + 2L) - 2R_1 L(L + Q_1)}{R_2 (R_1 - 2L) - 2(R_1 + R_2)(Q_1 + L) + 4L(Q_1 + L)} = Q_1 \quad (2.36)$$

Hence,

$$L R_1 (R_2 - L)(1/Q_1)^2 + 2L(R_2 - L)(1/Q_1) + (R_1 + R_2 - 2L) = 0 \qquad\qquad (2.37)$$

and

$$(1/Q_1) = -\frac{1}{R_1} \pm \frac{j}{R_1} \left[\frac{(R_1 + R_2 - 2L)R_1}{L(R_2 - L)} - 1 \right]^{\frac{1}{2}} \qquad (2.38)$$

The radius of curvature of the wavefront, corresponding to the real parts of (2.11) and (2.38), matches the radius of mirror 1 - as expected. The width of the laser beam at mirror 1 is determined by comparison of the imaginary parts of (2.11) and (2.38), giving

$$w_1^4 = \{\frac{\lambda R_1}{\pi}\}^2 \cdot \frac{(R_2 - L)L}{(R_1 - L)(R_1 + R_2 - L)} \qquad (2.39)$$

Similarly, the spot size at mirror 2 is determined from

$$w_2^4 = \{\frac{\lambda R_2}{\pi}\}^2 \cdot \frac{(R_2 - L)L}{(R_2 - L)(R_1 + R_2 - L)} \qquad (2.40)$$

It is of considerable practical importance to evaluate the position of the beam waist with respect to either mirror. Recalling that the wavefront is plane at the waist and that we arbitrarily located the origin of z at this position, so that $Q(o)$ was purely imaginary, it follows from (2.13) that the distance of mirror 1 from the waist (z_{w1}) is given by the real part of Q_1. Hence, from (2.38),

$$z_{w1} = - \frac{(R_2 - L)L}{(R_1 + R_2 - 2L)} \tag{2.41}$$

and

$$z_{w2} = (z_{w1} + L) = + \frac{(R_1 - L)L}{(R_1 + R_2 - 2L)} \tag{2.42}$$

The waist diameter then follows from (2.41) and (2.14):

$$w_o^4 = \{\frac{\lambda}{\pi}\}^2 \frac{L(R_1 - L)(R_2 - L)(R_1 + R_2 - L)}{(R_1 + R_2 - 2L)^2} \tag{2.43}$$

One further resonance condition must be considered: after the beam has travelled from one mirror to the other the phase shift must be an integral multiple of π, in order that a well-defined standing-wave phase-structure can be established within the resonator. From (2.24) and (2.25) it follows that

$$kL - (m+n+1) \tan^{-1}(\lambda L/\pi w_o^2) = \pi(q + 1) \tag{2.44}$$

where q is the (very large) number of modes along the axial standing wave pattern, i.e. the number of half wavelengths is $q + 1$. Defining the characteristic beat frequency between successive longitudinal resonances as $\nu_o = c/zL$, where c is the velocity of light, (2.44) can be algebraically manipulated into the form

$$\frac{\nu}{\nu_o} = (q + 1) + \frac{1}{\pi}(m+n+1) \cos^{-1} \surd(1 - L/R_1)(L - L/R_2) \tag{2.45}$$

where the square root is given the sign of either of its component terms. The cavity stability condition (2.1) therefore expresses the physical requirement that the square root term in (2.45) be real, with a modulus less than unity, in order that the resonant frequency ν_{mnq} itself is real. (The stability condition (2.1) similarly ensures that the beamwidth is non-zero, i.e. that the beam parameter Q has a non-zero imaginary part, cf. Maitland and Dunn 1969). The cylindrical analogue of (2.45) is given by

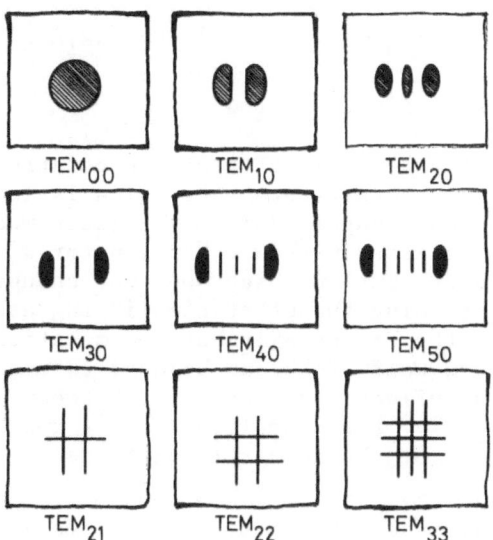

Fig 5. Mode patterns of a gas laser oscillator (rectangular symmetry).

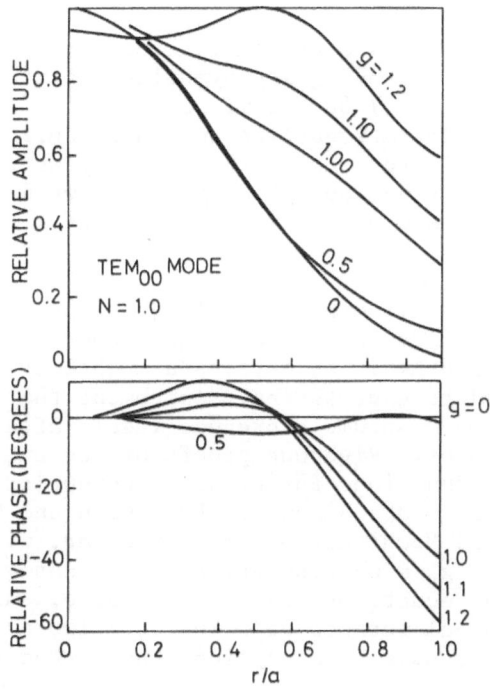

Fig 6. Relative field distributions of the TEM_{00} mode for a N = 1
resonator with circular mirrors (after Kogelnik and Li 1966).

$$\nu/\nu o = (q + 1) + \frac{1}{\pi} (2p + \ell + 1) \cos^{-1}(g_1 g_2)^{\frac{1}{2}} \qquad (2.46)$$

Stable Resonators of Finite Aperture

Resonators of practical interest are constructed of mirrors of finite size and are therefore intrinsically lossy. Unless energy is continuously supplied from the laser medium, a field configuration established within the resonator will slowly decay, with a relative distribution which does not change with time (Fox and Li 1960). Retaining the other simplifying assumptions discussed in 2.1, the Fresnel-Kirchoff formulation of Huygen's principle (Born and Wolf 1966) can be used to describe a travelling TEM wave which is reflected successively between spherical mirrors 1 and 2. The relative field amplitudes $E^{(1)}$ and $E^{(2)}$ at the two mirrors are related by the integral equations

$$\gamma^{(1)} E^{(1)} (s_1) = \int_{S_2} K^{(2)} (s_1, s_2) E^{(2)} (s_2) dS_2$$

$$\gamma^{(2)} E^{(2)} (s_2) = \int_{S_1} K^{(1)} (s_2, s_1) E^{(1)} (s_1) dS_1 \qquad (2.47)$$

where the integrations are taken over the mirror surfaces S_2 and S_1, respectively. s_1 and s_2 are symbolic notations for transverse coordinates (x,y or r,θ) on the mirror surfaces (1 and 2); $\gamma^{(1)}$ and $\gamma^{(2)}$ describe the attenuation and phase shift experienced by the wave in transit from one mirror to the other; the kernels $K^{(1)}$ and $K^{(2)}$ are known functions of distance between s_1 and s_2, since they are fully defined by the mirror geometry within the area of integration.

Each of the two linear integral equations (2.47) relates the field at one mirror to the field at the other; substituting one equation into the other, one obtains a double-transit equation which can be used to express the requirement that the field of a self-consistent mode should reproduce itself after a round trip within the resonator. Rigorous proofs of the existence of eigen values and eigen functions for kernals appropriate to common resonator geometries are discussed by Newman and Morgan (1964), Cochran (1965) and Hochstadt (1966). However, with the exception of confocal ($g_1 = g_2 = 0$) resonators, exact analytical solutions of these integral equations are not available, and recourse must be made to numerical methods of solution. Two alternative mathmatical techniques have been applied; the method of successive approximations (Fox and Li 1960, 1961, 1963 and Li 1965), and the method of kernal expansion (Streifer 1965, Heurtley and Streifer 1965). The first method follows the transient behaviour of the resonator after it has been excited by a wave of arbitrary spatial

distribution. This wave undergoes amplitude and phase changes as
it is reflected alternately and successively from each end of the
resonator; it also loses energy by diffraction at the mirror
apertures at each transit. After sufficient transits, it is found
that a quasi-static condition is achieved in which the field
changes only by a constant multiplicative factor at each further
reflection. The resultant field distribution is an eigen function
of the integral equations and represents, of course, the mode that
has the lowest diffraction loss for the assumed symmetry. The
constant multiplicative factor is the eigen value of that eigen
function, and thus gives the diffraction loss and phase shift of
the mode. Somewhat surprisingly, this iterative technique can be
modified (Fox and Li 1963) to yield higher order solutions of
(2.47); each of the resulting eigen functions are orthogonal over
their respective mirror surfaces, i.e.

$$\int_{S_1} E_m^{(1)}(s_1) \, E_n^{(1)}(s_1) \, d_1 S_1 = 0, \quad m \neq n$$

$$\int_{S_2} E_m^{(2)}(s_2) \, E_n^{(2)}(s_2) \, dS_2 = 0, \quad m \neq n \qquad (2.48)$$

where m and n denote the different mode orders. The second method,
involving kernel expansion techniques, yields eigen functions of
both high and low-order modes. We shall not pursue the mathematical
techniques in detail, but will summarize some of the more important
numerical results, since they are of considerable practical
significance. (The interested reader is referred to Maitland and
Dunn (1969) and references contained therein).

Figures 6 and 7 show the relative field distributions of the
TEM_{00} and TEM_{01} modes for a resonator having identical circular
mirrors, and a Fresnel number (N) of unity (i.e. $a_1^2 = a_2^2 = \lambda L = a^2$;
$g_1 = g_2 = g$). Curves are reproduced (after Kogelnik and Li 1966)
for various values of g, ranging from zero (confocal) through one
(plane-parallel) to 1.2 (unstable, convex). The field is
concentrated near the resonator axis in confocal resonators, but
tends to spread out as $|g|$ increases. (These curves are also
applicable to resonators with negative g values, provided the sign
of the phase distribution ordinate is reversed). When g = 0 an
exact analytic solution is given by the generalized prolate
spheroidal wave-functions (Boyd and Gordon 1961, Slepian 1964,
Heurtley 1964); for large Fresnel numbers these functions can be
approximated by the Laguerre-Gaussian functions discussed in the
section 'Higher Order Solutions of the Wave Equation'. Similarly,
for long rectangular mirrors the exact solutions are shown by the
same authors to be prolate spheroidal wave functions, which can be
approximated at large Fresnel number by the Hermite-Gaussian
functions.

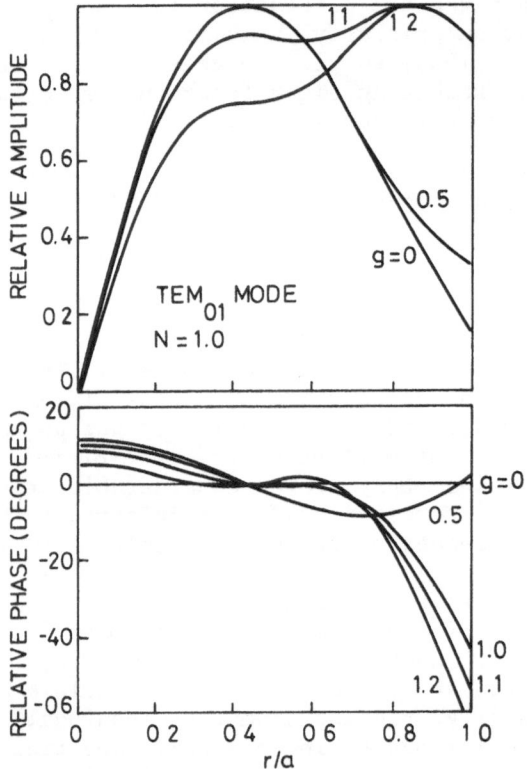

Fig 7.   Relative field distributions of the TEM_{01} mode for a N = 1
         resonator with circular mirrors (after Kogelnik & Li 1966).

Figure 8 shows the relative field distributions of some low-order modes of a Fabry-Perot resonator having identical plane-parallel mirrors of circular cross-section (N = 10, g = 1). The small wiggles on these curves (Schwering 1961) are related to the Fresnel number, and are characteristic of resonator geometries which are unstable, or nearly so. To first order, these field distributions are represented by Bessel functions (and by sine and cosine functions for long, rectangular, mirrors).

The diffraction loss α and phase shift β for particular modes are important quantities, since they determine the dominant mode, its Q factor and the resonant frequencies of the resonator. The diffracted energy loss per transit is related to (2.47) by

$$\alpha = 1 - |\gamma|^2$$

Typical values of α for a stable resonator having identical circular mirrors (Li, 1965) are plotted in Figures 9 and 10 as functions of the Fresnel number N, for various values of $|g|$. The angle of γ represents the phase shift (β) experienced by the wave over and above its geometrical phase shift $2\pi L/\lambda$, during transit from one mirror to the other. Curves of β, corresponding to the resonator losses of Figures 9 and 10, are shown in Figures 11 and 12, for positive g only. The limiting (horizontal) portions of the curves can be deduced from (2.46), from which it follows that

$$\beta_{(N \to \infty)} = (2p + \ell + 1) \cos^{-1} g \qquad\qquad (2.49)$$

Figures 9 and 10 show that the diffraction losses for the TEM_{01} mode always exceed those of the lowest order TEM_{00} mode, for given N and $|g|$. The power loss per transit is even higher for higher order modes, which have less of the total power confined near the axis (cf. Figures 6 and 7), and this fact offers the possibility of controlling those modes which lase, as we shall see later.

Explicit expressions for α and β can be obtained for the confocal (g = 0) geometry. For large N, with circular mirrors (Slepian 1964)

$$\alpha = \frac{2\pi (8\pi N)^{(2p+\ell+1)}}{p! \, (p+\ell+1)!} \; \exp(-4\pi N) \qquad\qquad (2.50)$$

$$\beta = (2p+\ell+1)(\pi/2)$$

Analogous approximations have been derived for small N, i.e. for large diffraction loss (Bergstein and Schachter 1965); Figure 13 shows the power loss per transit in a confocal cavity for several low-order modes, as a function of Fresnel number (McCumber 1965).

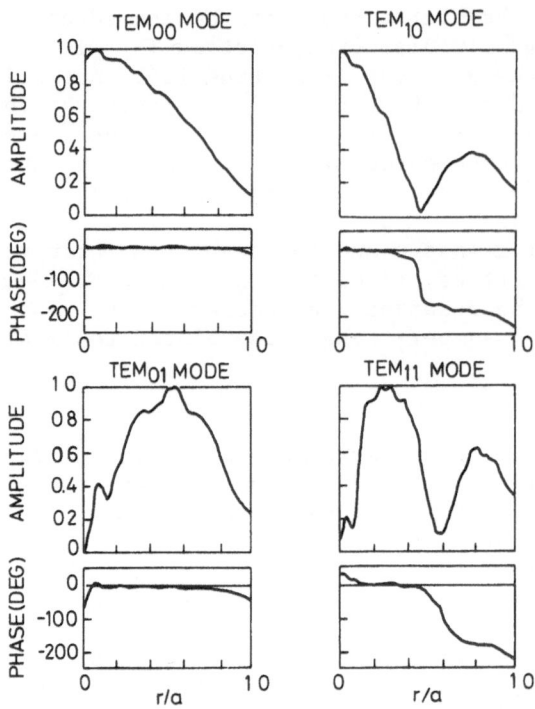

Fig 8. Field distribution of modes of a plane-parallel Fabry Perot
       N = 10 resonator having circular mirrors (after Schwering
       1961).

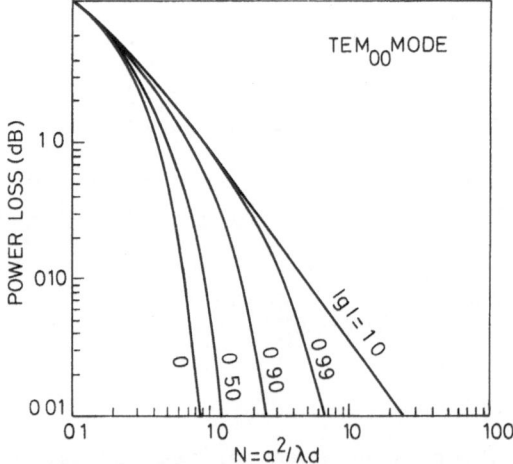

Fig 9. Diffraction loss per transit (in decibels) for TEM_{oo} mode
       of stable resonator with circular mirrors (after Li 1965).

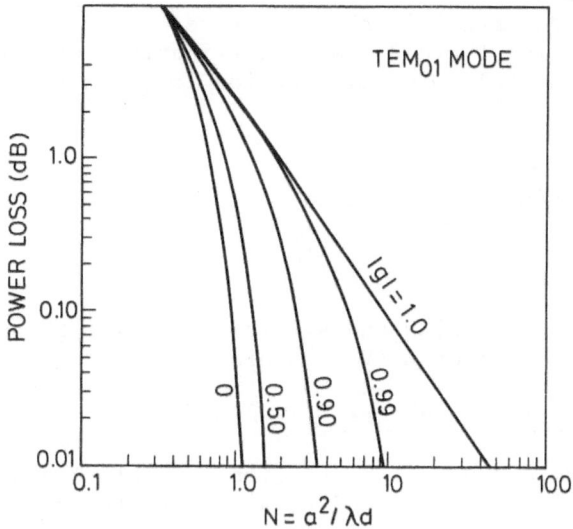

Fig 10. Diffraction loss per transit (in decibels) for TEM_{01} mode
of stable resonator with circular mirrors (after Li 1965).

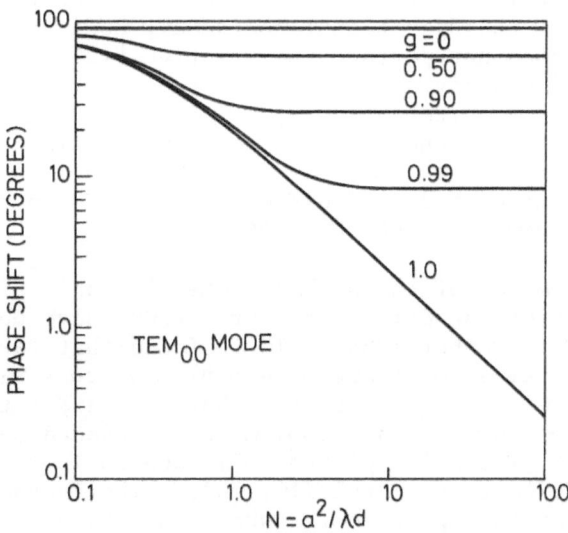

Fig 11. Phase shift per transit for TEM_{00} mode of stable
resonator with circular mirrors (after Li 1965).

Diffraction losses at rectangular confocal mirrors have been similarly discussed by Vainshtein (1963) and Slepian (1964). A symmetrical plane-parallel Fabry-Perot resonator of circular symmetry, for which typical field distributions were plotted in Figure 8, has the following diffraction loss, in the limit of large N:

$$\alpha = 8 \, K_{p\ell} \, \frac{\delta(M + \delta)}{[(M + \delta)^2 + \delta^2]^2} \tag{2.51}$$

$$\beta = \{\frac{M}{4\delta}\}\alpha$$

where $\delta = 0.2824$, $M = \sqrt{8\pi N}$ and $K_{p\ell}$ is the $(p + 1)$th zero of the Bessel function of order ℓ (Vainshtein 1963).

Hole-Coupled Stable Resonators

Much of the discussion in section ' Stable Resonators of Finite Aperture' was restricted to symmetric $(g_1=g_2=g)$ resonators, for reasons of mathematical simplicity. However, geometrically symmetric stable resonators also prove to be of considerable practical convenience, since the volume enclosed by the lowest order mode can roughly match that of any laser medium having cylindrical symmetry. If we wish to extract power preferentially at one end of the resonator, we may choose to make one of the mirrors partially transmitting, with a transmittance (t) greater than the round-trip diffraction loss for the lowest-order mode; the round-trip losses for higher modes may, however, exceed t. This approach does not perturb the relative field distributions of the various modes derived in section 'Stable Resonators of Large Cross Section'. An alternative approach is to drill a hole, of radius a_o, exactly at the centre of one of the mirrors; such an approach does, of course, perturb the symmetry of the resonator, and in general mixes modes of different p without perturbing the azimuthal (ℓ) symmetry (McCumber 1965).

The eigen modes of the stable symmetric confocal resonator formed by identical spherical annular mirrors have been examined in some detail (McCumber 1965). It is found that for a given p, modes with low values of ℓ are more sensitive to a small aperture on the axis than are the modes of higher ℓ; whereas the axial intensity of the lowest order (00) mode is reduced by the presence of the hole, the intensity of the (02) mode is increased. However, the strongest effect is on the (10) mode. The Fresnel number $(N_o = \alpha_o{}^2/\lambda L)$ of the aperture for which the losses of the (00) mode exceed that of the (10) mode is designated N_{oc} and is plotted in Figure 14 as a function of the mirror Fresnel number $N_M = a^2{}_M/\lambda L$, where a_M is the outer radius of the mirror. For $N_o < N_{oc}$, the laser will oscillate in a mode approximating to that of the lowest

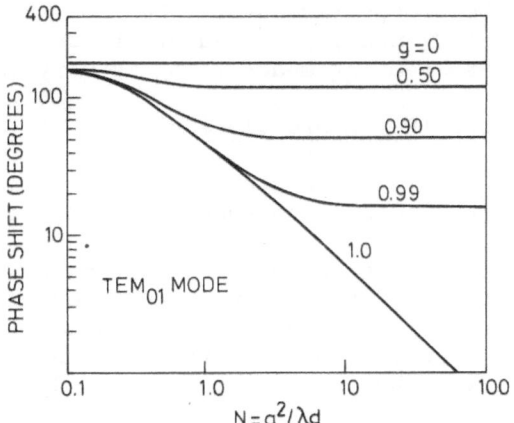

Fig 12. Phase shift per transit for TEM_{00} mode of stable resonator with circular mirrors (after Li 1965).

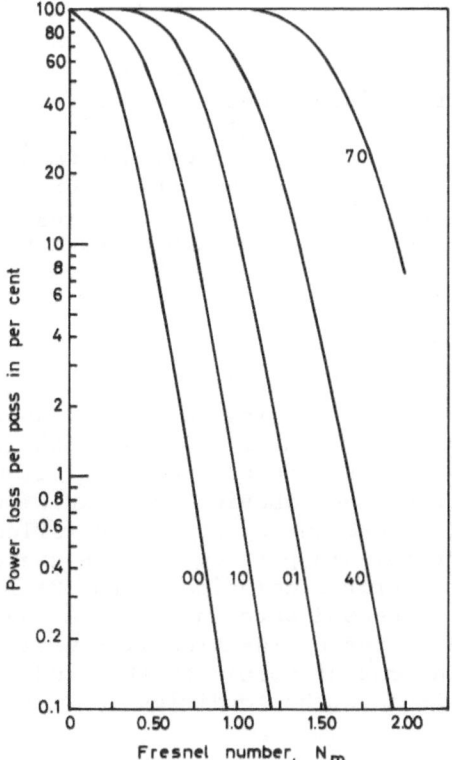

Fig 13. Power loss/pass for low-order modes in confocal cavity versus Fresnel number N_m (after McCumber 1965).

order mode of a hole-less resonator. For $N_o > N_{oc}$ the laser will
oscillate in the (10) mode, provided other conditions are not
changed, or in more complex modes if N_o is very large.

Unstable Resonators (Ray Approximation)

Theoretical studies of unstable resonators having mirrors
with sharply defined edges have shown characteristics significantly
different from those of the stable cavities discussed above. The
losses and the transverse variation of intensity within an unstable
resonator are influenced both by the length of the optical cavity
and by the mirror radii, in contrast to stable resonator modes
that change only slowly with the mirror parameters. In general
the mode intensity is not zero at the inner edge of an unstable
cavity, and so in the wave approximation a diffracted wavelet
whose source is the mirror's edge should be superposed on the
reflected wave (Born and Wolf 1968). Although the amplitude of
this additional wave decreases rapidly away from the direction of
specular reflection, some part of the wavelet may be scattered
through sufficiently large angles that it is retained as a
convergent wave within the resonator for many transits (Figure 15).
Since large-angle diffractive scattering depends strongly on the
nature of the mirror discontinuity (Sommerfeld 1954) wave theories
of unstable resonators are highly sensitive to strong gradients of
reflectivity, or phase variations which normally occur at the edges
of real mirrors. Alignment assymmetries, or fine-scale surface
irregularities at the edge of a mirror having a completely uniform
surface reflectivity can also scramble the net diffraction effect,
as noted in a review by Anan'ev (1972).

Ideally Aligned Resonators

With these practical considerations in mind, let us first
examine ideally-aligned cavities in the ray approximation.
Following Siegmann (1965) the field inside the resonator will be
regarded as the superposition of two spherical waves travelling
in opposite directions and obeying the laws of geometrical optics;
the waves transform into one another upon reflection at either of
the spherical mirrors. In particular, the centre of curvature of
the wave leaving, mirror 1 coincides with the image of the centre
of curvature of the second wave in mirror 1, and vice versa.
Figure 15(a) illustrates an identically equivalent resonator,
formed from a thin lens of focal length f and two plane mirrors
(1 and 2). In this case the condition for mutual imaging is
clearly given by

$$\frac{1}{\ell_1 + x} + \frac{1}{\ell_2 - y} = \frac{1}{f} \tag{2.52}$$

and,

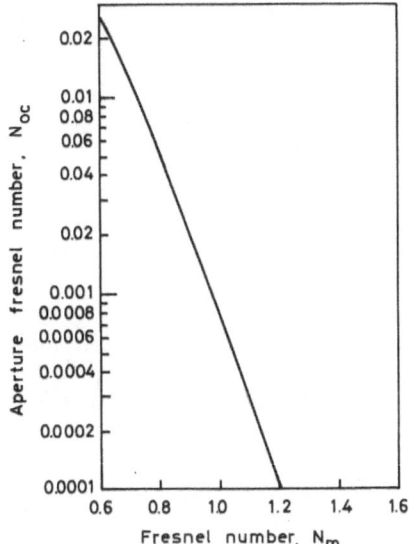

Fig 14. Critical aperture Fresnel number N_{oc} for which losses of (00) mode equal those of (10) mode versus mirror Fresnel number N_m (after McCumber 1965).

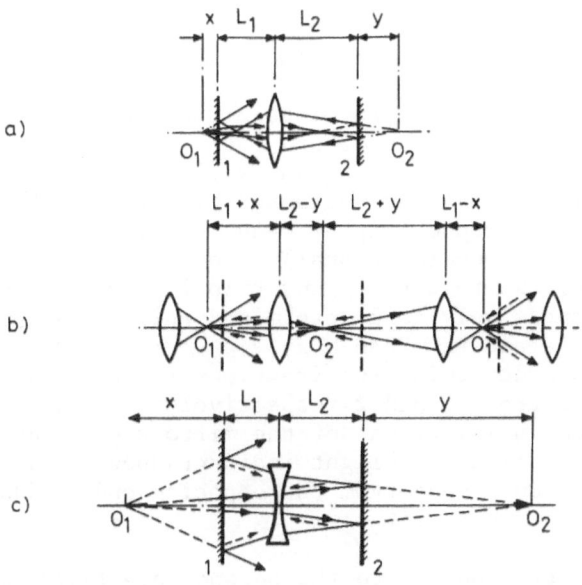

Fig 15. Paths of rays in an unstable resonator consisting of two plane mirrors (1 and 2) and a lens: a) resonator with a positive lens; b) system equivalent to that shown in (a); c) resonator with a negative lens.

$$\frac{1}{\ell_2 + y} + \frac{1}{\ell_1 - x} = \frac{1}{f} \tag{2.53}$$

The solution of these equations is

$$x_{1,2} = \pm\sqrt{\frac{\ell_1 - f}{\ell_2 - f}} \{\ell_1\ell_2 - f(\ell_1 + \ell_2)\}; \; y_{1,2} = \{\frac{\ell_2 - f}{\ell_1 - f}\} \cdot x_{1,2} \tag{2.54}$$

For convenience, let us choose $\ell_2 \geq \ell_1$, then the values of x and y are real if either $f < \ell_1\ell_2/(\ell_1 + \ell_2)$ or $\ell_1 < f < \ell_2$. Figure 15(a) shows the ray paths for a converging lens, $0 < f < \ell_1\ell_2/(\ell_1 + \ell_2)$ and Figure 15(c) for a diverging lens, $f < 0$.

The solution of (2.54) for x and y positive is represented by the continuous arrows, and that for x and y negative by the dashed ray paths. The first solution represents a situation in which the linear cross-section increases by the factor $M' = [(\ell_1 + x_1)/x_1]$ $[y_1/(\ell_2 - y_1)]$ between mirrors 1 and 2, and by the factor $M'' = [(\ell_2 + y_1)/y_1][x_1/(\ell - x_1)]$ between mirrors 2 and 1. Rearranging, the transverse dimension of the beam increases by a factor

$$M = M'M'' = \frac{1 + [1-f^2/(\ell_1 - f)(\ell_2 - f)]^{\frac{1}{2}}}{1 - [1-f^2/(\ell_1 - f)(\ell_2 - f)]^{\frac{1}{2}}} \tag{2.55}$$

during a round trip from mirror 1 to 2 and back to 1. Eventually, of course, radiation escapes around one or both of the mirrors, because of their finite radii a_1 and a_2. In a medium of zero gain the radiation intensity thus decreases by a factor M^2 during a (three-dimensional) round trip between spherical mirrors, and by a factor M during a (two-dimensional) round trip between cylindrical mirrors. It follows that the round trip losses, which are $(1-M^{-2})$ and $(1-M^{-1})$ respectively, are determined entirely by the resonator parameters ℓ_1, ℓ_2 and f. When $M'a_1 > a_2 > a_1/M''$, the mirror dimensions merely determine the fractions $(a_1/M''a_2)$ and $(a_2/M'a_1)$ reflected from mirrors 1 and 2 respectively. If this inequality is not satisfied, however, one of the mirrors will be sufficiently large to reflect all of the light incident upon it, and all of the radiation will then escape from the opposite end of the resonator, as illustrated in Figure 16.

The beam corresponding to the second (negative) solution of (2.54) decreases in cross-section by the identical factor M during its round trip within the resonator. This radiation cone would converge indefinitely in the ray approximation, but it can be shown (Siegmann 1965) that the wave is unstable to perturbations caused, for example, by diffraction. Thus a converging wave is

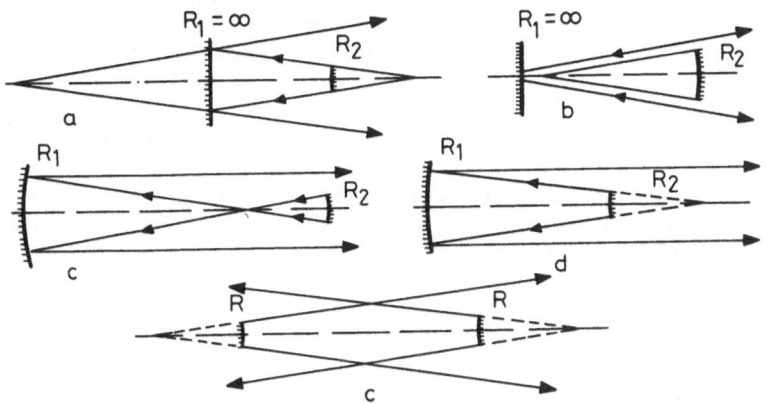

Fig 16. Unstable resonators: a) plane and convex mirror combination; b) plane and concave combination ($L > |R_2|$); c) asymmetrical confocal resonator formed by concave mirrors; d) confocal resonator formed by one concave and one convex mirror; e) symmetrical resonator formed by convex mirrors.

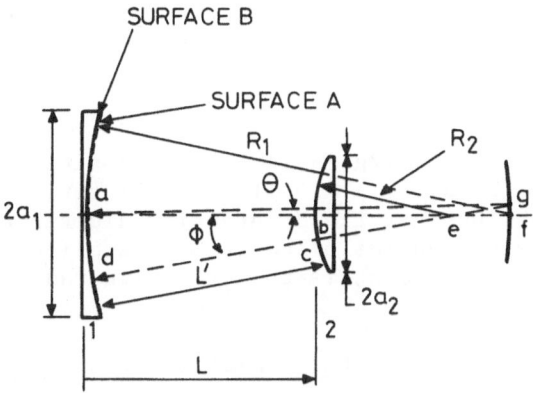

Fig 17. Geometry of (misaligned) spherical mirror resonator. R_1 (of aperture $2a_1$) is positive, and R_2 (of aperture $2a_2$) is negative, as shown.

eventually converted within the resonator into a stable diverging
wave of the type considered above. However, it should be
particularly noted that when the resonator has large transverse
dimensions, and contains a strongly amplifying medium, it is sometimes possible for a converging beam to grow at the expense of the
stable, diverging wave. (The source of the initial, converging,
perturbation may be reflection at an interface, volumetric scattering, or the diffraction at mirror edges discussed previously).

The power-loss for the generality of symmetric ($g_1 = g_2 = g$)
and half-symmetric ($g_1 = 1$, $g_2 = g$) resonators has been discussed
by Siegmann (1965). Of the various configurations illustrated in
Figure 16, probably the most widely used are the asymmetrical confocal resonators 16(c) and (d), since they both provide a one-sided,
and parallel, output beam. We shall therefore discuss these
exclusively, to avoid some tedious and cumbersome equations.
These resonators satisfy the confocality condition

$$R_1 + R_2 = 2L \qquad (2.56)$$

The positive-branch confocal resonator, for which g_1 and g_2 are
both positive, possesses mirrors of opposite curvature and has the
significant advantage of a quasi-cylindrical mode volume. The
negative-branch resonator on the other hand has two mirrors of
positive curvature, but either g_1 or g_2 is negative; it therefore
has a real internal focus, which can be a very real objection to
its use in lasers for which the cavity intensity becomes sufficiently high to give break-down problems. It follows from (2.55)
and (2.56) that the round-trip power loss of the lowest-order
geometric mode in unstable resonators formed from spherical mirrors
can be written

$$\Pi = 1 - M^{-2} \qquad (2.57)$$

where $M = (1 - 2L/R_1)^{-1} \equiv (2g_1 - 1)^{-1}$ for a positive branch
resonator and $M = (2L/R_1 - 1)^{-1} \equiv (1 - 2g_1)^{-1}$ for a negative
branch resonator. The near-field intensity distribution is a
uniformly illuminated annulus of constant phase (Barone 1967) having
an outer radius Ma_2 (where a_2 is the effective radius of the smaller
mirror).

Sensitivity to Variations in the Geometric Parameters

The dependence of the output coupling upon geometric parameters
of the resonator can be investigated very straightforwardly, using
first-order variational techniques. For example, Krupke and Sooy
(1969) find

$$\frac{d\Pi}{dL} = 0 \tag{2.58}$$

$$\frac{d\Pi}{dR_1}\bigg|_{\pm} = \pm(M \mp 1)(M^3 L)^{-1} \tag{2.59}$$

$$\frac{d\Pi}{dR_2}\bigg|_{\pm} = \pm(M \mp 1)(M^2 L)^{-1} \tag{2.60}$$

where the plus or minus sign hereafter denotes the positive or negative branch configuration respectively.

Similarly, if $(\delta i/\lambda)$ represents the extent of the departure from a plane between the centre and edge of a mirror of curvature R_i, in wavelengths (λ), and $\Delta(\delta i/\lambda)$ that change in $(\delta i/\lambda)$ which corresponds to a change ΔR_i in its radius, then

$$\Delta(\delta i/\lambda) = \frac{-a_i^2 \Delta R_i}{2\lambda R_i^2} , \tag{2.61}$$

$$\Delta\Pi = \Delta(\delta_1/\lambda)_{\pm} . \quad 8(M \mp 1)^{-1}(MN_1)^{-1} \tag{2.62}$$

and $\quad \Delta\Pi \quad \Delta(\delta_2/\lambda)_{\pm} . \quad 8(M \mp 1)^{-1}(M^2 N_2)^{-1} \tag{2.62}$

Thus the output coupling from a negative branch resonator is more tolerant to spherical distortion than is the positive branch, whilst all large Fresnel-number systems are less critical to the number of fringes of distortion across the mirrors. (Note that the negative-branch configurations normally use mirrors of greater curvature and that these can usually be fabricated with greater precision).

The focusing of laser beams will be discussed in detail in lecture 3. However, we note here that for resonators subject to time variations, such as might be produced by intermittent thermal loading with a high-power laser beam, the resultant fluctuations in far-field focal intensity may become a considerable practical embarrassment. Suppose that we wish to maintain the axial intensity constant within 20% of its ideal value, Krupke and Sooy (1969) demonstrate that the following inequalities must be maintained:

$$\frac{\Delta L}{L}\bigg|_{\pm} < \frac{M^2}{4N_1(M \mp 1)^2} \tag{2.63}$$

$$\Delta(\delta_1/\lambda) < 1/16 \quad \text{(both branches)}$$

$$\Delta(\delta_2/\lambda) \quad < 1/16 \quad \text{(both branches)} \tag{2.65}$$

It follows that the positive-branch configuration is far less length-sensitive, but that all mirror-figures must be maintained to 1/16 of a wavelength.

Effects if Mirror Misalignment

Figure 17 illustrates a misaligned mirror configuration of arbitrary geometry. When perfectly aligned the optic axis (which joins the centres of curvature of the two mirrors e,f) also intersects the physical centres a,b of the mirrors. If mirror 1 is now tilted around a by an angle θ, its centre of curvature moves to g, and the optic axis of the misaligned mirror system becomes g-e-b-c (making an angle Q with the original axis a.b.e.f.). The major effects of this misalignment are steering of the beam along the new direction g-e, and mode distortion. The angular sensitivity to beam steering, defined as (Q/θ), has been calculated by Krupke and Sooy (1969) to be

$$(Q/\theta)_\pm = \frac{2M}{(M \mp 1)} \tag{2.66}$$

Clearly positive-branch resonators having small output coupling $(M \sim 1)$ can be very sensitive to a small misalignment, θ. Figure 18 illustrates typical contours of constant (Q/θ) in the g_i plane; note that $Q/\theta \to \infty$ along the curved boundaries between the stable and unstable regions. These boundaries delineate concentric resonators $g_1 g_2 = 1$ for which there is no unique optic axis.

Two critical angles can be identified as θ is slowly increased. The first critical angle, θ_1 occurs when no power is coupled out of one side of mirror 2, and the second, θ_2 occurs when the optic axis passes beyond the edge of one of the mirrors. For confocal resonators,

$$\theta_1 = (a_1 - a_2)/R_1 = \frac{2a_1}{L} [\frac{M - 1}{2M}]^2 \quad \text{(+ve branch)} \tag{2.67}$$

$$= \frac{2a_1}{L} [\frac{M^2 - 1}{4M^2}] \quad \text{(-ve branch)}$$

and

$$\theta_2 = a_1 L/R_1 R_2 \tag{2.68}$$

$$= - \frac{a_1}{L} [\frac{M - 1}{2M}]^2 \quad \text{(+ve branch)}$$

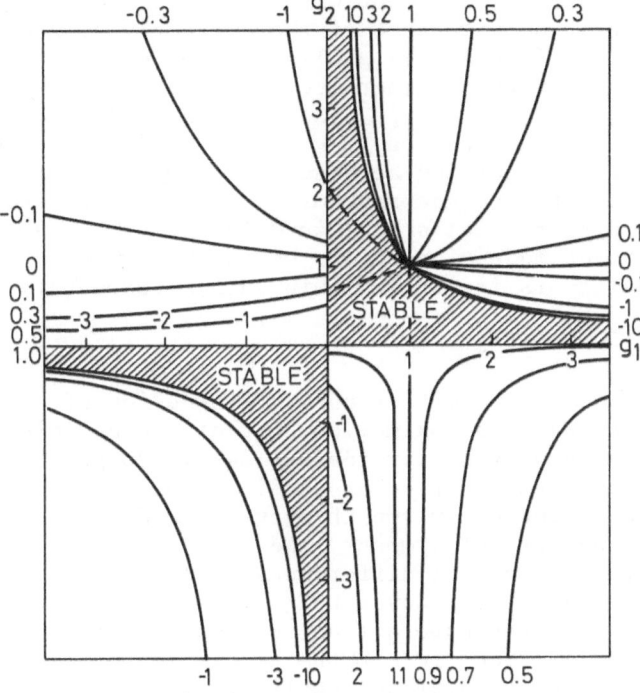

Parameter $M = (1 - g_2)/(1 - g_1 g_2)$

Fig 18. Contours of constant misalignment sensitivity (after Krupke and Sooy, 1969).

$$= + \frac{a_1}{L} [\frac{M + 1}{2M}]^2 \quad \text{(-ve branch)}$$

The positive branch is therefore more sensitive to misalignment than the negative branch, particularly for low coupling losses ($M \sim 1$). For angles exceeding θ_2 there exists no axis within the resonator having the original symmetry. If $\theta_2 > \lambda/2a_2$ the resonator will rapidly decrease in output power with increasing θ, but if $\theta_2 \ll \lambda/2a_2$ diffraction will ensure that the laser continues to oscillate when $\theta > \theta_2$, but with a mode structure exhibiting little correlation with that of the correctly aligned resonator. Krupke and Sooy (1969) have published some elegant experimental demonstrations of these general conclusions, and shown the need for high angular accuracy in practical mirror gimbal designs. In a typical low-magnification (M = 1.118) positive-branch resonator they found that a setting accuracy of < 100 μrad was necessary on the mirror mounts; an accuracy of $\sim$ 5 μrad was in fact achieved by the use of differential screw micrometers. (They also demonstrated the significant mode distortions which arise when a window of poor optical quality is inserted within such a resonator).

Summary

    Time does not permit extensive discussion of unstable resonator theory in the wave approximation, nor of unusual mirror geometries - such as those employing toroidal symmetry (Borghese 1980). Suffice it to note that the particular condition of the edge of the mirror (i.e. rough or optically smooth over the width of a Fresnel zone), and of its geometrical symmetry and accuracy of alignment, can make important practical differences to the coupling efficiency of an unstable resonator. Nevertheless, careful design of stable or unstable optical resonators (suitably matched to the geometry of the laser medium) does permit the generation of light sources having uniquely-low, diffraction-limited, divergence. How, then, can we optimize their transmission to a target?

FOCUSING THE BEAM

    Equations (2.14), (2.15) and (2.43) describe the propagation of the 'ideal' Gaussian beam; see also Figures 3 and 4 of § 2.

    Confusion often arises when the focusing of such 'single-mode'

TEM$_{00}$ beams is discussed, and particularly when comparisons are
made with annular (unstable cavity) profiles. Since neither the
laser medium nor the focusing lens is infinite in diameter, a
'Gaussian' profile is inevitably truncated in practise. We therefore propose to discuss this as a particular example of a more
general problem: Fraunhofer diffraction of a circular aperture
illuminated by a plane wave having an axisymmetric but radially nonuniform intensity distribution. The em field amplitude $\mu(\rho)$ of the
diffraction pattern at radius r' in the image plane is related to
the amplitude f(r) at radius r in the near field (the pupil
function) by the relation:

$$u(\rho) = \int_0^\infty f(r) \, Jo \, \frac{(\rho.r)r}{\sigma} \, dr \tag{3.1}$$

where $\rho = \frac{2\pi\sigma r'}{\lambda z} \{1 + 0(\frac{\lambda}{\sigma})^2\}$ is the (non-dimensional) radius in the
image plane, z = distance from aperture of outer radius σ to image
plane and Jo (x) is the Bessel function of order zero, vide.
standard texts on apodisation such as Born & Wolf (1970). Let us
consider (following Kaye, 1980):

i.   a uniformly illuminated aperture   f(r) = 1, r < σ

                                            = o, r > σ

ii.   an annular distribution   f(r) = 1, σM < r < σ

                                            = o, r < σM, r > σ

iii. truncated Gaussian   f(r) = c. exp$(-r/w)^2$, r < σ

                                            = o,   r > σ

Uniformly Illuminated Aperture

Here the intensity I(ρ) is given by:

$$I(\rho) = |u(\rho)|^2 = \sigma^4 \, [\frac{J_1 \, (\rho)}{(\rho)}]^2 \tag{3.2}$$

and $I(o) = \sigma^4/4$ \hfill (3.3)

The total power transmitted is:

$$P = 2\pi \int_0^\infty f^2(x).x.dx = \pi\sigma^2, \text{ as expected.}$$

Top-Hat Distribution

The total transmitted power

$$P = \pi\sigma^2 (1-M^{-2}) \tag{3.4}$$

From the Babinet principle, the far field amplitude (Selden 1979) is:

$$u(\rho) = \sigma^2 \frac{J_1(\rho)}{\rho} - \frac{\sigma^2}{M^2} \frac{J_1(\rho/M)}{(\rho/M)} \tag{3.5}$$

and hence the intensity on axis is given by;

$$I(o) = \frac{\sigma^4}{4} (1 - M^{-2})^2 \tag{3.6}$$

Kaye notes that an identical relationship exists for any function $f(r) = g(r) - g(Mr)$ e.g. for a Gaussian distribution obscured by an inverted Gaussian hole.

Truncated Gaussian Distribution

$$u(\rho) = c \int_0^\sigma \exp(-r^2/w^2) \, J_o \frac{(\rho r)}{\sigma} r.dr$$

Hence,

$$u(o) = c \frac{w^2}{2} [1 - \exp(-\sigma^2/w^2)] \tag{3.7}$$

The total transmitted power

$$P = 2\pi c^2 \int_0^\sigma \exp(-2r^2/w^2) r.dr$$

i.e. $$P = \pi \frac{c^2 w^2}{2} [1 - \exp(-2\sigma^2/w^2)] \tag{3.8}$$

Normalizing to unit power,

$$c = \{\frac{\pi w^2}{2} [1 - \exp(-2\sigma^2/w^2)]\} \tag{3.9}$$

And hence,

$$I_o \text{ (for unit power)} = \frac{w^2}{2\pi} \frac{[1 - \exp(-\sigma^2/w^2)]}{[1 + \exp(-\sigma^2/w^2)]} \qquad (3.10)$$

Comparison Between The Idealized Distributions

Normalizing to the same outer aperture and total transmitted power, it follows from (3.4) and (3.3) that the ratio of the axial intensity of an annular beam to one of uniform illumination is:

$$R_A = \frac{I(o) \text{ Top Hat}}{I(o) \text{ Uniform Aperture}} = (1 - M^{-2}) \qquad (3.11)$$

and from (3.10) and (3.4) that the corresponding ratio for a truncated Gaussian beam to one of uniform illuminations is:

$$R = \frac{I(o) \text{ Gaussian}}{I(o) \text{ Uniform Aperture}} = 2 \frac{w^2}{\sigma^2} \left[\frac{1 - \exp(-\sigma^2/w^2)}{1 + \exp(-\sigma^2/w^2)}\right] \qquad (3.12)$$

R_A and R are tabulated in Table 3.1, where it is seen that,

i.   For equal transmitted power, the far-field axial intensity increases with the severity of truncation of the Gaussian.
ii. A Gaussian beam truncated at the $(1/e^2)$ intensity points gives the same far-field axial intensity as an M=3 annular beam.

Table 3.1 Far-field axial intensity ratio of Gaussian to uniform aperture (R) and of top-hat to uniform distribution (R_A), both normalized to the same transmitted power and outside diameter (σ).

Gaussian near-field		Top-Hat near-field	
σ/a	R	M	R_A
0	1	∞	1
1	0.9	4	0.94
1.5	0.72	3	0.89
2	0.48	2	0.75
3	0.22	1.5	0.56
∞	0	1.0	0

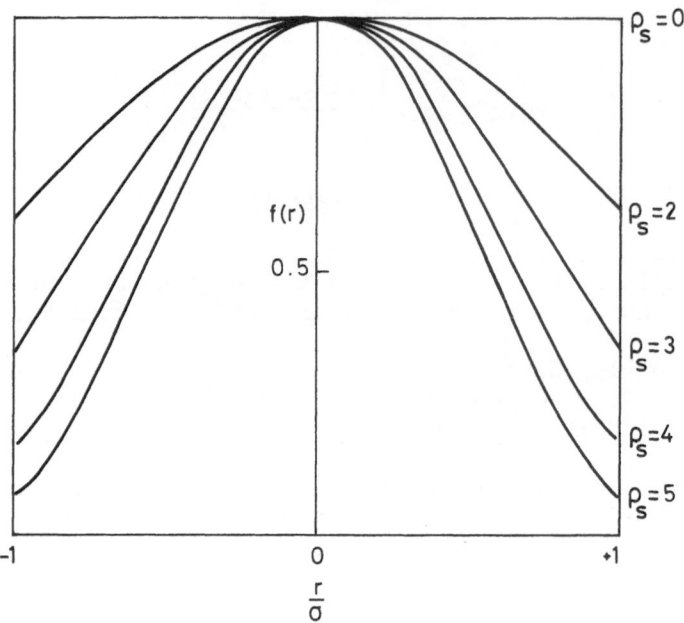

Fig 1. Pupil functions f(r) giving maximum power enclosed within ρs
at the focus (after Lansraux and Boivin 1961).

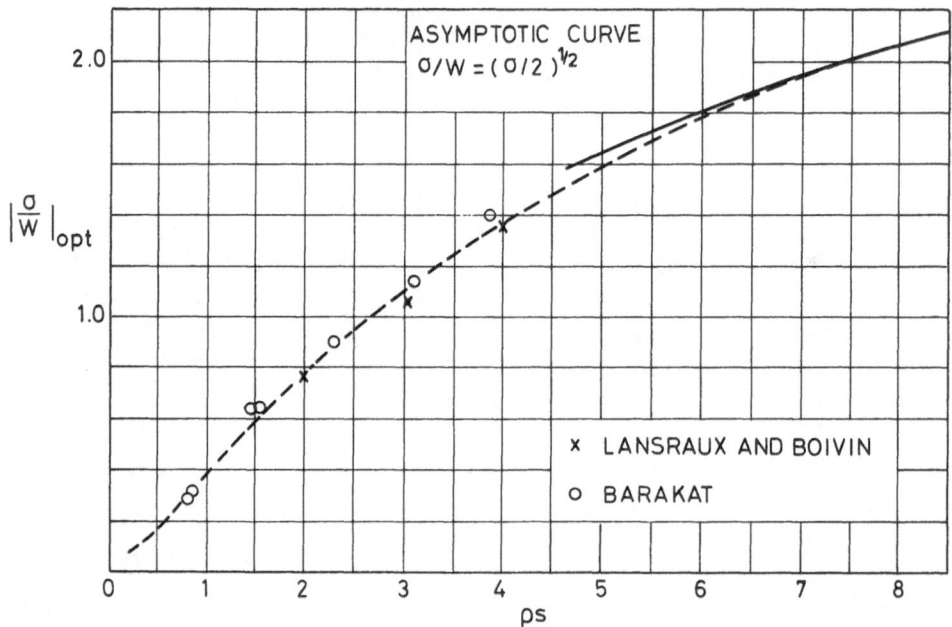

Fig 2. Optimum truncation (σ/W) opt. of a Gaussian beam as a
function of ρs (after Kaye 1980).

Alternative Optimising Criteria

It is not self-evident that peak intensity on axis is the most important criterion for optimisjng particular machining operations. As a specific example, for heat-treatment one might desire a uniform 'top-hat' distribution in the image-plane: this suggests a pupil-function

$$f(r) = \frac{J_1(r)}{r}$$

extending to infinity. However, for finite near-field apertures careful optimization of this pupil-function would then clearly be necessary, and a more convenient engineering solution might be to raster a tightly-focused spot across the area. This second approach utilizes the thermal capacity of the target to 'integrate' the incident flux. (A third approach is to use segmented focusing mirrors).

For cutting or welding one might wish to optimize the proportion $\varepsilon(\rho s)$ of the total power focused within a specified radius ρs in the image plane, i.e. to maximise:

$$\varepsilon(\rho s) = \int_0^{\rho s} 2\pi I(\rho)\rho d\rho / \int_0^{\infty} 2\pi I(\rho)\rho d\rho \qquad (3.13)$$

To determine the appropriate pupil-function it is usual to make a series-expansion approximation of the form

$$f(r) = \sum_{n=0}^{N} a_n g_n(r),$$

where $g_n(r)$ are functions whose coefficients (a_n) are then determined numerically (Lansraux and Boivin 1961, Barakat 1962). More recently, Slepian (1965) has presented analytical results. Typical results are summarized in Figure 1: no single pupil-function maximises the enclosed power for all values of ρs, but a truncated Gaussian (dotted curve) is a convenient approximation for large ρs. The equivalent e-folding radius, w, has been computed by Kaye (1980); it is plotted (as the function σ/w) against ρs in Figure 2. It will be seen that the pupil function maximising $\varepsilon(\rho s)$ changes continuously from an infinite Gaussian for large ρs, through an increasingly truncated Gaussian, to a uniform distribution for small ρs. The far-field intensity distributions corresponding to uniform near-field illumination, a truncated Gaussian optimized for $\rho s = 3.83$ ($\varepsilon = 97\%$ at $\sigma/w = 1.3$), and an M = 2 annular distribution are compared in Figure 3. Numerical integration (Ward 1980) of the power enclosed within a far-field of radius ρs for the annular beam are summarized in Figure 4, for comparison with (optimized) truncated Gaussians, and with the Airy function ($\sigma/w = 0$).

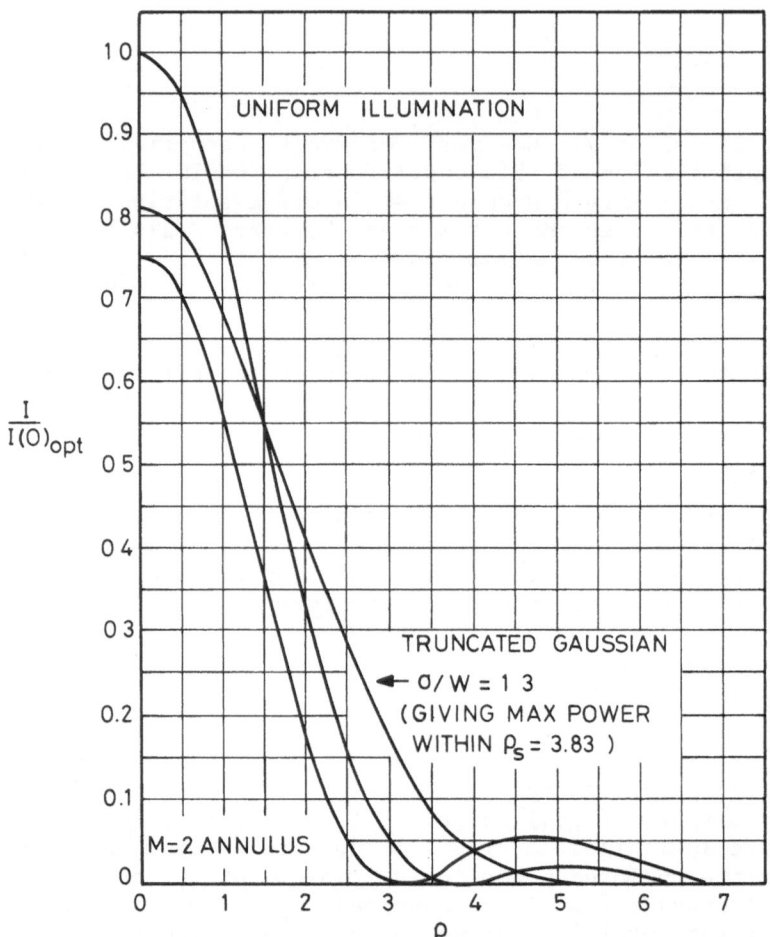

Fig 3. Focal intensity distribution for uniform illumination,
        truncated Gaussian (σ/W = 1.3), and annular (M = 2) beams
        as a function of ρ. (All beams have equal power and outer
        diameter.)

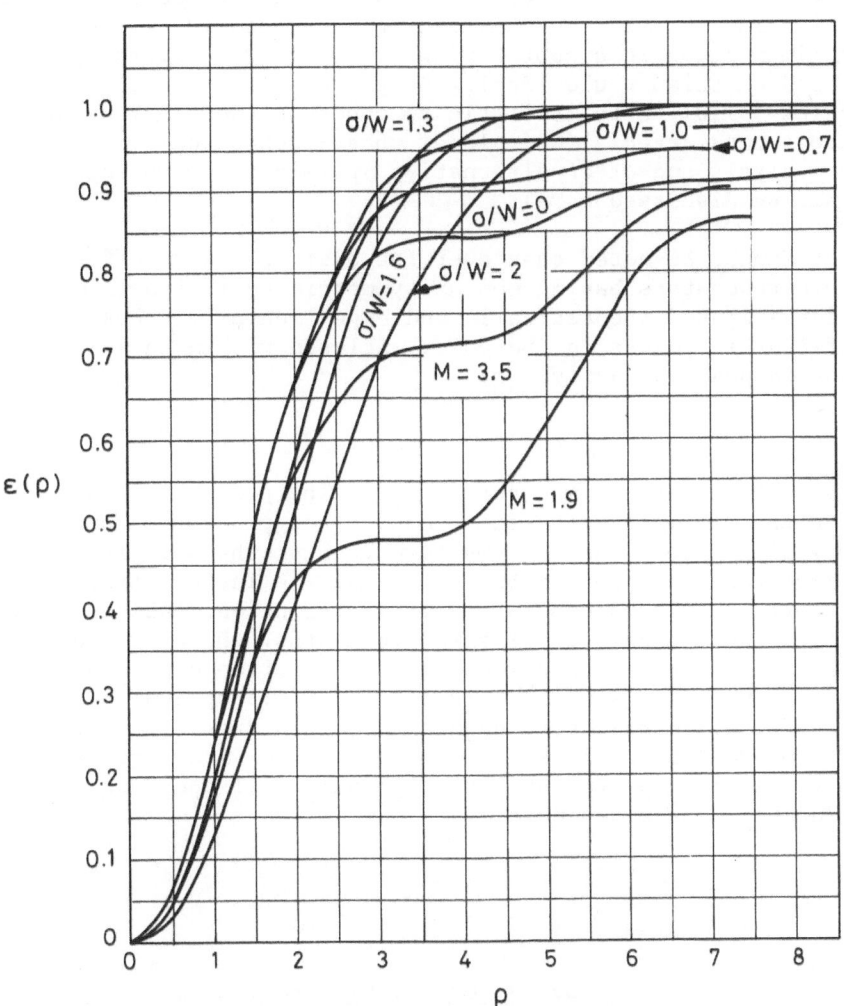

Fig 4.   Total power enclosed within radius ρ at the focus, for
         various pupil functions (Kaye 1980).

Optimization of the pupil function for a given application is clearly, then, dependent on the choice of ρs - a choice which is normally governed by experimental experience. However, it seems clear that for applications requiring maximum intensity within a small spot ($\rho s \lesssim 2$) a uniform near-field distribution is to be preferred, with little to choose between annular beams of modest magnification ($M \gtrsim 2$) and Gaussian beams truncated at $\sigma/w \lesssim 1.3$. For maximum power in a spot having $2 < \rho s < 4$, a carefully-optimized Gaussian would ideally be preferable. Finally for large spots ($\rho s > 10$) there is little to choose between any of the pupil functions considered. (Non-planar phase-fronts would modify the above analysis and offer alternative optimization procedures, but will not be discussed here).

It should be noted that over the past few years off-axis unstable resonators having non-axisymmetric pupil functions have been investigated theoretically and experimentally: these offer significant increases in the power enclosed within an Airy disc, but not in peak intensity.

Conclusion

For the practical man actual machining performance will always provide the ultimate test, since many other process parameters (e.g. the use of gas jets, and plasma formation at the workpiece) are also very important. It should be noted that the latter effect absorbs or distorts the in-vaccuo intensity distribution, and occurs more readily at the longer (IR) laser wavelengths. However, burn-patterns in perspex do provide a quick and often used qualitative test of a CO_2 laser's near-field and far-field optical performance. Such tests, of course, also demonstrate the adverse effects of any lens or mirror aberrations, and the departures from scalar theory which should be expected at very high numerical aperture (i.e. as the f/number approaches unity).

REFERENCES (Lectures 1-3)

Anan'ev, Yu.A., 1972, Sov. J. Quant. Electron. 1, 565.
Armandillo, E. and Kaye, A.S., 1980, J. Phys. D:Appl. Phys. 13, 321.
Barakat, R., 1962, J. Opt. Soc. Am. 52, 264.
Barone, S.R., 1967, Appl. Optics 6, 861.
Bergstein, L., 1968, Appl. Optics 7, 495.
Bergstein, L. and Schachter, H., 1965, J. Opt Soc. Am. 55, 1226.
Bertolotti, M., 1964, Nuovo Cimento 32, 1242.
Borghese, A., Canevari, R., Garifo, L., Pandarese, F., 1980, Paper
        WEE 6, CLEOS '80, Opt. Soc. America.
Born, M., and Wolf, E., 1970, Principles of Optics, (4th Edn.)
        Pergammon Press, Oxford.

Boyd, G.D., and Gordon, J.P., 1961, Bell Syst. Tech. J., 40, 489.
Boyd, G.D., and Kogelnik, H., 1962, Bell Syst. Tech. J., 41, 1347.
Brower, W., 1964, Matrix Methods in Optical Instrument Design,
        Bemjamin, New York.
Charschan, S.S., 1972, Lasers in Industry, (Ed) Van Nostrand
        Reinhold, New York.
Cochran, J.A., 1965, Bell Syst. Tech. J. 44, 77.
De Maria, A.J., 1973, Proc. IEEE 61, 731.
Duley, W.W., 1976, CO_2 Lasers, Effects and Applications, Academic
        Press, New York.
Fox, A.G., and Li, T., 1960, Proc. I.R.E. 48, 1904.
Fox, A.G., and Li, T., 1961, Bell Syst, Tech. J. 40, 453.
Fox, A.G., and Li, T., 1963, Proc. IEEE 51, 80.
Goubau, G., and Schwering, F., 1961, I.R.E. Trans. on Antennas and
        Propagation, Vol. AP-9, 248.
Heurtley, J.C., 1964, Proc. Symp. on Quasi Optics, pp367-375,
        Polytechnic Press, New York.
Heurtley, J.C., and Streifer, W., 1965, J. Opt. Soc. Am. 55, 1472.
Hochstadt, H., 1966, S.I.A.M. Review, 8, 62.
Jenkins, F.A., and White, H.E., 1976, Fundamentals of Optics,
        McGraw-Hill, New York.
Kaye, A.S., 1980, Private Communication, to appear as Culham Report.
Kogelnik, H., 1965, Appl. Optics, 4, 1562.
Kogelnik, H., 1965, Bell Syst. Tech. J. 44, 455.
Kogelnik, H., and Li, T., 1966, Appl. Optics, 5, 1550.
Krupke, W.F., and Sooy, W.R., 1969, IEEE J.Q.E. QE-5, 575.
Lansraux, G., and Boivin, G., 1961, Can. J. Phys. 39, 158.
Li, T., 1965, Bell Syst. Tech. J. 44, 917.
Luneberg, R.K., 1964, Mathematical Theory of Optics, U.C. Press,
        Berkeley.
McCumber, D.E., 1965, Bell Syst. Tech. J. 44, 333.
Maitland, A., and Dunn, M.H., 1969, Laser Physics, North Holland,
        Amsterdam.
Newman, and Morgan, 1964, Bell Syst. Tech. J. 43, 113.
Schwering, F., 1961, Arch. Elect. Ubertrag. 15, 555.
Schiff, L.I., 1955, Quantum Mechanics, (2nd Edn.) McGraw-Hill, New
        York.
Selden, A.C., 1979, Culham Laboratory Report CLM-R190.
Siegman, A.E., 1965, Proc. IEEE, 53, 277.
Slepian, D., 1964, Bell Syst. Tech. J. 43, 3009.
Slepian, D., 1965, J. Opt. Soc. Am. 55, 1110.
Somerfeld, A., 1954, Optics, Academic Press, New York.
Streifer, W., 1965, J. Opt. Soc. Am. 55, 868.
Svelto, O., 1976, Principles of Lasers, Plenum Press, New York.
Vainshtein, L.A., 1963, Sov. Phys. J.E.T.P., 17, 709.
Ward, B.A., 1980, Private Communication.
Weedon, T.M., 1980, Institute of Physics Meeting, Culham (15th May).
Willis, J.B., 1977, Laser Processing of Engineering Materials,
        (Seminar sponsored by Institute of Physics, London, 30th
        June 1977).

COUPLING OF LASER RADIATION TO METALS AND SEMICONDUCTORS

Martin F. von Allmen

Institute of Applied Physics
University of Bern
CH-3012 Bern, Switzerland

INTRODUCTION

Intense laser beams are currently being used in a whole
range of new applications - from growing crystals to forming
metallic glasses, and from depositing films to purifying
surfaces   . All of these actions are ultimately produced
by heat into which a smaller or larger part of the beam energy
is transformed by various coupling mechanisms. These mechanisms
determine not only the amount of heat created, but also its
spatial and temporal distribution.

Laser radiation [2] , in addition to its high intensity,
has some features not found in ordinary thermal radiation: It is
highly directional, and it is coherent, i.e. capable of self-
interference. The directionality enables the experimentalist to
concentrate the beam energy with the aid of lenses to a spot of
any size, down to little more than one wavelength. Self-
interference is the basis of Holography, but rather detrimental
in laser processing: Mutual interference of scattered parts of a
beam cause the intensity to vary from point to point on an
irradiated surface, creating problems of lateral nonuniformity.

This lecture reviews the fundamental mechanisms of interaction of laser beams with condensed matter from the point
of view of energy depostion. The starting point of our discussion
is classical optics which gives the "zero-order" approximation
to all coupling phenomena. We then outline various mechanisms
which lead to modifications of the optical properties at high
light intensities. Finally we discuss some practical aspects of
sample heating with laser beams.

1. Optical properties of materials

   1.1. General. Electromagnetic radiation with wavelengths
ranging from uv to ir interacts exclusively with electrons,
since atoms are too heavy to respond significantly to the high
frequencies ($\nu > 10^{13}$ Hz) of the electromagnetic field.
Therefore, the optical properties of matter are primarily
determined by the energy states of its valence electrons (bound
or free). Generally, bound electrons respond only weakly to the
external electromagnetic wave and mainly affect its phase
velocity. Free electrons are able to be accelerated and to
extract energy from the field more efficiently. However, since
the external field is periodically changing its direction, the
oscillating electrons re-radiate their kinetic energy unless
they undergo frequent collisions with the atoms. In this case,
energy is transmitted to the lattice and the external field is
weakened. Re-radiation of energy is the cause of reflection.

   1.2. Absorption and Dispersion. The general features of
linear interaction of light with solid matter can be obtained in
a rather elementary way by modeling the solid as a set of
harmonic oscillators under the action of an external, periodically
changing force field (Lorentz-model). Consider a bound electron
in a solid irradiated by a laser beam of angular frequency
$\omega = 2\pi c/\lambda$. The position r of the electron of mass m due to the
electrical field is described by

$$m\, \ddot{\vec{r}} + m\Gamma \dot{\vec{r}} + m\omega_o^2 \vec{r} = -e\, \vec{E}_{loc}. \tag{1.1}$$

Here $m\omega_o^2 \vec{r}$ is a Hooke's law restoring force and $m\Gamma \dot{\vec{r}}$ is a viscous
damping term. We have neglected the finite mass of the nuclei
(otherwise we would have written m_{eff} instead of m), as well
as the small force due to the magnetic field. E_{loc} denotes the
local electrical field as experienced by the electron. If we
take the local field to vary in time like $\exp(-i\omega t)$, the solution
of (1.1) is

$$\vec{r} = \frac{-e/m}{(\omega_o^2 - \omega^2) - i\Gamma\omega} \, \vec{E}_{loc} \tag{1.2}$$

It is important to realize, that the local field is due to
the external field $\vec{E}_{ext}$, as well as to the motion of the other
electrons in the solid. What we need to describe the optical
properties is a connection between $\vec{E}_{loc}$ (or some averaged
version of it) to the external field. To find this connection we
proceed as follows.

   The motion of the electron induces a dipole moment $\vec{p} =$

$- \vec{r} \ e \sqrt{4 \pi \varepsilon_o}$, ε_o being the dielectric constant[*]. For small excursions $\vec{r}$, $\vec{p}$ is a linear function of the field, i.e. $\vec{p} = a \ \vec{E}_{loc} \sqrt{4 \pi \varepsilon_o}$. This defines the atomic polarizability

$$a(\omega) = \frac{e^2/4 \pi \varepsilon_o \cdot m}{(\omega_o^2 - \omega^2) - i \ \Gamma \omega} \qquad (1.3)$$

Equation (1.3) gives the microscopic polarizability pertinent to the local field. One can define a macroscopic polarisation $\vec{P}$ by

$$\vec{P} = N \ a \ \varepsilon_o \langle \vec{E}_{loc} \rangle = \chi_e \varepsilon_o \vec{E}_{int} \qquad (1.4)$$

where N denotes the density of electrons and χ_e is the electric susceptibility of the material. $\langle \vec{E}_{loc} \rangle$ is obtained from the local field by averaging over the atomic site, while $\vec{E}_{int}$ is an average over the whole crystal, i. e. also over the spaces between atomic sites. The relation between $\langle \vec{E}_{loc} \rangle$ and $\vec{E}_{int}$ therefore depends on the structure of the material. For a free electron metal one may set $\langle \vec{E}_{loc} \rangle = \vec{E}_{int}$, but for insulators and semiconducors local field corrections have to be made which can be quite complicated [3]. For the present general discussion it is however sufficient to assume $\langle \vec{E}_{loc} \rangle = \vec{E}_{int}$.

As known from classical electrodynamics, one can write the external field $\vec{E}_{ext}$ as a superposition of the macroscopic field $\vec{E}_{int}$ plus the induced macroscopic polarisation,

$$\vec{E}_{ext} = \vec{E}_{int} + 4 \pi \ \vec{P} / \varepsilon_o = \vec{E}_{int} (1 + 4 \pi \ Na) = \varepsilon \ \vec{E}_{int}, \qquad (1.5)$$

where ε is the complex dielectric function, given by

$$\varepsilon = 1 + \frac{Ne^2/m \ \varepsilon_o}{(\omega_o^2 - \omega^2) - i\Gamma\omega} \equiv \varepsilon_1 + i \ \varepsilon_2 \qquad (1.6)$$

which provides the needed connection between $\vec{E}_{ext}$ and $\vec{E}_{int}$. Expression (1.6) pertains to a one-electron atom, but is easily generalized for atoms with many electrons, by replacing the polarizability a as given in (1.3) by a sum over all electrons, with N, ω_o and Γ replaced by the respective densities, resonance frequencies and damping constants. We note that a quantum mechanical treatment of light absorption leads to a very similar formula [3]

$$\varepsilon = 1 + \frac{Ne^2}{m \ \varepsilon_o} \sum_j \frac{f_j}{(\omega_j^2 - \omega^2) - i\Gamma_j \omega} \qquad (1.7)$$

---

[*]If not otherwise indicated, practical units are used.

In (1.7), the resonance frequencies correspond to differences $\hbar\omega_j$ between energy levels, and as additional factors the "oscillator strenghts" f_j are introduced, which account for the respective transition probabilities.

   Proceeding now with the simple formula (1.6), one obtains for the real and imaginary parts of the dielectric function

$$\varepsilon_1 = n^2 - k^2 = 1 + \omega_p^2 \; \frac{\omega_o^2 - \omega^2}{(\omega_o^2 - \omega^2) + \Gamma^2\omega^2} \tag{1.8}$$

$$\varepsilon_2 = 2nk = \omega_p^2 \; \frac{\Gamma\omega}{(\omega_o^2 - \omega^2) + \Gamma^2\omega^2}$$

where we have used

$$\omega_p^2 = Ne^2/m \, \varepsilon_o \tag{1.9}$$

For reasons to become more obvious later, ω_p is called the plasma frequency. Further, n and k are the complex refractive index and the extinction coefficient, respectively.

$$n^2 = \frac{1}{2} \; (\sqrt{\varepsilon_1^2 + \varepsilon_2^2} + \varepsilon_1) \tag{1.10}$$

$$k^2 = \frac{1}{2} \; (\sqrt{\varepsilon_1^2 + \varepsilon_2^2} - \varepsilon_2)$$

   We have thus obtained the macroscopic quantities that desribe the linear interaction of the light wave with the bound electrons in a solid. If I_o (W/cm^2) is the intensity of the light wave incident on the solid, then the power density Φ (W/cm^3) absorbed at depth z inside the material is given by

$$\Phi(z) = I_o \; (1-R) \; \alpha \; e^{-\alpha z} \tag{1.11}$$

where the reflectivity R and the absorption coefficient α are related to n and k by

$$R = \frac{(n-1)^2 + k^2}{(n+1)^2 + k^2} \tag{1.12}$$

$$\alpha = 2\omega k/c = 4\pi k/\lambda$$

Sometimes the absorption length $d = 1/\alpha$ is used instead of the absorption coefficient.

The pairs of parameters ε_1 and ε_2, n and k, and R and α are equivalent in describing the optical properties of the solid. Figures 1 - 3 show the frequency dependence of the three sets of quantities, as described by expressions (1.8), (1.10) and (1.12).

In metals, the optical properties are dominated by free electrons. An adequate description is easily obtained from the above Lorentz-model by setting the restoring force term and hence the resonance frequency ω_o equal to zero (Drude-model). Since the dominant damping mechanism in metals are electron-lattice collisions (an electron undergoing a collision, in the average, every t_c seconds), we also set $\Gamma = 1/t_c$. The real and imaginary parts of the dielectric function then become

$$\varepsilon_1 = n^2 - k^2 = 1 - \frac{\omega_p^2 t_c^2}{1 + \omega^2 t_c^2} \qquad (1.13)$$

$$\varepsilon_2 = 2nk = \frac{\omega_p^2 t_c}{\omega(1 + \omega^2 t_c^2)}$$

The dependence of ε_1 and ε_2, n and k, and R and α on light frequency for this case are illustrated in Fig. 4 - 6. Note that for $\omega = \omega_p$ and assuming $\omega^2 t_c^2 \gg 1$, (usually satisfied in the uv) we have $\varepsilon_1 \cong 0$, and $n^2 \cong k^2 \cong 0$. As evident from Fig. 6 and also Fig. 3, R and α start to decrease when the light frequency exceeds ω_p. On the other hand, the vanishing refractive index indicates that at ω_p the phase velocity $c = c_o/n$ of the light inside the material is infinite. This means that all electrons are oscillating in phase. At even higher frequencies, the interaction of the light with the electron plasma becomes weaker and the material becomes transparent.

In the other limit, at light frequencies small compared to the electron-lattice collision frequency $1/t_c$ we have $\omega^2 t_c^2 \ll 1$. By making use of the definition of the dc conductivity of a metal in the form

$$\sigma_o = Ne^2 t_c/m = \omega_p^2 t_c \varepsilon_o \qquad (1.14)$$

we can write $\varepsilon_1 \cong 1 - \sigma_o t_c/\varepsilon_o$ and $\varepsilon_\ell = \sigma_o/\varepsilon_o\omega$. By using (1.10) and (1.12) we obtain finally after some algebra

$$R = 1 - 2 \sqrt{2 \varepsilon_o \omega/\sigma_o} \qquad (1.15)$$

(1.15) is known as the Hagen-Rubens formula[4] and allows to calculate the ir reflectivity of a metal from its dc conductivity. It is fairly accurate for $\lambda \geq 10$ μm. The simple models discussed so far describe quantitatively all essential features of light coupling to solids at low intensities. In order to get

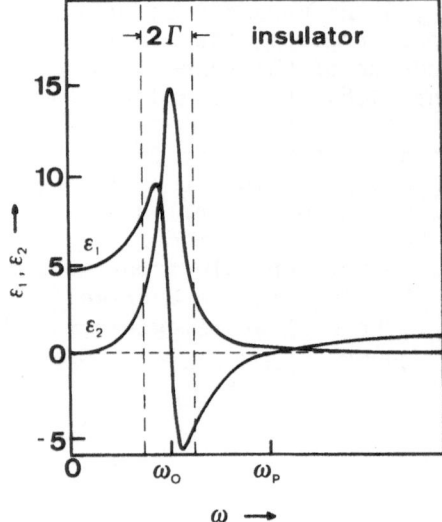

Fig. 1   Real- and imaginary
         parts of the dielectric
         function as a function
         of light frequency
         according to equ. (1.8)
         (schematic).

Fig. 2   n and k values
         corresponding to
         the data of Fig. 1.

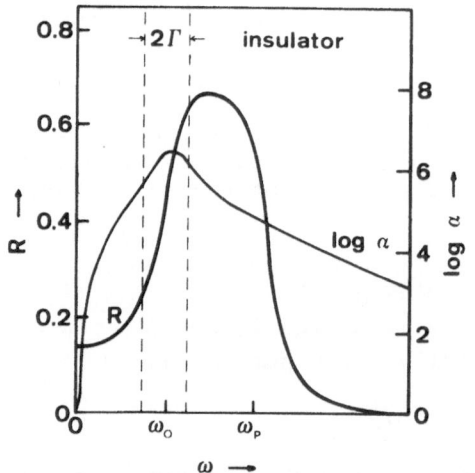

Fig. 3   R and α values corresponding to the data of Fig. 1.

(adapted, in part, from Ref. 3)

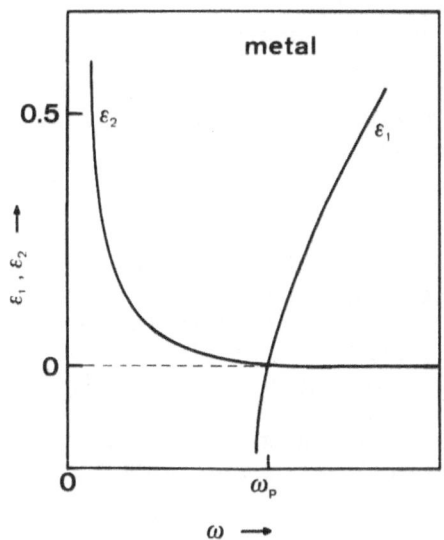

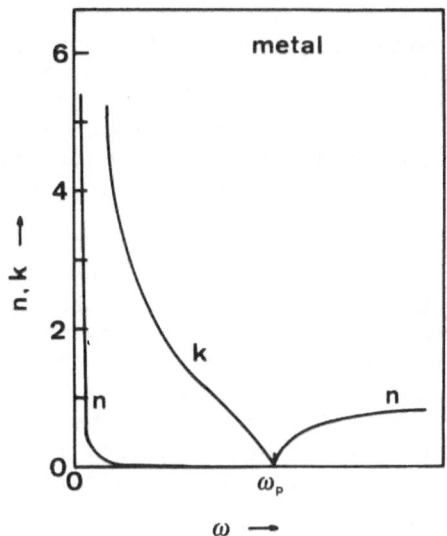

Fig. 4   Real- and imaginary
         parts of the dielectric
         function as a function
         of light frequency
         according to equ. (1.13)
         (schematic).

Fig. 5   n and k values
         corresponding to
         the data of Fig. 4.

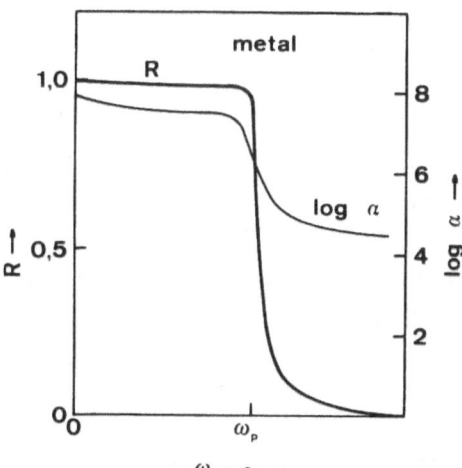

Fig. 6   R and α values corresponding to the data of Fig. 4.

(adapted, in part, from Ref. 3)

a quantitatively correct description, however, one has to use a
quantum mechanical formalism. Moreover, local field correction
have to be made in all but the simplest free electron metals.
Since such a general approach is far beyond the scope of this
introduction, we proceed by briefly looking at some empirical
data of real materials of interest.

   1.3. Metals. Metals, due to their large density of free
electrons have very small absorption lengths (in the order of
10 nm) over the whole optical spectrum. The reflectivity is high
above a critical wavelength which corresponds to the plasma
frequency ω_p (1.9). Below the critical wavelength λ_{cr}, which is in
the visible or uv part of the spectrum, the reflectivity decreases
sharply. As an example, Fig. 7 shows R and α for Al and Au at
room temperature. The critical wavelength of Al is below 100 nm,

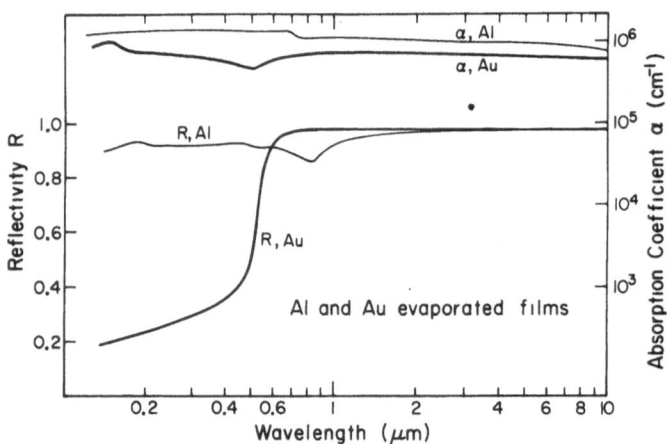

Fig. 7.   Absorption coefficient and reflectivity for Al and Au
          at room temperature

while that of Au is at about 600 nm (the yellow color of Au
is due to its low reflectivity in the green and blue).

   At longer wavelengths the reflectivity of all metals tends
towards unity, as predicted by (1.15). Values for R and d_5 of various
metals at selected laser wavelength are given in Table I [5] .

Table I: Optical absorption lengths and reflectivities of evaporated
metal films and crystalline semiconductors and insulators
at room temperature for various wavelengths [5].

Material	$\lambda = 0.25 \ \mu m$		$\lambda = 0.5 \ \mu m$		$\lambda = 1.06 \ \mu m$		$\lambda = 10.6 \ \mu m$	
	d	R	d	R	d	R	d	R
Al	8 nm	0.92	7 nm	0.92	10 nm	0.94	12 nm	0.98
Ag	0 nm	0.30	14 nm	0.98	12 nm	0.99	12 nm	0.99
Au	18 nm	0.33	22 nm	0.48	13 nm	0.98	14 nm	0.98
GaAs	6 nm	0.63	100 nm	0.39	67 um	0.31	> 1 m	0.28
Ge	7 nm	0.42	16 nm	0.49	200 nm	0.38	> 1 m	0.36
Si	6 nm	0.61	500 nm	0.36	200 um	0.33	1 mm	0.30
KCl	> 1 m	0.05	> 1 m	0.04	> 1 m	0.04	> 1 m	0.03
SiO_2	> 1 m	0.06	> 1 m	0.04	> 1 m	0.04	40 um	0.2

1.4. Semiconductors and insulators. In materials with a
completely filled valence band, photons can induce two kinds of
electronic transitions, namely intraband- and band-to-band
transitions.In principle, each of these transitions is described
by an expression of the form (1.6). Further, direct coupling to
high-frequency phonons is observed. If the photon energy $\hbar\omega \ll E_g$
(generally true for semiconductors in the infrared and for
insulators in the visible region) only intraband- and phonon-
absorption is possible. Coupling in this regime is weak. The
absorption lengths d range typically from cm to m, but may
decrease sharply as a function of intensity or temperature,
as discussed in section 2. Basically, this is an unfavorable
regime for beam heating.

When the photon energy approaches the gap energy, the
degree of absorption increases because band-to-band transitions
become available (corresponding to a resonance at $\hbar\omega_0 = E_g$
in (1.6)) At $\hbar\omega > E_g$ the optical response of insulators
and semiconductors tends towards that of metals. Fig. 8 shows
the situation for crystalline Si: For $\lambda < 1 \ \mu m$ the absorption
coefficient α is low, while R is still relatively high, due to

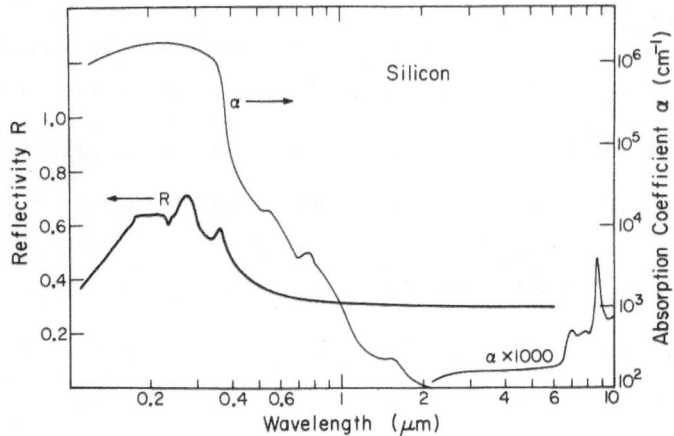

Fig. 8   Absorption coefficient and reflectivity of
crystalline Si at room temperature

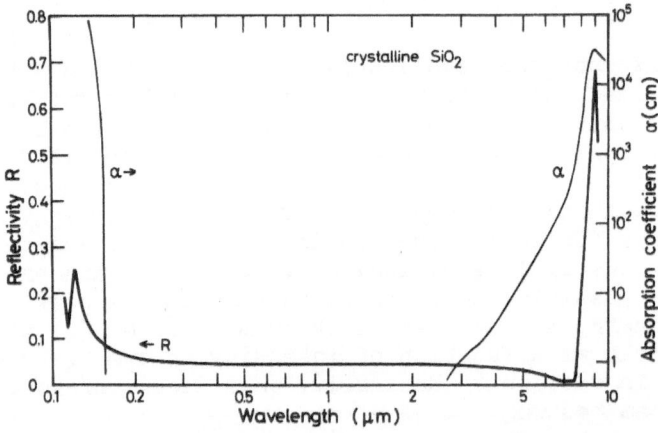

Fig. 9   Absorption coefficient and reflectivity of Quartz at room temperature

the large refractive index (n = 3.5 at 1 um). The weak
absorption band above λ = 6 μm is due to phonon absorption [6].
Below λ= 1 um both R and α increase because of free carrier
generation. In the near uv the reflectivity decreases again, as
the light frequency exceeds the free carrier plasma frequency,
in analogy to the behavior of metals.

Insulators such as oxides or ionic salts behave in similar
ways, except that the gap energies correspond to wavelengths
in the vacuum uv. The refractive indices and consequently
the reflectivities in the visible region are lower than those of
Si or Ge in the infrared, due to a smaller lattice polarizability.
As an example, Fig. 9 shows R and α for crystalline Quartz
(SiO_2). Some typical values for R and d can also be found in
Table I.

2. Modified optical properties

2.1. General. So far, we have briefly reviewed optical
properties at low light intensities. At sufficiently high
intensities, the response of matter to light can be modified by
a number of mechanisms of which we can only outline some basic
features. It is convenient to subdivide the various effects into
two categories, namely effects due to carrier excitation and
effects due to lattice heating. While the former category is
of practical significance only for non-metals, the latter
applies for any material.

2.2. Carrier excitation When a beam of photons of energy
$\hbar\omega > E_g$ is absorbed in a semiconductor or an insulator,
excited carriers are generated with a frequency $1/t_g = \Phi\sigma/\hbar\omega$,
where σ is the pertinent absorption cross section of the
carriers. At high absorbed power densities Φ, the rate of
carrier generation may exceed the rate of carrier relaxation. As
a result, the energy distribution of the carriers is significantly
changed as compared to that defining the optical properties of
the material at low light intensities.

Relaxation of hot carriers is a complex phenomenon and
still a subject of active research. However, for the sake of
simplicity we may assume that only two mutually independent
mechanisms are important, namely carrier-lattice collisions
(at a rate $1/t_c$), and recombination (at a rate $1/t_r$).
Usually, $1/t_c \gg 1/t_r$. Thus, in the order of increasing
power density two regimes can be distinguished: In the regime
$1/t_g > 1/t_r$ significant densities of free carriers are built
up, leading to optically induced free-carrier absorption. If
even $1/t_g > 1/t_c$, a hot carrier plasma is created, the
temperature of which exceeds the lattice temperature.

2.2.1. Optically induced free-carrier absorption (FCA).

Absorption of photons with $\hbar\omega > E_g$ in a non-metal always
leads to free carriers. However, only above a certain intensity
becomes their density sufficient to produce a measurable contribution to total absorption [7]. Under some simplifying assumptions,
an effective absorption coefficient α_{eff} can be defined as the sum
of the absorption coefficient α_o for normal band-to-band
absorption and a coefficient for FCA. The latter is the product of
the density N_c of carrier pairs and a FCA cross-section σ_{FC}, thus

$$\alpha_{eff} = \alpha_o + N_c\,\sigma_{FC} \tag{2.1}$$

The value of σ_{FC} is independent of N_c for not too high carrier
densities [8], but it depends on the nature of the dominant
scattering mechanism (phonon scattering or ionized impurity
scattering [9]), and it increases with the wavelength. For $\lambda =$
1.06 um, $\sigma_{FC} = 5.1 \cdot 10^{-18}\ cm^2$ was found in high-purity Si at RT [10].
Between 200 and 400° K, σ_{FC} was proportional to absolute
temperature [11].

The attenuation of a light beam inside the material is
described by α_{eff}, while the carrier density created is equal
to the density of photons absorbed by band-to-band absorption
only. An expression for the resulting carrier density for short
times, when recombination or diffusion of carriers can be
neglected, was given by Gauster and Bushnell [7]. Using their
result, α_{eff} for $z \ll 1/\alpha_o$ can be expressed as

$$\alpha_{eff} \cong \alpha_o + \frac{\sigma_{FC}}{\hbar\omega} \int_0^t \Phi\,(0,t')\,dt' \tag{2.2}$$

where Φ is given by (1.11) with $d = 1/\alpha_o$. For Si, (2.2) is
expected to be a reasonable approximation for ns laser pulses (for
longer pulses optically induced FCA does usually not play a role).

An absorption length that increases as a function of time
and pulse intensity makes beam heating strongly nonlinear. It is
found that the contribution from FCA can reduce the pulse energy
necessary for surface melting significantly[12].

An additional complication in the description of heating in
the present regime arises from the fact that only photons absorbed
by FCA are available for lattice heating instantaneously (on a
ns time scale), while band-to-band absorption releases heat only
via recombination, thus with a delay t_r that again depends on
N_c [13].

2.2.2. Carrier avalanche multiplication. Above a certain threshold intensity, free carriers created by excitation across the band gap (or by ionization of localized impurity states within the gap) can acquire enough energy to create additional carriers by impact ionisation. The result is an avalanche - multiplication of free carriers, similar to the one occuring in the dc-breakdown of dielectrics [14]. Hereby the absorption coefficient increases suddenly by orders of magnitude above a certain threshold intensity. The threshold intensity is generally found to scale like $1/\lambda^2$ and decreases with pulse duration [15]. This mechanism is often responsible for damage in optical materials [16].

2.2.3. Hot plasma effects. At sufficiently high power densities $1/t_g > 1/t_c$, i. e. the rate of energy gain by the carriers exceeds the rate of energy loss to the lattice. The result is a splitting-up of carrier- and lattice-temperatures [8]. Fig. 11 shows a schematic energy diagram for the electrons in a semiconductor, illustrating this regime. A typical collision rate $1/t_c$ is 10^{12} s^{-1}, whereas $1/t_c$ for beam intensities of 10^8 W/cm^2 may be comparable or even higher. A large density of hot carriers therefore builds up before the lattice is appreciably heated. Collisions between carriers establish thermal equilibrium between them on a time scale of 10^{-14} s. Auger processes and other recombination mechanisms establish equilibrium between electrons and holes, but do not affect the total carrier density. After the pulse ends, the hot carriers thermalize with a time constant t_r and recombine within typically $t_r = 10^{-9}$ s [13]. The effects of the presence of a dense, hot plasma on the lattice are not yet completely understood [17]. As far as coupling is concerned, however, the effect of the plasma is to ultimately limit the rate of local energy deposition in two ways:
- at a sufficiently high carrier density, light is reflected from the plasma (for reflection of visible light a density of about 10^{20} cm^{-3} is necessary);
- diffusion of hot carriers away from the absorbing surface effectively increases the heated volume. Yoffa [18] has shown that for Si this diffusion can be characterized by an effective diffusivity of about 100 cm^2/s for typical laser annealing parameters. Under these circumstances the expression (1.11) for the absorbed power density is not a good approximation for the distribution of the produced heat flux.

2.3. Lattice heating

The phenomena discussed now can play a role already at moderate light intensities, if only the total pulse energy is sufficient to raise the temperature of the solid. The optical

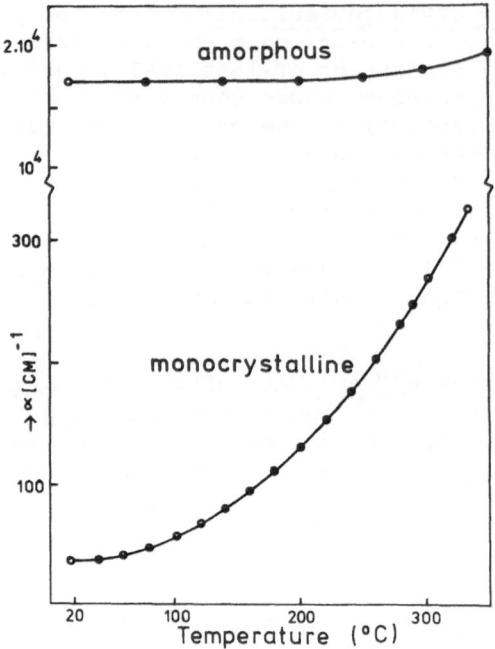

Fig. 10   Measured absorption coefficient in Si at λ = 1.06 μm as
a function of sample temperature.   From Ref. 19.

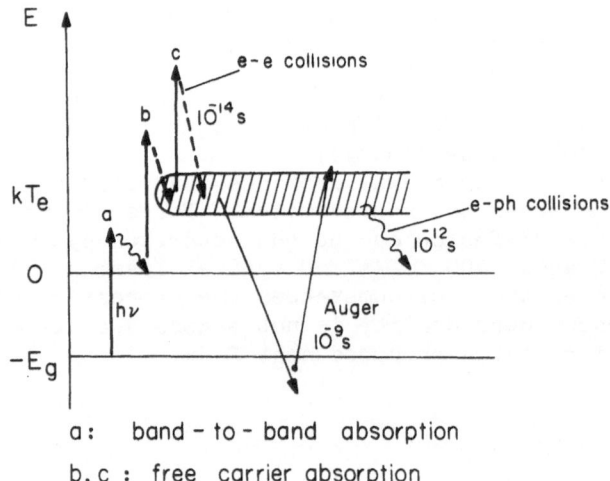

a :   band – to – band   absorption

b, c : free carrier absorption

Fig. 11   Schematic energy diagram of the electrons in a semiconductor under intense irradiation.

properties can then change for two basic reasons: 1) at elevated
temperatures more phonons are present which can interact either
with the photons or with the carriers; and 2) the band structure
of the solid may change with temperature.

2.3.1. Phonon assisted transitions. Interaction of phonons
with photons are necessary for light absorption in indirect
semiconductors. In such materials, the absorption of a photon
with $\hbar\omega \gtrsim E_g$ requires the simultaneous absorption of a phonon
for momentum conservation. At higher temperatures more phonons
are present, and the absorption coefficient increases. An
example is the absorption of Nd-laser light ($\hbar\omega$ = 1.17 eV) in
cristalline Si. Fig. 10 shows the absorption coefficient as a
function of sample temperature for crystalline and amorphous
Si [19]. Note, that for amorphous Si α changes only slightly
with temperature, because momentum conservation plays no role in
the disordered lattice.

2.3.2. Phonon scattering. Phonons are also required to
transfer energy from the carrier system to the lattice. Therefore,
the carrierlattice collision frequency $1/t_c$ increases with the
lattice temperature. A practical consequence is, that the
reflectivity of most metals decreases with temperature. As an
example, Fig. 12 shows the quantity (1-R) versus temperature for
Al and Au. The curves were calculated from electrical dc-resisti-
vity data using classical formulae [4]. In addition to the
steady increase of (1-R) with temperature, an abrupt increase is
observed for most metals at the melting point [20].

2.3.3. Thermally induced free-carrier absorption. This effect
is similar to the one discussed in 2.2.1., except that the
carriers are thermally instead of optically excited. An effective
absorption coefficient formally identical to the one in (2.1)
can be defined, where α_0 now stands for the residual lattice
absorption (due to intraband- or phonon coupling), and N_c is a
function of temperature [21]

$$N_c = N_o + 4 \ (2\pi kT/h^2)^{3/2} \cdot (m_e \ m_h)^{3/4} \cdot e^{-E_g(T)/2kT} \tag{2.3}$$

Here m_e and m_h are the electron- and hole effective mass,
respectively, and N_o is the concentration of ionized impurities.
Thermally induced FCA governs the absorption of CO_2-laser
radiation (λ = 10.6 um) in Ge and Si and is often referred to as
"thermal runaway". Fig. 13 shows a measured transmission curve
in Si at λ = 10.6 um, showing the typical features of thermally
induced FCA [22].

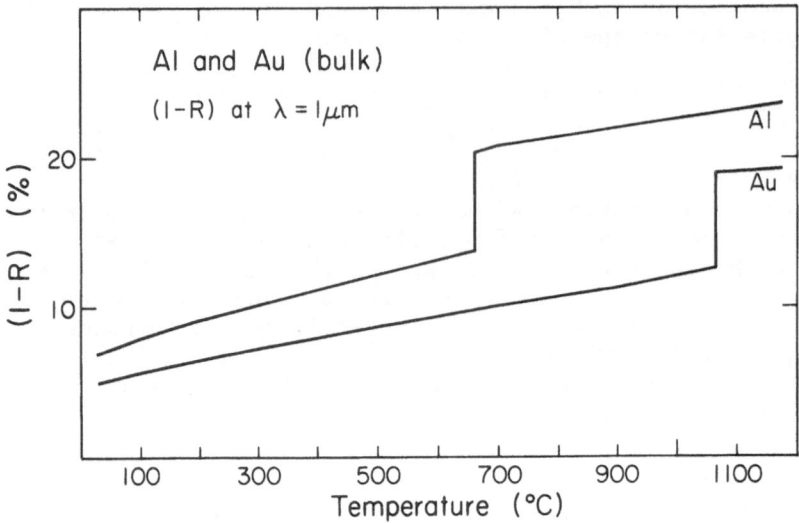

Fig. 12   (1-R) as a function of temperature for Al and Au, cal-
         culated from dc-resistivity data.

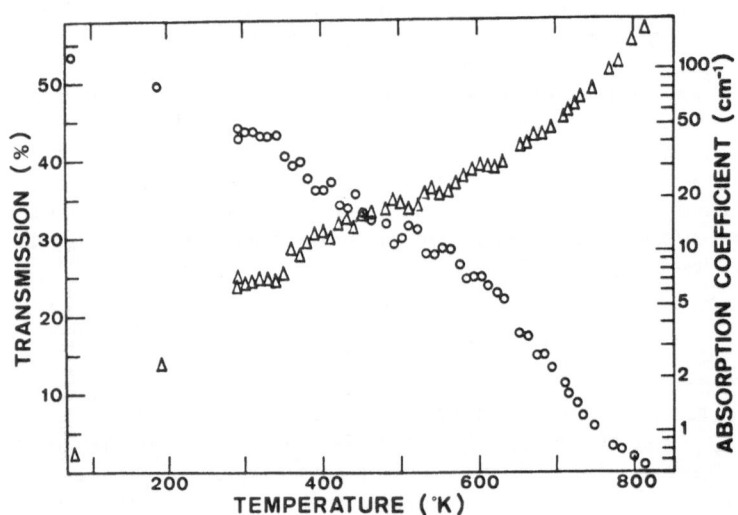

Fig. 13   Measured transmission (o) and absorption coefficient
         (Δ) as a function of sample temperature in a 290 μm
         thick Si wafer, at 10.6 μm.   From Ref. 22.

   2.3.4. Thermally induced changes in band structure. Under
this heading a number of phenomena can be summarized that lead
either to an increase or a reduction in the degree of absorption.
A few examples may illustrate the point.
- The gap energy of most semiconductors decreases with temperature.
  This affects primarily the absorption coefficient at photon
  energies close to the gap energy.
- A number of semiconducting materials, including Si, become
  metallic upon melting. This, of course, induces a strong
  increase in α and R.
- The reverse is observed when amorphous semiconductors crystallize.
  For Si, the ratio of the room temperature absorption coefficients
  of the amorphous and crystalline phase is about 10^4 (see
  Fig. 10).

   2.3.5. Evaporation effects. Yet, another large class of
phenomena is observed if the sample reaches temperatures where
intense evaporation takes place. Vapors, due to the localisation
of the electrons, are basically transparent insulators. However,
in dense vapors optical breakdown by avalanche multiplication of
carriers is observed at light intensities above 10^7 W/cm^2 [23].
The process is very similar to the one described in 2.2.2. The
threshold intensity is again proportional to $1/\lambda^2$, but decreases
strongly in the presence of easily ionizable impurities such as
dust particles [24]. The presence of a strongly absorbing vapor
can lead to shielding of the target surface [25]. Alternately,
a drastic decrease of the apparent reflectivity of metals has
been observed in the presence of a vapor plasma [26]. This
regime cannot be properly discussed without referral to the
dynamics of evaporation and gas expansion and it is not treated
further here.

3. Practical aspects

   After having discussed basic principles of light-solid
coupling we now turn to a few aspects of practical importance
in the use of high power beams for sample heating.

   3.1. Interference. Laser-irradiated samples often reveal a
microscopic surface structure consisting of rings, ripples,
corrugations and the like, particularly if surface melting is
induced during irradiation [27]. Some of these structures
result from interference effects [28], while others are due to
materialrelated thermal or mechanical phenomena [29]. Only
the first category of effects will be briefly discussed here,
because it directly related to the coupling problem.

Laser light has a very narrow bandwidth, since it results
from an undamped oscillation. Radiation with a bandwith $\Delta\nu$
is said to be coherent over a length $l_{coh} = c/\Delta\nu$, the
coherence length. Two parts split off a coherent beam that
travel along different paths can produce a persisting interference
pattern if their path lengths differ by less than l_{coh}. For
typical cw gas lasers l_{coh} can be several meters, whereas for
solid-state Q-switch lasers it is usually in the order of 1 cm.
This means, that the conditions for interference are practically
always given.

Interference patterns result if part of the laser beam is
scattered from an obstacle and interferes with the remainder of
the beam. If the obstacle is a simply-shaped macroscopical
object such as an aperture or an edge, interference leads to the
familiar diffraction patterns described in standard textbooks on
optics [30] . Apart from diffraction effects (which are easily
eliminated) interference phenomena may result from the following
reasons:

Scattering from small particles, e. g. dust particles in
the atmosphere or small protrusions on the sample surface. If
the scatterer is small enough for scattering to be isotropic,
we have interference of a spherical wave with a plane wave.
The resulting pattern consists of concentric circles (for
normal incidence) or ellipses (for an incidence angle $\varphi \neq 0$),
and is described by [31]

$$r_n = n\lambda \cdot \frac{G + \sqrt{1 + H(1-G^2)}}{1 - G^2} \qquad (3.1)$$

with $G = \cos\iota \sin\varphi$, and $H = 2h/n\lambda \cos\varphi$. Here r and ι are the
polar coordinates of the interference maxima, h is the distance
of the scattering center from the sample surface, and n is an
integer. Some typical interference patterns described by this
formula are shown in Fig. 14.

Scattering from a diffusor, such as a diffusing screen or a
diffusely reflecting surface. The patterns of this type are
called speckling patterns and consist of a random distribution of bright spots whith a mean diameter of [32]

$$\Lambda = \lambda \ a/D \qquad (for \ a \gg D) \qquad (3.2)$$

where D is the beam diameter and a is the optical distance
between the diffusor and the sample.

Multimode interference. This effect is not related to
scattering of any kind, but depends on the properties of the
laser source. Whenever a laser oscillates simultaneously in

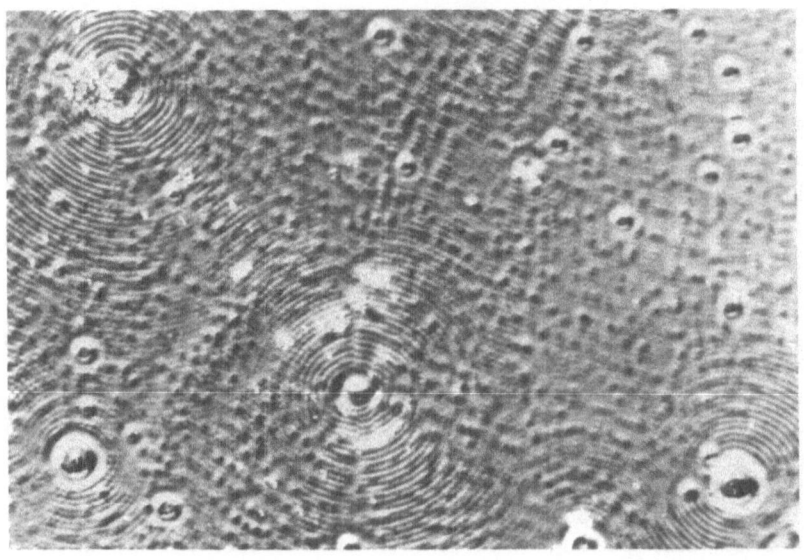

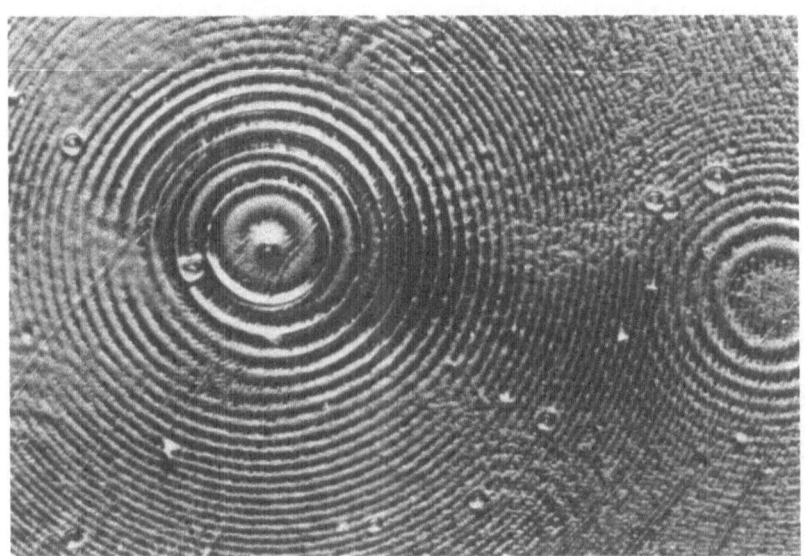

Fig. 14   Interference fringes on the surface of laser-irradiated
Palladium silicide (normal incidence). The equidistant
fringes in the upper micrograph are produced by scattering
centers at the surface (h = 0 in equ. (3.1)); the fringes
in the lower figure are due to scattering from particles
in the atmosphere in front of the sample surface[28].

several transversal resonator modes, then the intensity on a
sample is found to fluctuate in space and time because of the
interfering modes. The maximum intensity reached in a "hot spot"
can be estimated from [33]

$$I_{max}/<I> = \pi\ r_m^2\ /L\ \lambda \tag{3.3}$$

where r_m is the useful output mirror radius, L is the mirror
separation and $<I>$ is the mean intensity on the beam axis. An
upper limit to the "lifetime" of the hot spot is given by the
inverse of the frequency separation of adjacent resonator
modes [34] .

There are two obious ways to reduce the detrimental effects
of interference on beam heating:
- The physical size of the fringes or hot spots can be made small
  compared to the thermal diffusion lenght $2\sqrt{\varkappa t}$, where $\varkappa$ is the
  thermal diffusivity; or
- the lifetime of the hot spots can be reduced such that their
  energy is too small to produce a material effect. This can be
  achieved by "scrambling" the phases in the laser beam with a
  suitable device [35] .
A more general discussion of the sensitivity of the sample
temperature to lateral variations of intensity is given
in the following.

3.2. Temperature control

In most annealing procedures the crucial parameter is the
sample temperature, which has to be kept within narrow limits to
assure the desired result. Here is not the place to give a
general discussion of heat flow in laser-irradiated samples;
what we want to do is just demonstrate how the discussed coupling
phenomena influence the temperature distributions produced.

Fig. 15 shows instantaneous temperature profiles, produced
by a beam of constant and uniform intensity I_o, assuming all
material parameters to be constants. The three cases shown
pertain to the following situations.
a) Pure surface heating ($\alpha = \infty$). The temperature in this
case is given by [36]

$$T(z,t) = \frac{I_o\ (1-R)\ \delta}{\rho\ c_p\ \varkappa}\ ierfc\ \{\frac{z}{\delta}\} \tag{3.4}$$

Here ρ, c_p and $\varkappa$ are the density, specific heat and thermal
diffusivity of the material, respectively, and $\delta = 2\sqrt{\varkappa t}$.
b) The light penetrates the sample surface and is dissipated

uniformly within a layer of thickness d. The resulting temperature is

$$T(z,t) = \frac{I_o(1-R)\delta}{\rho\, c_p\, \mathcal{æ}} \left[i^2 \mathrm{erfc}\,\{\tfrac{z-d}{\delta}\} - i^2\mathrm{erfc}\,\{\tfrac{z+d}{\delta}\} - (\tfrac{z-d}{\delta})^2\, p(z) \right] \qquad (3.5)$$

where p = 1 for z < d and p = 0 otherwise.
c) Finally, the light is absorbed according to equation (1.11) with d = 1/α . In this case [37]

$$T(z,t) = \frac{I_o(1-R)\delta}{\rho\, c_p\, \mathcal{æ}} \left[\mathrm{ierfc}\,\{\tfrac{z}{\delta}\} - \right.$$

$$\left. \frac{d}{\delta}\,(e^{-z/d} - \tfrac{1}{2}\, e^{\delta^2/4d^2}(e^{-z/d}\cdot \mathrm{erfc}\,\tfrac{\delta}{2d} - \tfrac{z}{\delta} + e^{z/d}\cdot \mathrm{erfc}\{\tfrac{\delta}{2d} + \tfrac{z}{\delta}\})) \right. \qquad (3.6)$$

Now, real-world laser beams have a finite reproducibility and contain lateral variations of intensity. Scattering and interference effects also tend to create large lateral variation of intensity. The result is a variation in sample temperature, which one would like to keep as small as possible.

The amount of these variations depends on the dynamics of the heating process. Fig. 16 shows schematic temperature-vs -fluence curves illustrating the point: in curve b) the temperature rises linearly with the pulse fluence, in curve a) the rate of temperature rise decreases with fluence, and in curve c) the reverse is true. The final temperature reached is the same in all cases, but obviously a given fluctuation in beam fluence (either due to lateral variations of intensity or to variations from shot to shot) causes a large variation in temperature in case c), whereas in case a) it does not. From a practical point of view, clearly case a) is the desirable one. Which of the cases in Fig. 16 represents a given experimental situation, depends on coupling phenomena as well as on heat conduction. This is demonstrated by the following simple argument.

A beam of fluence $F(J/cm^2)$, incident on the surface of a sample of thickness L, creates heat in a surface layer of thickness d = 1/α . During the heat pulse of duration t (thought to be fixed), heat diffuses over a dimension $\delta = 2\sqrt{\mathcal{æ}t}$. (If the sample thickness L is smaller than δ , the following discussion applies with δ replaced by L). Two limiting cases can be distinguished: If d >> δ , then heat is created practically only at the surface. The surface temperature at the end of the pulse is, from (3.4) with z = 0

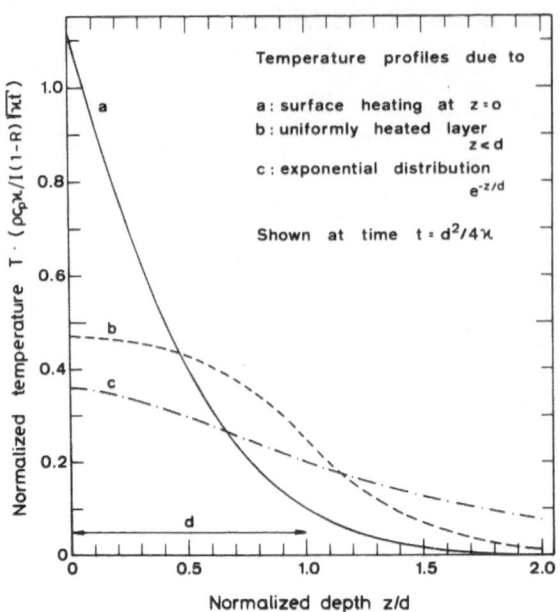

Fig. 15   Normalized temperature profiles calculated from equations
          a) (3.4), b) (3.5) and c) (3.6).

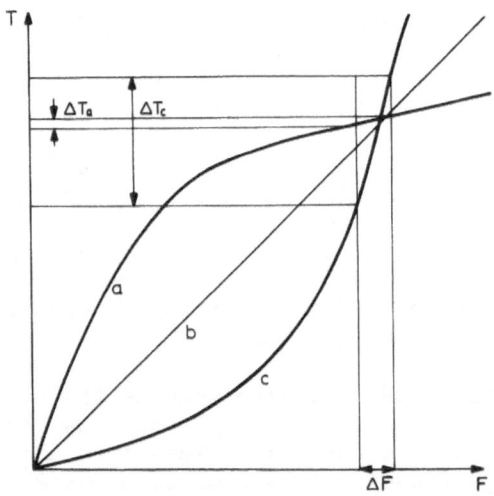

Fig. 16   Schematic temperature - versus - fluence curves, showing
          the temperature fluctuation ΔT caused by a fluctuation
          in fluence ΔF for various heating characteristics.

$$T(0,t) = 1.13 \, F(1 - R)/\rho \, c_p \, \delta \qquad\qquad (3.7)$$

The slope of $T(F)$ now depends mainly on the behavior of $R(T)$ and $c_p(T)$. Since c_p usually increases with temperature, $T(F)$ tends to saturate, corresponding to the case a) above, if $R(T)$ is constant. If, however, R strongly decreases with temperature (as in the case of light reflection of metals), then the slope of $T(F)$ is shifted towards case b) or c).

If, on the other hand, $\underline{d \gg \delta}$, then heat is created over a large depth. From (3.6), by setting $z = 0$, expanding into a series and neglecting higher order terms in δ/d, we find

$$T(0,t) = F(1 - R)/\rho \, c_p \, d \qquad\qquad (3.8)$$

In this case, the behavior of d as a function of temperature or fluence is crucial. We have discussed several mechanisms which lead to a strong decrease of the absorption length for light during irradiation. A decrease of d produces a temperature rise like case c) above, at least as long as the condition $d \gg \delta$ is fulfilled [38].

Two examples may illustrate how the discussed concepts work out in practical laser annealing experiments. Consider a sample consisting of several alternating thin layers of Au and Si on a glass substrate, to be reacted by a 30 ns Nd laser pulse [39]. If the top layer is Au, then $d \ll \delta$, and the decrease of $R(T)$ leads to an unfavorable situation. If the top layer is chosen to be, say, 30 nm of Si (acting also as an antireflection coating), then R even increases with temperature (Si turns metallic upon melting) and the heating curve resembles curve a) in Fig. 16. The unfavorable decrease of the absorption length of Si plays no role here, since the Si layer in only 30 nm thick, corresponding to a case with $d \ll \delta$. The experiment confirmed, that the latter configuration could be annealed easily, while in the former the control of pulse intensity was extremely critical.

Similar arguments apply for the case of pure crystalline Si heated by a short Nd-laser pulse. Here the strong decrease of d during irradiation results in an extreme type-c heating curve [38]. However, by simply preheating the sample to, say $250°C$ (thus decreasing the initial absorption length), a sufficiently stable heating curve is achieved [40]. The same effect can be obtained if part of the pulse energy is converted to a higher photon energy by frequency-doubling [41], in order to create an appropriate density of free carriers.

Table 2. Beam-Induced Optical Coupling Phenomena

Effect	applies to material	spectral range	results in
Optically induced FCA	semiconductors	$\hbar\omega \geq E_g$	$\alpha\,(F)\nearrow$
Carrier avalanche	semiconductors insulators	$\hbar\omega > E_g$	$\alpha\,(F)\uparrow$
	(with imperfections)	$\hbar\omega < E_g$	$\alpha\,(F)$
hot plasma	semiconductors	$\hbar\omega \gtrsim E_g$	$D\nearrow,\ R\nearrow$
photon assisted transitions	indirect semiconductors	$\hbar\omega \sim E_g$	$\alpha\,(T)\nearrow$
thermally induced FCA	semiconductors	$\hbar\omega < E_g$	$\alpha\,(T)\nearrow$
variations of E_g	semiconductors (Si, Ge)	$\hbar\omega \sim E_g$	$\alpha\,(T)\nearrow$ for $E_g(T)\searrow$
semicond. to metal trans.	some semiconductors (Si, Ge)	any	$R\nearrow,\ \alpha\nearrow$ upon melting
crystallization	amorphous Si, Ge	$\hbar\omega \sim E_g$	$\alpha\searrow$
vapor breakdown	metals	any	$R\downarrow$, or shielding

4. Concluding remarks

The remarkable results obtained in materials processing with beam energy these days are all more or less directly related to heat - or more specifically, to the high degree of temporal and spatial concentration of thermal energy available with high-power beams. We have therefore limited our inroduction into the concepts of absorption of laser- and electron beams to those aspects that are relevant for beam heating. We would like to point out, however, that, notably for laser beams, a wealth of "nonthermal" material effects can be observed in irradiated materials. Such effects are
- the resonant and highly selective excitation of atoms with a tunable light source (which is the basis of photoche-mistry),
- the stimulated scattering from coherent phonons (mainly used for analysis of crystal structure),
- nonlinear optical effects (such as optical harmonic generation and optical mixing),

only to mention a few examples. All these effects are well known, but have as yet practically not been utilized in the field of material processing. Here seems to be a large field for future developments.

References

1. a - Laser-Solid Interactions and Laser Processing-1978, ed. by S. D. Ferris, H. J. Leamy, J. M. Poate, AIP Proceedings No. 50 (1979),
   b - Laser and Electron Beam Processing of Electronic Materials, ed. by C. L. Anderson, G. K. Celler, G. A. Rozgonyi, The Electrochem. Society, 1980,
   c - Laser- and Electron Beam Processing of Materials - 1979, ed. by C. W. White, P. S. Peercy, Academic Press 1980. (in press).
2. See e. g. Laser Handbook Vol. 1, ed. by F. T. Arecchi, E. O. Schulz-Dubois, North Holland 1972.
3. See, e. g. F. Wooten, Optical Properties of Solids, Academic Press 1972.
4. N. F. Mott, H. Jones, The Theory of the Properties of Metals and Alloys, Dover 1958.
5. American Instiute of Physics Handbook, 3rd ed., McGraw-Hill 1972, and other sources.
6. W. G. Spitzer in Semiconductors and Semimetals Vol. 3, ed. by R. K. Willardson, A. C. Beer, Academic Press 1967, chapter 2.
7. W. B. Gauster, J. C. Bushnell, J. Appl. Phys. 41, 3850 (1970).
8. E. Yoffa, Phys. Rev.B 21, 2451 (1980).
9. R. Rosenberg, M. Lax, Phys. Rev. 112, 843 (1958).

10. K. G. Svantesson, J. Phys. D. Appl. Phys. $\underline{12}$, 425 (1979).
11. K. G. Svantesson, N. G. Nilsson, J. Phys. C. Solid State
    Phys. $\underline{12}$, 3837 (1979).
12. A. Lietoila, J. F. Gibbons, Appl. Phys. Lett. $\underline{34}$, 332 (1979).
13. J. Dziewior, W. Schmid, Appl. Phys. Lett. $\underline{31}$, 346 (1977)
14. L. H. Holway, D. W. Fradin, J. Appl. Phys. $\underline{46}$, 279 (1975).
15. D. W. Fradin, N. Bloembergen, J. P. Letellier, Appl. Phys.
    Lett. $\underline{22}$, 635 (1973).
16. N. Bloembergen, IEEE J. Quantum Electron. $\underline{QE-10}$, 375 (1974).
17. J. A. Van Vechten, D. Hoonhout, F. W. Saris, Phys. Lett. A
    (in press).
18. E. Yoffa, Appl. Phys. Lett. (in press).
19. M. R. T. Siregar, M. von Allmen, W. Lüthy, Helv. Phys. Acta
    $\underline{52}$, 45 (1979).
20. A. J. Gubanov, Quantum Electron Theory of Amorphous Con-
    ductors, Consultants Bureau 1965.
21. C. Kittel, Introduction to Solid State Physics, 5th. ed.,
    Wiley 1976, p. 228 ff.
22. M. R. T. Siregar, W. Lüthy, K. Affolter, Appl. Phys. Lett.
    $\underline{36}$, 787 (1980).
23. V. I. Bergelson, A. P. Golub, I. V. Nemchinov, S.P. Popov,
    Sov. Phys. Quantum Electron. $\underline{3}$, 288 (1974).
24. D. C. Smith, J. Appl. Phys. $\underline{48}$, 2217 (1977).
25. E. Stürmer, M. von Allmen, J. Appl. Phys. $\underline{49}$, 5648 (1978).
26. M. von Allmen, P. Blaser, K. Affolter, E. Stürmer, IEEE
    J. Quantum Electron. $\underline{QE-14}$, 85 (1978).
27. C. Hill in Ref. 1c.
28. K. Affolter, W. Lüthy, M. Wittmer, Appl. Phys. Lett. $\underline{36}$,
    559 (1980), and references therein.
29. G. Vitali, M. Bertolotti, L. Stagni in Ref. 1a, p.111,
    and references therein.
30. M. Born, E. Wolf, Principles of Optics, 5th ed., Pergamon
    Press 1975.
31. M. von Allmen, unpublished.
32. L. I. Goldfisher, J. Opt. Soc. Am. $\underline{55}$, 247 (1965).
33. D. M. Ryter, Maximale und mittlere Intensität bei der Ueber-
    lagerung von transversalen Moden, University of Bern (1975).
34. A. G. Fox, T. Li, Proc. IEEE $\underline{51}$, 80 (1963).
35. A. G. Cullis, H. C. Webber in Ref. 1b, p. 220.
36. H. S. Carlslaw and J. C. Jaeger, Conduction of Heat in
    Solids, 2nd. ed., Oxford University Press 1959.
37. This solution was also given by F. W. Dabby, U.-C. Paek,
    IEEE J. Quantum Electron. $\underline{QE-8}$, 106 (1972).
38. M. von Allmen in Ref. 1a, p. 43.
39. M. von Allmen, S. S. Lau, M. Mäenpää, B. Y. Tsaur, Appl.
    Phys. Lett. $\underline{36}$, 207 (1980).
40. M. von Allmen, W. Lüthy, J-P. Thomas, M. Fallavier, J.M.
    Mackowsky, R. Kirsch, M.-A. Nicolet, M. E. Roulet, Appl.
    Phys. Lett. $\underline{34}$, 82 (1979).

41. D. H. Auston, J. A. Golovchenko, A. L. Simons, R. E. Slusher,
    P. R. Smith, C. M. Murko, T. N. C. Venkatesan, in Ref. 1a,
    p. 11.

LASER HEATING OF SOLIDS

Michael Bass

Center for Laser Studies
University of Southern California
Los Angeles, CA 90007

I. INTRODUCTION AND OUTLINE

We are concerned with heating metals and semiconductors to
produce a change in their properties in the heated region only.
The change may be a solid to solid conversion as in heat treating
steels, a solid to liquid to solid conversion as in semiconductor
annealing or a solid-to-liquid-to vapor conversion as in cutting
or drilling. The questions which we might wish to answer before
designing a processing procedure include

1. What sort of laser do we need?
2. How must it be delivered to the sample?
3. How long should it dwell on the irradiated site? and
4. Will there be any unwanted side effects?

In order to answer these questions, we can try to solve the
thermodynamical problem of laser heating. This can be attempted
in a number of nearly equivalent ways but all require a knowledge
of, 1, the target's optical and thermal properties during the
irradiation, 2, the laser beam distribution, 3, the dynamics of
the irradiation process, and 4, the processes of phase change in
the target material. As could be expected, it is extremely difficult to solve this problem exactly in the general case and so
we resort to some reasonable approximations and to the usual procedure of solving the problems that are easy to solve. These are
then useful as guides to the solutions of the problems we cannot
solve.

We will examine the equation of heat conduction in a solid
with the laser energy absorbed on the irradiated surface as a heat

source. By judiciously selecting the beam geometry, dwell time and
sample configuration, we will reduce the problem to solvable one
and two dimensional heat flow analyses. We will also include phase
transitions and calculate the temperature distributions that one
may produce. Examples will be selected to provide specific guid-
ance in the choice of lasers and materials. The result of all this
will be a notion of the effects that one may produce by laser
heating solids.

II. THE PROBLEM OF HEAT FLOW

II.A. The equation for heat flow in a three dimensional solid is

$$\rho C \frac{\partial T}{\partial t} = \frac{\partial}{\partial x}\left(K\frac{\partial T}{\partial x}\right) + \frac{\partial}{\partial y}\left(K\frac{\partial T}{\partial y}\right) + \frac{\partial}{\partial z}\left(K\frac{\partial T}{\partial z}\right) + A(x,y,z,t) \quad (1)$$

where

    ρ = the material density, gm/cm^3

    K = the thermal conductivity, W/cm^oC, and

    C = the heat capacity, J/gm^oC

are material properties which will depend upon temperature and
position.[*] The quantity $A(x,y,z,t)$ is the rate at which heat is supplied to the solid per unit time per unit volume in $J/sec\ cm^3$ and
$T = T(x,y,z,t)$ is the resulting temperature distribution in the
material.

The temperature dependence of the thermal properties results
in a nonlinear equation which is very hard to solve exactly. Where
the functional dependence of these quantities on temperature is
known, it is sometimes possible to employ numerical integration
techniques to obtain a solution. A further complication arises
from the temperature dependence of $A(x,y,z,t)$ through that of the
material's absorbtivity. When phase transitions occur one can
attempt a solution of the problem by solving for each phase separately and including the heat required for the transition where
appropriate.

For most materials the thermal properties do not vary greatly
with temperature and can be assigned an average value for the temperature range to be studied. In this case it is possible to solve
the heat flow problem. A further simplification is obtained by
assuming that the material is homogeneous and isotropic. Under
these conditions Eq. (1) reduces to

---

[*] A list of the symbols used appears at the end of this paper.

$$\nabla^2 T - \frac{1}{\kappa} \frac{\partial T}{\partial t} = - \frac{A(x,y,z,t)}{K} \tag{2}$$

where $\kappa = K/\rho C =$ the thermal diffusivity. In the steady state $\partial T/\partial t = 0$ and we have

$$\nabla^2 T = - \frac{A(x,y,z,t)}{K} \tag{3}$$

If there is no heat source as in the case of cooling of heated material the temperature distribution will be given by

$$\nabla^2 T = \frac{1}{\kappa} \frac{\partial T}{\partial t} \tag{4}$$

in the time dependent case and

$$\nabla^2 T = 0 \tag{5}$$

in the steady state.

II.B. The Thermal Diffusivity

We first would like to examine Eq. (4) to obtain a physical insight into the meaning of the quantity κ. First a simple analysis of the units show that

$$(\kappa t)^{\frac{1}{2}} = \text{distance.}$$

This distance or a multiple thereof is generally known as the thermal diffusion distance for the particular problem and is a handy quantity to use when scaling the effects of laser heating. It is often very useful to know how the optical absorption depth compares with the thermal diffusion distance during a laser pulse. To see the meaning of the thermal diffusion depth more clearly, let us consider some particular solutions of Eq. (4). For example:

(A) $\quad T = T_o \, e^{\, t/T \, \pm \, z/z_D}$

and so

$$\frac{\partial T}{\partial t} = \frac{1}{T} T \text{ and } \frac{\partial^2 T}{\partial z^2} = \left(\frac{1}{z_D} \right)^2 T \quad .$$

Thus

$$\left[\left(\frac{1}{z_D}\right)^2 - \frac{1}{\kappa T}\right] T = 0$$

and we have a solution if

$$z_D = (\kappa T)^{\frac{1}{2}} \quad .$$

Therefore a characterisitic distance is related to a characteristic time by $(\kappa T)^{\frac{1}{2}}$

For a more realistic case consider:

(B) $T = T_0 \, t^{-\frac{1}{2}} \, e^{-z^2/4\kappa t}$ for $t > 0$.

Then we have

$$\frac{\partial T}{\partial t} = T_0 \left(-\frac{1}{2t^{3/2}} + \frac{z^2}{4\kappa T^{5/2}} \right) e^{-z^2/4\kappa t}$$

and

$$\frac{\partial^2 T}{\partial z^2} = T_0 \left(-\frac{1}{2\kappa t^{3/2}} + \frac{z^2}{4\kappa t^{5/2}} \right) e^{-z^2/4\kappa t}$$

which satisfies

$$\frac{\partial^2 T}{\partial z^2} - \frac{1}{\kappa} \frac{\partial T}{\partial z} = 0 \quad .$$

At $t = 0$ we will set $T = T_0$ at $z = 0$ and $T = 0$ for $z > 0$.
Now consider at any time t_p the temperature at

$$z_D = 2 \, (\kappa t_p)^{\frac{1}{2}}$$

is at $1/e$ of the value at $z = 0$. Carslaw and Jaeger[1] point out
that this solution corresponds to the release of the quantity of
heat $2\rho C T_0 \, (\pi\kappa)^{\frac{1}{2}}$ per unit area over the plane $z = 0$ at time
$t = 0$.

This could be a decent approximation to the laser heating
case of a semi-infinite metal irradiated by a short pulse (defined
in terms of the relative values of beam radius and $(\kappa t_p)^{\frac{1}{2}}$) in a
uniform beam and asking for the temperature on the beam axis. The
quantity of heat per unit area would be the laser intensity I

times the pulse duration, T_p, times the fraction absorbed or

$$IT_p \alpha = 2\rho CT_0 \, (\pi\kappa)^{\frac{1}{2}}$$

where α = absorbtance of the metal. Then we have

$$T_0 = \frac{\alpha IT_p}{2\rho C(\pi\kappa)^{\frac{1}{2}}}$$

and

$$T(z,t) = \frac{\alpha IT_p}{2\rho C(\pi\kappa)^{\frac{1}{2}}} \; t^{-\frac{1}{2}} \; e^{-z^2/4\kappa t}$$

or

$$T(z,t) = \frac{\alpha IT_p}{\rho C\pi^{\frac{1}{2}}} \frac{e^{-z^2/4\kappa t}}{(4\kappa t)^{\frac{1}{2}}} \qquad \text{for } t > 0 \quad .$$

We see the important role of the quantity $z_D = 2(\kappa t)^{\frac{1}{2}}$ in describing this process.

Now we will turn to some other examples.

III. CASES OF ABSORPTION IN A VERY THIN SURFACE LAYER

(i.e.: $\alpha^{-1} \ll (\kappa T_{laser})^{\frac{1}{2}}$ where T_{laser} = duration of the irradiation)

III.A. Continuous Radiation Turned On At t = 0 and the Slab is Infinitely thick

The incident intensity is given by

$$I_0(t) = \begin{matrix} 0 & t < 0 \\ I_0 & t \geq 0 \end{matrix}$$

and is assumed to be uniformly distributed. Again we are considering a special case which is only useful as an approximation to the
temperature distribution on axis. We do not allow for heat diffusion in the x and y directions in solving this problem and this
may be important in certain applications. However, the results of
this special case is very useful since most often the on-axis temperature is the desired quantity.

The boundary condition is that

$$T(z,t) = 0 \quad \text{for} \quad t < 0 \text{ and all } z$$

Rather than go through the tedious process of solving the diffusion equation in these examples, we will use the results derived so elegantly by Carslaw and Jaeger. [1]   For this problem they show

$$T(z,t) = (2\alpha I_0/K)(\kappa t)^{\frac{1}{2}} \text{ ierfc} \left[\frac{z}{2(\kappa t)^{\frac{1}{2}}} \right] \quad (6)$$

where

$$\text{ierfc}(X) = \int_X^\infty \text{erfc}(X')dX' \quad (7)$$

and

$$\text{erfc}(X) = 1 - \text{erf}(X) = \frac{2}{\pi} \int_X^\infty e^{-(X')^2} dX' \quad (8)$$

and

$$\text{erf}(X) = \frac{2}{\pi} \int_0^X e^{-(X')^2} dX' \quad . \quad (9)$$

[The error function, erf X, has the following properties

$$\text{erf } 0 = 0 \quad \text{erf } \infty = 1 \quad \text{erf}(-X) = -\text{erf } X$$

erfc 0 = 1 and erfc ∞ = 0.

In order to find the limiting value of ierfc X we integrate by parts as follows:

Let,

$$B = \text{ierfc } 0 = \int_0^\infty \text{erfc } xdx$$

then dV = dx and U = erfc x

This gives

$$B = x \text{ erfc } x \Big|_0^\infty - \int_0^\infty x \frac{d(\text{erfc } x)}{dx} dx = -\int_0^\infty x \frac{d(\text{erfc} x)}{dx} dx .$$

Now $\frac{d}{dx}(\text{erfc} x) = \frac{d}{dx}\left(1 - \frac{2}{\sqrt{\pi}}\int_0^x e^{-\mu^2} d\mu\right) = -\frac{2}{\sqrt{\pi}} e^{-x^2} .$

Therefore

$$B = +\frac{2}{\sqrt{\pi}} \int_0^\infty x e^{-2} d = \frac{1}{\sqrt{\pi}} = \text{ierfc } 0 .$$

Also ierfc $= 0.\Big]$

Thus the surface temperature is given by

$$T(0,t) = \frac{2\alpha I_0}{K} (\kappa t)^{\frac{1}{2}} \left(\frac{1}{\pi}\right)^{\frac{1}{2}} . \tag{10}$$

We see that the surface temperature is proportional to $t^{\frac{1}{2}}$ and if there were no phase changes (i.e.: melting and vaporization) the temperature would continue to increase. Note that the energy flux absorbed by the surface is

$$E = \int_0^t \alpha I dt = I_0 t \alpha_0$$

but because of conduction the surface temperature increases more slowly, as $t^{\frac{1}{2}}$. This means that a greater surface temperature can be achieved for a given laser pulse energy by shortening the pulse and increasing I_0 !

Since we have mentioned pulses, let us model a simple pulse as one where

$$I = I_0 \quad \begin{array}{ll} 0 & t < 0 \\ & 0 \le t \le T \\ 0 & T < t \end{array}$$

During the interval 0 to T the temperature is as given above with the maximum temperature obtained at t = T. For t > T

$$T(z,t) = \frac{2\alpha I_0}{K} \left[(\kappa t)^{\frac{1}{2}} \text{ ierfc } \left(z/2(\kappa t)^{\frac{1}{2}} \right) - \right.$$

$$\left. \left(\kappa(t-T) \right)^{\frac{1}{2}} \text{ ierfc} \left(z/2 \, \kappa(t-T) \right)^{\frac{1}{2}} \right] \quad (11)$$

Remembering that this problem is only a one dimensional approximation to the actual case and that the assumption of an infinitely thick medium requires that the slab thickness, L, be

$$L > 2 \ (\kappa T)^{\frac{1}{2}}$$

allows one to use these results to calculate the temperature distribution achieved. As pointed out above the result will only be correct near the beam axis. The cooling rate following the irradiation in this example is obtained by taking the time derivative of T(z,t) in Eq. (11).

III.B.   Same as III.A but the beam is uniformly distributed in a circle of radius A.

The power is

$$P = \begin{array}{cc} 0 & t < 0 \\ P_0 & t \geq 0 \end{array}$$

Using cylindrical coordinates (See Fig. 1) Carslaw and Jaeger [1] find for this case

$$T(r,z,t) = \frac{\alpha P_0}{2\pi AK} \int_0^\infty \frac{d\lambda}{\lambda} \ J_0 \ (\lambda r) \ J_1 \ (\lambda A) \ \times$$

$$\left\{ e^{-\lambda z} \text{ erfc} \left[\frac{z}{2(\kappa t)^{\frac{1}{2}}} - \lambda(\kappa t)^{\frac{1}{2}} \right] - \right.$$

$$\left. e^{\lambda z} \text{ erfc} \left[\frac{z}{2(\kappa t)^{\frac{1}{2}}} + (\kappa t)^{\frac{1}{2}} \right] \right\} \quad (12)$$

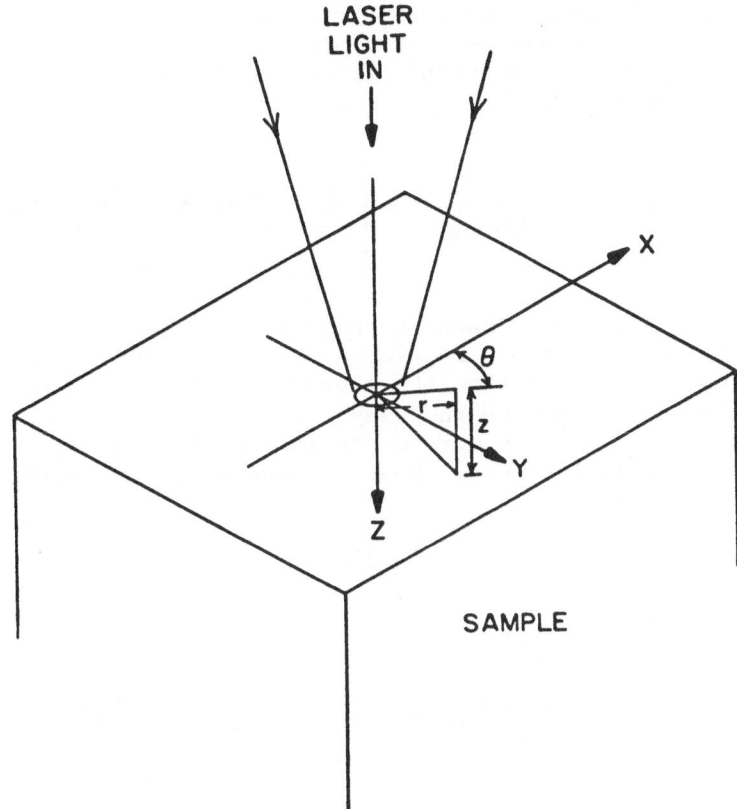

Fig. 1.   Irradiation Geometry and Coordinate Axes.

where J_0 and J_1 are Bessel functions of the first kind.   If one so
desires, he may evaluate this numerically.   It is far more useful
in planning experiments to evaluate the result directly under the
beam since this will be the hottest place.   In other words we want
the temperature at r = 0.   This is given as

$$T(0,z,t) = \frac{2\alpha P_0}{A^2 K} (\kappa t)^{\frac{1}{2}} \left[ierfc \left(\frac{z}{2(\kappa t)^{\frac{1}{2}}} \right) \right.$$

$$\left. - \, ierfc \left(\frac{(z^2 + A^2)^{\frac{1}{2}}}{2(\kappa t)^{\frac{1}{2}}} \right) \right] \qquad (13)$$

which is clearly the contribution one would have if the entire z = 0 plane were irradiated, as in case A above, less a term which accounts for the finite extent of the irradiation. On axis and at the surface z = 0,

$$T(0,0,t) = \frac{2\alpha P_0}{\pi A^2} \frac{(\kappa t)^{\frac{1}{2}}}{K} \left[\frac{1}{\pi^{\frac{1}{2}}} - \text{ierfc} \left(\frac{A}{2(\kappa t)^{\frac{1}{2}}} \right) \right] \quad (14)$$

.

Note that if $A \gg 2(\kappa t)^{\frac{1}{2}}$ then at those values of t one can treat the problem as the simple one in case A. In other words while the beam radius is greater than the thermal diffusion distance, the problem can be reduced to the simple case for the on axis temperature. However, as time marches on this will break down and we will find some differences. Most obviously we will find that $T \not\to \infty$ but instead $T \to T_{max}$.

In the limit as $t \to \infty$ we find

$$T(0,0,\infty) = \frac{\alpha P_0}{\pi AK} = T_{max} \quad . \quad (15)$$

This reveals a curious point – the quantity P_0/A determines the maximum achievable temperature. If, for example, we wish to heat the material without melting the surface then we must choose P_0/A such that $T_{max} < T_{melt}$.

For example Duley [2] lists the following numbers assuming a Drude like behavior for the absorption of metals at 10.6 μm.

Minimum P_0/A at 10.6 μm to Melt the Surface of the Listed Metal with a Beam Focused to a Circular Spot of Radius A

Material	P_0/A (x 10^4 W/cm)
Al	13
Au	26
Cu	34
Sn	0.85
Ni	4.4
W	8.9
Mo	7.3

Carslaw and Jaeger [1] also computed the average surface temperature that can be achieved in this case. In the steady state (i.e. $\partial T / \partial t = 0$) they find

$$T_{av} = \frac{8 \alpha P_0}{3 \pi^2 AK} \simeq 0.85 \; T_{max} \; .$$

We note that $T_{av}/T_{max} \simeq 1$ and that this ratio is independent of material!!

III.C. The Noncircular Beam

As is often the case, the focused laser beam will not be circular. Distortion in the beam handling optics or imperfect laser media will lead to noncircular focal distributions. Again assuming that the light distribution is uniform we can examine the distorted case by approximating it by a rectangular distribution. Again using Carslaw and Jaeger's [1] results we find

$$T(0,0,\infty) = \alpha P/2K\pi\ell a \; (a \sinh^{-1} (\ell/a) + \sinh^{-1}(a/\ell)) \tag{16}$$

for the maximum equilibrium temperature, a is the width and ℓ the length of the irradiated rectangle.

If for example a, the width of the rectangle, and A, the radius in the circular case, were set equal and we asked what power would be needed to reach a specific temperature in each case we would have

$$\frac{P_{circ}}{P_{rect}} = \frac{1}{2\ell} \left(a \sinh^{-1} \left(\frac{\ell}{a}\right) + \ell \sinh^{-1} \left(\frac{a}{\ell}\right) \right) . \tag{17}$$

This is plotted in Fig. 2.

Even a square beam requires more power to produce a specific temperature than a circular beam. (Of course it does, a larger area is irradiated and the absorbed energy must heat up more material.) Thus to achieve equal maximum temperature, we must provide more input. If $\ell/a = 3$ then we see from this expression (Fig. 2) that the power must be increased by a factor of ~ 2. This simple example demonstrates the importance of good beam quality and focusing optics in order to obtain maximum heating from laser irradiation.

III.D.   Gaussian Beam Irradiation

The Gaussian beam intensity distribution is given as

$$I = I_0 \, e^{-r^2/w^2} \qquad \qquad (18)$$

.

[Note: It is common to describe a Gaussian mode of a laser by the electric field distribution

$$E = E_0 \, e^{-r^2/w_0^2}$$

and then give

$$I = I_0 \, e^{-2r^2/w_0^2} = I_0 \, e^{-r^2/(w_0/\sqrt{2})^2} \qquad .$$

The beam parameter w used in this treatment is the radius at which the <u>intensity</u> has fallen to $1/e$ of its on axis value. It is the radius at which the electric field has fallen to $e^{-\frac{1}{2}}$ of <u>its</u> on axis value. If we used w_0 to describe the Gaussian, then it corresponds to the radius at which the intensity has fallen to e^{-2} of its on axis value.]

Carslaw and Jaeger [1] give the following expression for the temperature distribution in a semi infinite solid due to irradiation at the surface by an instantaneous ring source of radius r' and total deposited energy Q

$$T_{\substack{inst \\ ring}}(r,z,t) = \frac{Q}{4\rho C\,(\pi \kappa t)^{3/2}} \, \exp\left(\frac{-r^2-r'^2-z^2}{4\kappa t}\right) I_0\,(r\,r'/2\kappa t) \quad .$$

In this expression I_0 is the modified Bessel function of order zero.

For a Gaussian source we have

$$Q = q_0 \, e^{\left(-r'^2/w^2\right)} \, 2\pi r'dr'$$

where q_0 = energy per unit area at the origin.

We insert this and integrate. However Ready [3] has already done this integration and has obtained

$$T_{inst\ Gaussian}(r,z,t) = \left(\frac{q_0 w^2}{\rho C (\pi \kappa t)^{\frac{1}{2}} (4\kappa t + w^2)} \right) \times$$

$$\exp \left(-\frac{z^2}{4\kappa t} - \frac{r^2}{4\kappa t + w^2} \right) \ . \tag{19}$$

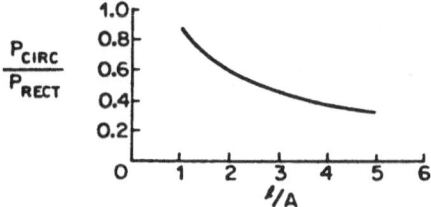

Fig. 2.   Ratio of Power Required to Heat the
          Surface to a Specific Temperature
          for a Circular Beam of Radius A to
          that for a Rectangular Beam of Dim-
          ensions A by ℓ Plotted Versus ℓ/A.

III.D.1 The Short Pulse Gaussian Beam

If we replace q_0 by $\alpha I_0 T_p$ where T_p is the duration of a pulse and

$$w >> (4\kappa T_p)^{\frac{1}{2}}$$

we can use equation 19 for the temperature distribution following irradiation by a short pulse. NOTE: SHORT PULSE IS DEFINED IN TERMS OF "DOES HEAT DIFUSE SIGNIFICANTLY WITH RESPECT TO THE BEAM DIMENSION DURING THE PULSE? IF IT DOES NOT, THEN THE PULSE IS "SHORT".

We rewrite

$$T_{inst.\ Gaussian}(r,z,t) = \frac{\alpha I_0 T_p}{\rho C(\pi\kappa t)^{\frac{1}{2}}}\left(\frac{1}{\frac{4\kappa t}{w^2}+1}\right)$$

$$\exp\left(-\frac{(z/w)^2}{\frac{4\kappa t}{w^2}} - \frac{(r/w)^2}{\frac{4\kappa t}{w^2}+1}\right) \qquad (20)$$

and define the unitless quantities

$$T' = \frac{w\rho C\pi^{\frac{1}{2}}}{2\alpha I_0 T_p}T$$

$$t' = 4\kappa t/w^2$$

$$z' = z/w$$

$$r' = r/w$$

so that

$$T' = \left(\frac{1}{t'}\right)^{\frac{1}{2}}\frac{1}{t'+1}\ e^{\left(-(z')^2/t' - (r')^2/(t'+1)\right)} \qquad (21)$$

For any z' and t' we see that T' decreases with r' as $e^{-(r')^2/t'+1}$. Similarly for any t' and r', T' decreases as $e^{-(z')^2/t'}$. Thus having found the on-axis surface temperature one can easily find it at any point in the material!!

Again, remember where this simple form holds! We should consider some numbers now to get a feeling for the scale of things.

For copper $\kappa \simeq 1$ cm^2/sec. Thus for $w = 0.5$ mm we would require

$$T \ll \frac{.0025}{4} \text{ sec} \simeq 6 \times 10^{-4} \text{ sec}\quad.$$

For steel $\kappa \simeq .15$ cm^2/sec and $w = 0.5$ mm we would require

$$T_p \ll 4 \times 10^{-3} \text{ sec}.$$

Thus for pulsed Nd:YAG lasers where $T_p \approx 10^{-5}$ sec this approximation is quite adequate. For CO_2 TEA laser pulses with $T_p \approx 10^{-6}$ sec again we can use this form. However, for discharge pulsed CO_2 lasers with $T_p \approx 10^{-3}$ sec or for CW lasers (or for smaller w) we must use the non-instantaneous form.

III.D.2 CW Gaussian Irradiation

This case can be treated by setting $q_0 = \alpha I_0 (\mu) d\mu$ where μ is a dummy variable for time) and integrating $T_{inst. Gaussian}$ from 0 to t.

This gives

$$T_{non-inst. Gaussian} (r,z,t) = \frac{\alpha I_0^{max} w^2}{K} \left(\frac{\kappa}{\pi}\right)^{\frac{1}{2}} \times$$

$$\int_0^t \frac{p(t-\mu)d\mu}{(\mu)^{\frac{1}{2}}(4\kappa\mu + w^2)} \exp\left(-\frac{z^2}{4\kappa\mu} - \frac{r^2}{4\kappa\mu + w^2}\right)$$

where $I_0 (\mu) = I_0^{max} p(\mu)$

now setting

$$\mu' = \frac{4\kappa\mu}{w^2} \qquad z' = z/w \qquad r' = r/w$$

we have

$$T_{non-inst. Gaussian} (r',z't') = \frac{\alpha I^{max} w}{K 2\pi^{\frac{1}{2}}} \times$$

$$\int_0^{t'} \frac{p(t'-\mu') d\mu'}{(\mu')^{\frac{1}{2}}(\mu't1)} \exp\left(- (z')^2/\mu - (r')^2/(\mu +1)\right)$$

and so for a dimensionless $T' = \dfrac{2\pi^{\frac{1}{2}} K T}{\alpha I_0^{max} w}$

we have

$$T' = \int_0^{t'} \frac{p(t'-\mu')d\mu'}{(\mu')^{\frac{1}{2}}(\mu'+1)} \; \exp\left((z')^2/\mu - (r')^2/(\mu+1)\right) \quad (22)$$

.

At the surface, $z' = 0$ and at the center, $r' = 0$ and so we have

$$T' = \int_0^{t'} \frac{p(t'-\mu')d\mu'}{(\mu')^{\frac{1}{2}}(\mu'+1)}$$

.

For a CW laser or one where $T_P \gg \frac{w^2}{4\kappa}$

we have $p(t' - \mu') = 1$ and

$$T' = 2 \tan^{-1}(t')^{\frac{1}{2}}$$

or

$$T(0,0,t) = \frac{\alpha I_0^{max} \, w}{K\pi^{\frac{1}{2}}} \; \tan^{-1}\left(\frac{4\kappa t}{w^2}\right)^{\frac{1}{2}} \quad (23)$$

.

III.E.   Comparison of Results

We have several expressions for the induced surface tempera-ture.

For the uniformly irradiated surface of a semi-infinte sample (i.e.: III.A.)

$$T_A(0,0,t) = \frac{2\alpha I_0}{K} \left(\frac{\kappa t}{\pi}\right)^{\frac{1}{2}}$$

.

For the uniformly irradiated beam of radius A (i.e., III.B)

$$T_B(0,0,t) = \frac{2\alpha P_0}{\pi A^2 K} (\kappa t)^{\frac{1}{2}} \left(\frac{1}{\pi^{\frac{1}{2}}} - ierfc \; \frac{A}{2(\kappa t)^{\frac{1}{2}}}\right)$$

and

$$T_B(0,0,\infty) = \frac{\alpha P_0}{\pi A K}$$

.

For the noncircular uniform beam (i.e., III.C)

$$T_C(0,0,\infty) = \frac{\alpha P_0}{2K\pi\ell a}\left(a\ \sinh^{-1}(\ell/a) + \ell\ \sinh^{-1}(a/\ell)\right)$$

For the Gaussian beam (i.e., III.D.2) with $I = I_0 e^{-r^2/w^2}$,

$$T_D(0,0,t) = \frac{\alpha I_0 w}{K\pi^{\frac{1}{2}}}\ \tan^{-1}(4\kappa t/w^2)^{\frac{1}{2}}$$

and

$$T_D(0,0,\infty) = \frac{\alpha I_0 w \pi^{\frac{1}{2}}}{2K}\ .$$

We may ask how much difference there is between these results. For example if we used a Gaussian beam but treated it as a circular beam of radius w with $I = I_0$, what would happen?

Thus we compare cases D and B. At equilibrium we have

$$\frac{T_D(0,0,\infty)}{T_B(0,0,\infty)} = \left.\frac{\alpha I_0\ w\ \pi^{\frac{1}{2}}}{2K}\middle/ \frac{\alpha P_0}{\pi w K}\right.$$

$$= \left.\frac{\alpha I_0\ w\ \pi^{\frac{1}{2}}}{2K}\middle/ \frac{\alpha I_0\ \pi w^2}{\pi w K}\right.$$

$$= \frac{\pi^{\frac{1}{2}}}{2} = 0.886$$

Thus $T_D(0,0,\infty) = 0.886\ \ T_B(0,0,\infty)$

and we see that the error in approximating the Gaussian by a uniform circle with average intensity equal to the peak on-axis
intensity of the Gaussian is only 11%. Considering the approximations involved in setting the thermal and optical properties
equal to some average value, this is acceptable. A similar conclusion could be reached when comparing the other on-axis, surface
temperature at other times. THUS, AN ACCEPTABLE ESTIMATE OF THE
ON-AXIS SURFACE TEMPERATURE PRODUCED BY A BEAM WITH $I(r = 0) \neq 0$
IS OBTAINED FROM THE SIMPLEST CASE WITH A FINITE BEAM RADIUS, THAT
IS CASE III.B.

To estimate the off-axis temperatures produced by a Gaussian
beam irradiation one could scale the on-axis temperature by a
Gaussian fall off in I_0. This approximation will be examined in
subsequent sections when we examine different approaches to calculating temperature distributions.

III.F   Some Numerical Examples

Now we shall try out some numbers to get a feeling for the
scale of things and to have in mind in the later discussions
where we include phase changes. We will use aluminum, 304 stainless steel and carbon phenolic as interesting examples. Their
properties are as listed:

Material	$K(W/cm\,^{o}C)$ Solid	Liquid	C $(J/g\,^{o}C)$	ρ (g/cm^3)	κ (cm^2/sec)
Al	2.0	1.0	1.0	2.7	0.74
304 S.S.	0.26	0.26	0.6	8.0	0.054
Carbon Phenolic	0.01	--	1.7	1.45	0.004

Material	$L_m(\frac{kJ}{g})$	$L_v(\frac{kJ}{g})$	$T_m^*(^{o}C)$	$T_v^*(^{o}C)$
Al	0.4	11	640	2430
304 S.S.	0.27	4.65	1430	2980
Carbon Phenolic		14.6		4000

*T_m and T_v are measured as the increase from $20\,^{o}C$ need to melt
or vaporize respectively.

The other property that we need to estimate the temperatures
which can be obtained is the material's absorbtivity for the
laser in question. The carbon phenolic can be assumed with

reasonable accuracy to absorb 100% of incident 1.06 and 10.6 μm
laser light. The Al and 304 S.S. absorbtivities can be questioned
as they will depend on surface finish and the presence of any
molten material. Also, the absorbtivity of a metal is expected
to increase with temperature. Recent measurements in our laboratory for Al and for 1018 steel show that at 10.6 μm the Drude
model gives accurate values for both the absorptivity and its
temperature coefficient. [4] At 1.06 μm the experimental results
do not agree with the Drude Model. The absorbtivity is higher
than predicted. While we are working further on this question,
we have still further data to add to the confusion. Whenever we
melt even a little bit of any metal the heat affected zone that
results corresponds to an absorption of ∿20% of the incident
laser light. While this may be due to an unexpected increase in
absorbtivity at the melting temperature or due to geometrical
considerations upon melting we feel that a fair estimate for the
absorbtivity of almost any metal exposed to 1.06 and 10.6 μm
laser beams is ∿10%. (You can't go too far wrong with this estimate particularly if a little melting occurs.)

For a Gaussian beam at the surface and on axis

$$T(0,0,t) = \frac{\alpha I_0 w}{K \pi^{\frac{1}{2}}} \tan^{-1} \left(\frac{4 \kappa t}{w^2} \right)^{\frac{1}{2}}$$

$$= \frac{\alpha I_0}{K \pi^{\frac{1}{2}}} \; 2(\kappa t)^{\frac{1}{2}} \quad \text{for small } t \text{ or}$$

$$\text{for } w >> 2(\kappa t)^{\frac{1}{2}}$$

$$= \frac{\alpha P_0}{K \pi w^2} \; 2 \left(\frac{\kappa t}{\pi} \right)^{\frac{1}{2}}$$

and

$$T(0,0,\infty) = \frac{\alpha I_0 w \pi^{\frac{1}{2}}}{2K}$$

$$= \frac{\alpha P_0}{2K \pi^{\frac{1}{2}} w} \quad .$$

Let us assume a 500W beam focused to a spot with w = 0.02 cm
(200 μm). Then we have:

	Al	304 S.S.	Carbon Phenolic
$T(0,0,\infty)$	$353^\circ C$	$2712^\circ C$	$705,237^\circ C$

Obviously: Al gets warm, 304 S.S. melts, and Carbon Phenolic
           vaporizes.

We can see from this the role of K and α! Note that more power
or smaller w enter the answer linearly!

Now

$$\frac{T(0,0,t)}{T(0,0,\infty)} = \frac{4(\kappa t)^{\frac{1}{2}}}{w\pi} \quad .$$

Let us assume that the approximation $w \gg 2(\kappa t)^{\frac{1}{2}}$ holds if

$$\frac{w}{2(\kappa t)^{\frac{1}{2}}} = 5 \quad .$$

For times such that this inequality holds we have

$$\frac{T(0,0,t)}{T(0,0,\infty)} \approx 0.3 \quad .$$

In other words we reached one third of the final temperature in
the following times:

	Al	304 S.S.	Carbon Phenolic
$T_{1/3}$	5.4×10^{-6} sec	7.4×10^{-5} sec	1×10^{-3} sec

III.G. The Time Required to Achieve Melting or Vaporization

    It is interesting to consider the time required to achieve
melting or vaporization for a given intensity. Assuming that
this can be accomplished while we are in the "short time" or
diffusion free range we have

$$T(0,0,t) = \frac{2 \alpha P_0}{K\pi^{3/2}w^2} (\kappa t)^{\frac{1}{2}}$$

and so

$$t_m = \pi^3 w^4 K^2 T_m^2 / 4 \alpha^2 P_0^2 \kappa \qquad (24)$$

$$\text{or} \qquad \qquad \text{or}$$

$$v \qquad \qquad v \qquad \qquad \quad .$$

This expression is quite useful in designing a process for a given laser. For example, if your laser is fixed in output capability and you must work a certain material the only variable you can adjust to obtain melting is $\underline{w}$. Thus you had best spend your time preparing a proper beam handling optics. (Of course it helps to have $t_{m\ or\ v} \propto w^4$.) If you work out the numbers for some typical materials and irradiation conditions, you will find out that these times can be very short. It is of interest to consider how things which were so heated then cool!

III.H. Rates of Heating and Cooling

Some elementary considerations concerning pulsed heating and cooling (These matter a great deal when considering studies of rapid resolidification, special alloying, annealing and metastable phases.) We are concerned with order of magnitude estimates of the heating and cooling reates and for this purpose will assume no latent heat due to phase transitions. There are two limiting cases:

A: When α is very large or the optical absorption depth α^{-1} is very small compared to the thermal diffusion depth $(\kappa t_p)^{\frac{1}{2}}$. Here t_p is the duration of the irradiation.

In this case the energy absorbed per unit area (assuming a uniform beam) is

$$\alpha I_0\ t_p$$

and it is used to heat a layer

$$(\kappa t_p)^{\frac{1}{2}}$$

thick. Thus

$$\Delta T = \alpha I_0\ t_p / C\rho (\kappa t_p)^{\frac{1}{2}}.$$

Note: [This differs by $2/\pi^{\frac{1}{2}} = 1.13$ from the result of Carslaw and Jaeger.] and the heating rate is

$$\frac{\Delta T}{t_p} = \alpha I_0 / C\rho (\kappa t_p)^{\frac{1}{2}}.$$

After the laser pulse it takes about the same time, t_p, for the heat from the layer to diffuse a distance $(\kappa t_p)^{\frac{1}{2}}$ into the material. During this time the temperature at the surface drops by an amount which is of order of magnitude equal to ΔT. Thus the cooling rate is also

$$\frac{\Delta T}{t_p} = \alpha I_0 / C\rho (\kappa t_p)^{\frac{1}{2}} \quad .$$

B: If α^{-1} is greater than $(\kappa t)^{\frac{1}{2}}$ then the light is absorbed in the medium according to

$$I = I_0 \, e^{-\alpha z}$$

and a temperature distribution is created which is roughly given as

$$\Delta T(z) = \alpha(1-R)I_0(e^{-\alpha z}) \, t_p / C\rho$$

where R = reflectivity at the surface.

After the pulse cooling will occur if heat diffuses a distance of $\approx 1/\alpha$ and so the cooling time is approximately given as

$$t_c = \frac{(\alpha^{-1})^2}{\kappa} \quad .$$

Thus the cooling rate at the surface is

$$\frac{\Delta T(z)}{t_c} = (1-R) \alpha^3 \, I_0 \, \kappa t_p / \rho C \quad .$$

In most applications (i.e., annealing) one only wishes to obtain $T = T_m$ and so by comparison it is clear that the cooling rate for Case A will always be very high. In fact

$$\frac{dT}{dt} = \frac{T_m}{t_p}$$

and if one has sufficient energy to obtain T_m in say a 10^{-9} sec pulse one will have

$$\frac{dT}{dt} \approx 10^{12} \, ^\circ C/sec \; !!$$

P. L. Liu et.al [5] have annealed Si with 3×10^{-11} sec pulses for a record $\frac{dT}{dt} = 4.7 \times 10^{13} \, ^\circ C/sec$.

IV. LASER HEATING INCLUDING MELTING, VAPORIZATION AND BURN-
     THROUGH

IV.A Formulation of the Problem

     This section is based on the work of R. J. Harrach [6] in
which he used the "heat balance integral" method to calculate the
space and time dependent thermal response of a slab of arbitrary
thickness to a time-varying laser source. One could continue the
Carslaw and Jaeger approach but it often helps to try out different procedures.

     In this scenario laser radiation impinges on the front sur-
face of the material at time t = 0. A fraction α is absorbed
and a fraction $1 - \alpha$ is reflected. As a result of the absorption
of energy, the surface temperature rises and a distribution $T(z,t)$
arises in the material due to thermal conduction. If the laser
is sufficiently intense, the surface will melt and finally vaporize. A crater forms and the material ablates as the vapor-solid
boundary moves into the medium. There are the following important
times in this problem:

  t = 0   Irradiation begins

  t = t_ℓ the back surface (z = ℓ) undergoes an appreciable
             temperature rise

  t = t_v the front surface (z = 0) begins to vaporize

  t = t_{BT} the back surface (z = ℓ) begins to melt or burn-
             through occurs.

In general t_ℓ can precede or follow t_v as can t_{BT}. The exact
relations depend on the irradiation and the properties of the
target.

The problem then is to solve the heat flow equation

$$\kappa \, \frac{d^2 T(z,t)}{dz^2} \;=\; \frac{\partial T(z,t)}{\partial t} \tag{25}$$

subject to the boundary conditions:

$$-K \frac{\partial T}{\partial z} = 0 \qquad \text{for } z = \ell \text{ and } 0 \leq t \leq t_{BT}$$

assuming that the back surface is insulated

$$= \alpha I(t) \qquad \text{for } z = 0 \text{ and } 0 \leq t \leq t_v$$

$$= \alpha I(t) - \rho L \dot{z}_s(t) \text{ for } z = z_s(t) \geq 0 \text{ and } t_v \leq t \leq t_{BT}$$

L is the total latent heat and $z_s(t)$ describes the moving surace of the vapor-solid boundary

$$T(z,t_i) = 0 \qquad \text{for } t_i = 0 \text{ and } 0 \leq z \leq \ell$$

$$= U(z) \qquad \text{for } t_i = t_\ell \text{ or } t_v \text{ and } z_s \leq z \leq \ell$$

The unknown position of the ablating surface is to be determined along with $T(z,t)$. Since this additional unknown enters the problem after $t = t_v$ we must have one additional condition in order to solve for it. This is taken to be the balance between laser beam absorption and surface vaporization to maintain the surface at T_v or

$$T(z = z_s(t), t) = T_v \text{ for } t \geq t_v \quad .$$

We note that we are still solving one dimensional heat flow problems.

The calculations can include a time varying pulse shape $I(t)$, and $\alpha = \alpha(T)$. However, it is assumed that the thermal parameters are constants and that the temperature variations in a single condensed phase are to be found. The presence of the liquid phase is taken into account by the definition of the effective melting temperature and the total latent heat. That is

$$T'_M = T_M (1 + (L_M/CT_M))$$

and

$$L = L_M + L_V \quad .$$

Where T_M = actual melting temperature in degrees above the ambient. The only heat loss mechanism included is LONGITUDINAL HEAT CONDUCTION. Convection, radiation and radial heat conduction are neglected. Also, it is assumed that the absorption occurs in an

infinitesinally small layer. Any vapor blow-off plume for $t > t_v$
is assumed to be sufficiently cool, nonionized and tenuous to be
transparent to the incoming laser beam.

The last assumption is generally all right except under two
conditions. In one case it is possible to be using a sufficiently high laser intensity that following t_v a laser supported
air detonation (LSD) wave is produced. This wave effectively
decouples the incoming radiation from the target. If the intensity is less than that needed for an LSD wave but sufficient to
obtain vaporization, one might obtain a laser supported combustion
(LSC) wave. There is rapidly growing evidence that in the presence of an LSC wave the coupling of the incoming energy to the
target is significantly enhanced. [7]   We shall use Harrach's
assumption but we should remember that α may need to be changed
when an LSC wave is formed.

The crux of the heat balance integral method is that the
heat flow equation must only be satisfied in an average sense
throughout the medium and not at each point z. We integrate
Eq. (25) from z_s to ℓ and write

$$\int_{z_s}^{\ell} \kappa \, \frac{\partial^2 T}{\partial z^2} \, dz \;=\; \int_{z_s}^{\ell} \frac{\partial T}{\partial t} \, dz \quad .$$

In general the derivative of an integral is

$$\frac{d}{dt}\left(\int_{z_s}^{\ell} T \, dz \right) = \int_{z_s}^{\ell} \frac{\partial T}{\partial t} \, dz + T(z=\ell,t)\, \frac{\partial \ell}{\partial t} - T(z=z_s,t)\, \frac{\partial z_s}{\partial t} \quad .$$

Thus

$$\kappa\left(\frac{\partial T}{\partial z}\bigg|_{z=\ell} - \frac{\partial T}{\partial z}\bigg|_{z-z_s} \right) = \frac{d}{dt}\left(\int_{z_s}^{\ell} T \, dz \right) + T(z_s,t)\, \dot{z}_s \quad .$$

For the post vaporization state we insert the boundary conditions and have

$$\frac{d}{dt}\left(\int_{z_s}^{\ell} T \, dz\right) = \frac{\kappa}{K}\left[\alpha I(t) - \rho L \dot{z}_s(t)\right] - T_v \, \dot{z}_s(t) \tag{26}$$

For prevaporization we can set $z_s = \dot{z}_s = 0$ to obtain

$$\frac{\partial}{\partial t}\left(\int_0^{\ell} T \, dz\right) = \frac{\kappa}{K} \alpha I(t) \tag{27}$$

IV.B The Prevaporization Stage

This case, for $0 \leq t \leq t_\ell$ and $t \leq t$ and α = constant, is heating without a phase change as before in III.B.1! In other words, we have the case of laser heating a very thick slab.
Harrach selects the following solution to Eq. (37)

$$T(z,t) = T(0,t)\left(1 - z/\delta(t)\right)^2 e^{-z/\delta(t)} \text{ for } 0 \leq z \leq \delta(t) \leq \ell$$

$$= 0 \text{ for } z > \delta(t) \quad . \tag{28}$$

Here $\delta(t)$ is a thermal penetration depth. When
$z = 0.213$ then $T(z,t) = T(0,t)/2$. At
$z = 0.5$ then $T(z,t) = 0.152 \, T(0,t)$

We define the diffusion time or the back surface heating time t according to

$$\delta(t_\ell) = \ell \quad . \tag{29}$$

Integration of this solution from 0 to ℓ and using it in Eq. (27) gives the differential equation

$$\frac{d}{dt}\left(T(0,t)\right)\delta(t) = \frac{\kappa\alpha I(t)}{\left(1 - \frac{2}{e}\right)K} \quad .$$

Differentiating the solution with respect to t and using the boundary conditions gives

$$8KT(0,t)/\delta(t) = \alpha I(t)$$

and so

$$T(0,t) = \left(\frac{\kappa}{3(1 - \frac{2}{e})K^2}\right)^{\frac{1}{2}} \left(\alpha I(t) \int_0^t \alpha I(t')dt'\right)^{\frac{1}{2}}$$

and

$$\delta(t) = \left(\frac{3\kappa}{1 - 2/e}\right)^{\frac{1}{2}} \left(\int_0^t \frac{\alpha I(t')dt'}{\alpha I(t)}\right)^{\frac{1}{2}} .$$

IV.B.1 The Step Function Irradiation

The simplest case is that of a step function $I(t)$ and constant α.

Then

$$T(0,t) = \left(\frac{\kappa t}{3\left(1 - \frac{2}{e}\right)}\right)^{\frac{1}{2}} \frac{\alpha_0 I_0}{K} = 1.123 \frac{\alpha_0 I_0}{K} (\kappa t)^{\frac{1}{2}}$$

(30)

$$\delta(t) = \left(\frac{3\kappa t}{(1 - \frac{2}{e})}\right)^{\frac{1}{2}} = 3.37 (\kappa t)^{\frac{1}{2}}$$

In section III.B we derived the following expression for $T(0,t)$ at early times

$$T(0,t) = \frac{2 \alpha_0 I_0}{K} (\kappa t)^{\frac{1}{2}} \frac{1}{\pi^{\frac{1}{2}}}$$

$$= 1.1287 \frac{\alpha_0 I_0}{K} (\kappa t)^{\frac{1}{2}}$$

which is nearly the same as Harrach's solution.

By inserting Eq. (30) into Eq. (28) we then obtain the solution for $T(z,t)$.

Now we wish to find t_ℓ and t_v in this example

By definition

$$\delta(t_\ell) = \ell$$

$$\text{or} \quad 3.37 (\kappa t_\ell)^{\frac{1}{2}} = \ell$$

whereupon

$$t_\ell = \frac{\ell^2}{(3.37)^2 \kappa} \quad \text{for } \alpha_0 I_0 \ell / 3KT_v \lesssim 1$$

.

Again by definition

$$T(0, t_v) = T_v = 1.123 \frac{\alpha_0 I_0}{K} (\kappa t_v)^{\frac{1}{2}}$$

and so

$$t_v = \frac{T_v^2 \, K^2}{(1.123 \alpha_0 I_0)^2 \kappa} \quad \alpha_0 I_0 \ell / 3KT_v \gtrsim 1$$

.

However the first result holds only if $\alpha_0 I_0 \ell$ is such that $t_\ell < t_v$. That is our solution holds only if the front surface does not vaporize. This means that the inequality must hold. On the other hand the equation for t_v holds only so long as $t_v < t_\ell$ or else back surface heating must be taken into account. In this case the reverse inequality must hold. (Note that in Section III-G we found t_v from the Carslaw and Jaeger approach. The ratio of Harrach's value to that in III-G is 1.0096.)

One could, if one knew the form of $\alpha = \alpha(t)$ include it in the integration. However, in most cases it will be sufficient to include an appropriate average value. In general it is better to weigh the average more heavily towards the high temperature values since at high temperatures α for metals can be 5 to 10 times greater than at room temperature and during laser irradiation the material reaches high temperatures quite early on.

IV-C For the Time Interval $t_\ell \lesssim t \lesssim t_v$

If $t_\ell < t_v$ and we use a step function $I = I_0$ and $\alpha = \alpha_0$ is a constant or the inequality

$$\alpha_0 I_0 \ell / 3KT_v < 1$$

holds then up to the time t_ℓ the preceding expression applies.

For the interval $t_\ell \lesssim t \lesssim t_v$ Harrach uses the solution

$$T(z,t) = (1 - \frac{z}{\ell})^2 \left(e^{-z/\ell}\right) \left(T(0,t) - T(\ell,t)\right) + T(\ell,t) \quad .$$

Substituting into the differential equation results in

$$T(0,t) - T(\ell,t) = \ell \alpha I(t)/3K$$

and

$$T(0,t) = \left(\frac{\kappa}{\ell K}\right) \int_{t_\ell}^{t} \alpha I(t') \, dt' + \left((\frac{2}{e}) \ell/3K\right) \quad x$$

$$\left(\alpha I(t) + \frac{(1-2/e)}{2/e} (\alpha I(t))\Big|_{t=t_\ell}\right) \quad .$$

We see from the first of these that during the time interval in question the front to back temperature difference depends only

on $\ell \alpha I(t)$!!

For $I = I_0$ for $t \geq 0$

$= 0$ for $t \leq 0$

and $\alpha = \alpha_0$, a constant we find

$$T(0,t) = \frac{(2/e)\alpha_0 I_0}{3K} \left(\frac{3\kappa t}{(2/e)\ell^2} + 1\right) \tag{31}$$

and

$$T(\ell,t) = \frac{(1 - 2/e)\alpha_0 I_0 \ell}{3K} \left(\frac{3\kappa t}{(1-2/e)\ell^2} - 1\right) \tag{32}$$

where $t_\ell < t \leq t_v$ and t_ℓ is as derived earlier.

It is interesting to find the front surface temperature as a function of the back surface temperature because often one will want to know the former but can only measure the latter. We take the ratio of $T(0,t)$ and $T(\ell,t)$ and find

$$\frac{T(0,t)}{T(\ell,t)} = \frac{1}{1 - \left(\frac{\ell^2}{\left(\frac{2}{e}\right)\ell^2 + 3\kappa t}\right)} .$$

From this equation it is possible to find $T(0,t)$. We might note here that $T(\ell,t)$ should be a linear function of t. If during the measurements deviations from linearity are found, they may be attributed to changes in α_0 or the thermal properties with T. Since the latter are usually not strongly dependent on T it is possible to infer some of the properties of $\alpha(T)$ from such measurements.

Setting $T(0,t) = T_v$ gives the vaporization time

$$t_v = \left(\frac{KT_v\ell}{\kappa\alpha_0 I_0}\right) - \frac{2}{e}\frac{\ell^2}{3\kappa} . \qquad (33)$$

Similarly from the condition $T(\ell,t_{BT}) = T'_m - T_m(1 + L_m/CT_m)$ we obtain the burn through time

$$t_{BT} = \frac{(1 - 2/e)}{3}\frac{\ell^2}{\kappa} + \frac{\ell KT'_m}{\kappa\alpha_0 I_0} \qquad (34)$$

which holds for

$$\frac{\alpha_0 I_0 \ell}{3K(T_v - T'_m)} < 1 .$$

This latter inequality insures that the front surface does not vaporize before melt through occurs.

IV-C Post Vaporization $t_v \lesssim t \lesssim t_\ell$

This is the case that the sample behaves as a semi-infinite solid. An accurate trial solution can be written in terms of the instantaneous position $z_s(t)$ of the front surface and a thermal penetration depth $d(t)$ measured from $z_s(t)$ (See Fig. 3). Thus for positions in the interval

$$0 \lesssim (z - z_s(t)) \lesssim d(t)$$

and times

$$t_v \lesssim t \lesssim t$$

Harrach gives:

$$T(z,\ell) = T_v \left(1 - \frac{z - z_s(t)}{d(t)}\right)^2 e^{-(z - z_s(t))/d(t)}$$

For $z - z_s(t) > d(t)$, $T(z,\ell) = 0$.

Note the similarity to the case of prevaporization heating.

The time t_ℓ is now given by

$$d(t_\ell) = \ell - z_s(t_\ell) \qquad .$$

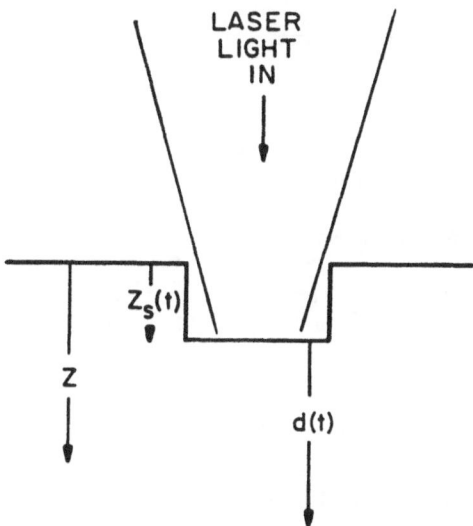

Fig. 3.   Definitions for the heat flow problem in the
          post vaporization stage.

Inserting the assumed solution into the differential equations
gives

$$3KT_v/d(t) = \alpha I(t) - L\dot{z}_s(t)$$

and

$$\dot{d}(t) = \frac{\kappa(1-(KT_v/\kappa\rho L))}{(1-2/e)\,KT_v}\,(\alpha I(t) - L\dot{z}_s(t)) - \frac{\alpha I(t)}{(1-2/e)\rho L}\,.$$

At this point Harrach defines a number of dimensionless variables, rewrites the equations, and manipulates them to obtain solutions in specific cases. The point is that while he can set down certain results one of the most important is the case where no back surface heating occurs (i.e.: the piece appears as if it were a semi-infinite sheet).

Instead of Harrach's treatment we will do this one more simply. We have shown (Section III.B) that

$$T(0,0,t) = \frac{2\alpha P_0}{\pi^{3/2} A^2 K} (\kappa t)^{\frac{1}{2}}$$

is a very good estimate of the surface temperature and that the time required for vaporization is

$$t_v = \pi^3 A^4 K^2 T_v^2 / 4 P_0^2 \alpha^2 \kappa \quad .$$

When vaporization occurs a volume

$$V = \pi A^2 v$$

is vaporized per unit time where $v = \dot{z}_s(t)$ in Harrach's notation. The energy required to do this is

$$Q = \pi A^2 v (\rho C T_v + L) \quad .$$

This amount of energy must be absorbed per unit time in order to vaporize material and, for example, drill a hole. The energy absorbed per second per unit area is

$$\alpha I_0$$

and that absorbed in the irradiated area is

$$\alpha I_0 A^2 \pi \quad .$$

Thus the speed of the vaporization surface is obtained by setting

$$\alpha I_0 A^2 = A^2 v (\rho C T_v + L)$$

and

$$v = \frac{\alpha I_0}{\rho C T_v + L} \quad . \tag{35}$$

{Note: In order to be physically real $v \lesssim v_{max}$ where

$$v_{max} = C' \exp(-Lz/\rho NkT_v)$$

is the speed with which particles can leave a surface at T

Here C' = speed of sound in the material

z = atomic number of the material

N = Avogadros Number

k = Planck's constant.}

We can ask a practical question.   How deeply will a certain laser drill a hole and would it be better to use a cw or a pulsed laser?

The depth drilled would be

$$\ell_{cw} = v \, t_{dw}$$

where t_{dw} = the dwell time of the laser on the site.

Thus

$$\ell_{cw} = \frac{\alpha I_0^{cw} t_{dw}}{\rho CT_v + L} = \frac{\alpha E_0^{cw}}{\rho CT_v + L}$$

where $E_0^{cw} = I_0^{cw} t_{dw}$ = the total energy incident upon the site during the irradiation!

If we used a repetitively pulsed laser operating at a pulse repetition frequency of f Hz, we would have to sum up the incremental hole depths drilled by each pulse.   That is

$$\ell_p = \sum_{t_{dw}} v \, t_p \qquad \text{where } t_p = \text{the duration of each laser pulse} .$$

Here we are making the assumption that for the laser pulse intensity and dwell time the material reaches T_v and $t_v \ll t_p$.

The number of pulses used is ft_{dw} and so

$$\ell_p = ft_{dw} \frac{\alpha(I_0^P)t_p}{\rho CT_v + L} .$$

Here, $E_p = I_0^P t_p$ and the total energy incident on the sample is

$$E_0^P = ft_{dw} \, I_0^P t_p .$$

again

$$\ell_p = \frac{\alpha E_0^{\ P}}{\rho C T_v + L}$$

suggesting that if all else were equal, equal average power cw or repetitively pulsed lasers would drill equal holes. However, all things are not equal. The intensity in the pulses will be significantly greater than I_0^{cw}. As a result the material's vapor may react with the incoming light to produce a hot plasma which can under certain circumstances greatly increase the fractional coupling of energy into the surface.[7] Also as a hole is formed the coupling may increase due to the altered geometry. Finally when driven by an intense pulse, the vapor pressure can be very high and it can act to support the walls of the hole so that the light can penetrate within the medium.

V. The Moving Laser Beam

In this section we will briefly touch upon the question of thermal analysis of the problem of laser heating when the irradiated site moves through the medium. That is, what differences, if any, are there when the sample and the light spot are translated with respect to each other. To do this we use the approach of Cline and Anthony.[8]

A coordinate frame fixed in the medium is used. The light beam impinges parallel to the z axis and strikes the sample (the plane z = 0) at time t = 0 (See Fig. 1). The laser moves in the +x direction with a speed v. In order to be more realistic a Gaussian intensity distribution is assumed. A moving Gaussian normalized to give a total power P for a spot of radius R is

$$Q = \frac{P\ e^{-\left(\left((x-vt)^2 + y^2\right)/(2R^2)\right)}}{2\pi R^2} \quad \frac{h(z)}{\lambda} \tag{36}$$

where λ = absorption depth
and $h(z) = 0$ for $z > \lambda$. The heat diffusion equation

$$\frac{\partial T}{\partial t} - \kappa \nabla^2 T = \frac{Q}{C} \tag{37}$$

is solved by superposition of the known solution for the thermal distribution of a unit point source G at the coordinates x' y' t' on the surface. Heat flow from this source affects the temperature at (x,y,z) and a later time t. The solution technique is that of using a moving Green's function. Cline and Anthony give the Green's function for the diffusion equation at the surface as

$$G(x'y'z't') = \frac{e^{-r^2/4\kappa(t-t')}}{4\{\pi\kappa(t-t')\}^{3/2}}$$

where $r^2 = (x - x')^2 + (y - y')^2 + (z - z')^2$ (38)

and at the surface $z' = 0$. At $t = 0$, the temperature distribution of the laser beam is a superposition of Gaussians at earlier times t' when the laser was at x' y'. The superposition is obtained from the integral expression

$$T = \int_{-\infty}^{t} \int_{-\infty}^{\infty} \int_{-\infty}^{\infty} \int_{-\infty}^{\infty} \frac{Q(x'\,y'\,z't')}{C}$$

$$G(x'y'z't'|xyzt) \, dx'dy'dz'dt' \quad (39)$$

Now fortunately the x' and y' integrals result in a known Gaussian form. Thus we find

$$T = \frac{P}{C} \int_{0}^{\infty} \frac{\exp\left(-\left(\left((x + vt'')^2 + y^2\right)(2R^2 + 4\kappa t')^{-1} + z^2/4\kappa t'\right)\right)dt'}{(\pi^3 \kappa t')^{\frac{1}{2}}(2R^2 + 4\kappa t')}$$

(40)

In the limit of $R \to 0$ we have a moving point source. With this simplification we can integrate the equation to find

$$T = \frac{P}{C\kappa 2\pi r} \exp\left(-\frac{v(r + x)}{2\kappa}\right) \quad (41)$$

where $r^2 = x^2 + y^2 + z^2$. We note here the onset of answers that are not one dimensional. In most cases this is more than is necessary but it is included for completeness sake. We see from the last equation that the temperature of a moving point source is infinite at $r = 0$ and falls off with distance from the source: For a beam of finite radius the following unit changes lead to a form of T which can be evaluated numerically:

Let
$$\mu^2 = \frac{2\kappa t}{R^2}, \rho = (R/D)v, X = \frac{x}{R}, Y = y/R \text{ and } Z = z/R$$

Eq. (35) then gives

$$T(X,Y,Z) = \frac{P}{CR} \ f(X,Y,Z,v) \tag{42}$$

where

$$f = \int_0^\infty \frac{\exp(-H)}{(2\pi^3)^{\frac{1}{2}}(1+\mu^2)} \ d\mu \tag{43}$$

where

$$H = \frac{\left(X + \frac{\rho}{2}\mu^2\right)^2 + Y^2}{2(1+\mu^2)} + \frac{Z^2}{2\mu^2} \quad . \tag{44}$$

Those who wish to examine the numerical solutions of Cline and Anthony are welcome to do so. We will consider a useful special case.

We want the surface temperature on-axis. Thus

  z = 0
  x = 0
  y = 0

and

$$H = \frac{\rho^2}{4}\mu^4/2(1+\mu^2) \tag{45}$$

In the limit that v = 0, ρ = 0 and so H = 0. This gives

$$f = \int_0^\infty \frac{d\mu}{(2\pi^3)^{\frac{1}{2}}(1+\mu^2)} = \frac{1}{(2\pi^3)^{\frac{1}{2}}} \ \tan^{-1}\mu \Big|_0^\infty$$

$$= \frac{1}{(2\pi^3)^{\frac{1}{2}}}(\frac{\pi}{2} - 0) = \frac{1}{2^{3/2}\pi^{\frac{1}{2}}}$$

and

$$T(0,0,0) = \frac{P}{\kappa CR} \ \frac{1}{2^{3/2}\pi^{\frac{1}{2}}} \tag{46}$$

For a Gaussian beam we found in Section III-E that the final temperature on-axis and on the surface is given by

$$T(0,0,\infty) = \frac{\alpha \, I_0 w \pi^{\frac{1}{2}}}{2K}$$

$$= \frac{\alpha \, P_0}{2\pi^{\frac{1}{2}} w \rho C \kappa}.$$

In order to compare this Carslaw and Jaeger result with Eq. (46) we must re-examine things a bit. In Cline and Anthony the specific heat, thermal conductivity and thermal diffusivity are related by

$$K = C\kappa$$

where we have been using

$$K = \rho C \kappa .$$

Thus we can replace Cline and Anthony's κ with our $\rho \kappa$. Next we note that Cline and Anthony use a value of R such that at $r = R$

$$Q(r) = Q(0)e^{-\frac{1}{2}}$$

We used w such that at $r = w$

$$Q(r) = Q(0)e^{-1}$$

Thus $w = \sqrt{2} \, R$. With this we have from Eq. (44)

$$T(0,0,0) = \frac{\alpha P_0}{\kappa \rho C w} \, \frac{1}{2\pi^{\frac{1}{2}}}$$

where Cline and Anthony's "total power" is related to the incident power by the absorbtivity . In other words the results are identical!! Of course, we have used a very special case to check out Cline and Anthony's approach. Nevertheless it gives one some confidence in the results and where necessary the Cline and Anthony result can be used for detailed analyses. Again, remember that Cline and Anthony make no provision for temperature varying absorptions and thermal properties. Also they do not show how to use their approach if a phase change occurs. As a result it is more often than not most valuable to make the best guesses one can and proceed to use the simplest analysis for calculating laser induced thermal distributions.

Definitions Used Throughout

   T = temperature, $^\circ$C

   t = time, sec

   x = space coordinate, cm

   y = space coordinate, cm

   z = space coordinate, cm

   ρ = material density, gm/cm^3

   K = thermal conductivity, W/cm$^\circ$C

   C = heat capacity, J/gm$^\circ$C

   κ = thermal diffusivity, cm^2/sec

   α = absorbtivity, cm^{-1}

   I = intensity, W/cm^2

   P = power, W

   A = radius of uniformly irradiated circle

   w = Gaussian beam radius at which the intensity is 1/e of its
        on axis value

   t_p or T_p = pulse duration

   L_m = latent heat of melting, J/g

   L_v = latent heat of vaporization, J/g

   T_m = melting temperature, $^\circ$C

   T_v = vaporization temperature, $^\circ$C

   t_v = time required to initiate front surface vaporization

   t_m = time required to initiate front surface melting

Additional Definitions Used in Section IV:

   t_ℓ = time at which back surface (at $z = \ell$) undergoes appreci-
        ble heating

   t_v = time at which front surface (at $z = 0$) undergoes vapori-
        zation

   t_{BT} = time at which back surface begins to melt

   $\delta(t)$ = a thermal penetration depth

   $z_s(t)$ = instantaneious position of the front surface.

References

1. H. S. Carslaw and J. C. Jaeger, "Conduction of Heat in
    Solids", 2nd ed. (Oxford University Press (Clarendon)
    London and New York, 1959).
2. W. W. Duley, "CO_2 Lasers: Effects and Applications"
    (Academic Press, New York, 1976).
3. J.F. Ready, "Effects of High Power Laser Radiation"
    (Academic Press, New York, 1971).
4. M. Bass, D. Gallant and S. D. Allen, "The Temperature
    Dependence of Optical Absorption in Metals" NBS
    Special Publication No. 574, p. 48, May 1980,
    and D. Gallant, M.S. Thesis, University of Southern
    California, June 1980.
5. P. L. Liu, R. Yen, N. Bloembergen and R. T. Hodgson,
    Appl. Phys. Letters 34 : 864 (1979).
6. R. J. Harrach, J. Appl. Phys. 48 : 2370 (1977).
7. J. A. McKay and J. T. Schriempf, Appl. Phys. Letters
    33 : 877 (1978).
8. H. E. Cline and T. R. Anthony, J. of Appl. Phys. 48 : 3895
    (1977).

THERMODYNAMICS AND KINETICS OF LASER-INDUCED STRUCTURE CHANGES

David Turnbull

Division of Applied Science
Harvard University, Cambridge, Mass.

1. THERMODYNAMICS OF SOLIDIFICATION AND MELTING

A. Modes of Solidification

We suppose that a body is solid when its shear viscosity, η, exceeds the somewhat arbitrarily chosen value of 10^{15} poise, at which the time constant for shear relaxation would be about one day. At temperatures below its thermodynamic crystallization temperature, T_m, a fluid will solidify heterogeneously by growth from seed crystals, or "nuclei", to a crystalline body. In this process η is increased discontinuously, by many orders of magnitude, with the advance of the crystal-fluid interface. Alternatively, the fluid may, under certain conditions, solidify homogeneously by the continuous increase of η with falling temperature (or increasing pressure) into the solid range. Experience shows that the temperature, T_g, at which this solidification occurs is always less than T_m and that the solid body formed is amorphous in structure. Amorphous solids can sometimes also form discontinuously by condensation of material from fluids onto substrates held at temperatures below T_g.

B. Configurational Freezing

Amorphous solids, however formed, are not in internal equilibrium but, rather, are slowly undergoing internal rearrangement ("relaxation") to configurations with lower energies. The rate of this relaxation falls sharply as temperature decreases so that at temperatures well below T_g the solid is virtually frozen into a single configuration. The particular configuration and energy

117

of the frozen state will be determined by the rate of quenching
through the temperature range of configurational freezing. Thus an
amorphous solid may, depending on the conditions of its formation,
exist in any one of a continuum of configurationally frozen states,
each state characterized, perhaps, by a particular set of positions
about which the atoms oscillate.

The configurational freezing of melts (melt - glass transition) is generally marked by sharp drops in heat capacity and
thermal expansion coefficient to levels quite close to those of the
crystal. These drops occur at temperatures which are lower the
lesser is the cooling rate and so, as Kauzman [2] explained, they
do not reflect any thermodynamic transition but only the configurational freezing.

The phenomenon of configurational relaxation and freezing is
not confined to amorphous solids but is very widespread. For
example, the equilibrium degree of compositional short range order
(SRO) in a disordered crystalline alloy is reached by interposi-
tional exchanges of atoms. As temperature is decreased the time
constant for these exchanges will increase and below some temperature become larger than the periods within which observations are
convenient. Thus the alloy becomes frozen into an arrangement in
which the SRO is quite different from what it would be at thermo-
dinamic equilibrium. Actually most of the solids of common
experience are in configurationally frozen states which are, to
some degree, out of thermodynamic equilibrium. Nearly all the
solids produced by the new microprocessing techniques, e.g. ion
implantation, ion beam mixing, laser pulsing, rapid melt quenching,
etc. are likely to be in frozen states far from configurational
equilibrium.

C. Metastability, Instability and Impurity Trapping

When in internal equilibrium a body may freely pass through
the large number of configurational microstates which partly
constitute its macrostate. A metastable body, e.g. an undercooled
liquid, is one in internal equilibrium but with a higher free
energy than the same mass in some different macrostate. Generally
a metastable body transforms to a more stable form by the macroscopically discontinuous process of nucleation and growth. A
configurationally frozen solid, while "kinetically" stable over
a long time period, is thermodynamically unstable in the sense
that its free energy may be decreased by highly localized atomic
rearrangements which are macroscopically homogeneous.

The isobaric courses of the chemical potential, μ of the
solid and molten states of a glass forming body are shown schematically in Figure 1.The chemical potential of the liquid in its

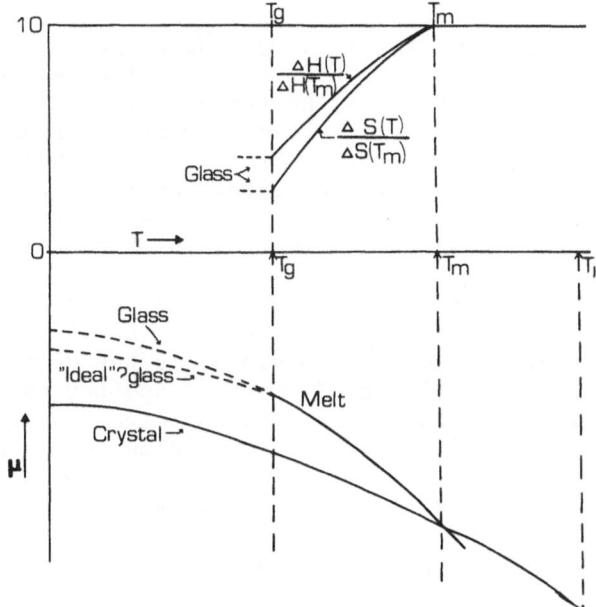

Fig. 1   Chemical potential μ of solid and
molten states of glass forming body

undercooled metastable state is well defined from T_m to T_g. At $T < T_g$ μ of the amorphous phase is no longer uniquely defined but rather exhibits at a given T a dispersion of values corresponding to the continuum of configurationally frozen states which may exist at that T.

The possible existence of a crystal in a state which is metastable relative to its melt, as depicted in Figure 1, is a much debated problem. However, it is well established [3] that crystals, such as quartz, cristobalite, albite or P_2O_5, which melt to very viscous fluids can be superheated quite far into their metastable range without internal melting. Thus the problem has mainly to do with the nature of melting to fluids with low viscosity.

As observed, such melting is always heterogeneous, i.e., it is effected by the motion of crystal-melt interfaces. The thermodynamic crystallization point is the temperature at which two highly dynamic opposing processes, the forward and reverse motion of a planar crystal-melt interface, are exactly in balance so that the interface is stationary. If the interface moves forward the melt in its vicinity must be undercooled while if it moves backward the crystal at the interface must be superheated to some degree. Presumably the regime of crystal metastability will be terminated at some temperature, T_I, where the shear modulus goes continuously to zero. Calculation of this instability point, at which homogeneous melting may be possible, is the object of most microscopic theories of melting. However, since the shear modulus is generally quite large and slowly falling with T at T_m, the metastability range should extend to very large superheatings.

The general procedure in microprocessing is to energize a material and then quench or condense it. In the quenching or condensation the material passes through metastable or unstable intermediate states of high atomic mobility which are far removed from the stable equilibrium state. It is then configurationally frozen into an unstable state derived from an intermediate state. The outcome of such procedures is determined by what is kinetically preferred within the range of thermodynamic possibilities.

Thermodynamically the given mass may, of course, reconstitute itself into any state having lower Gibbs free energy than that of the intermediate state. For example, supersaturated crystalline solutions can be produced by the crystallization of highly undercooled liquid solutions. The thermodynamic constraints on this process have been set forth by Baker and Cahn [4]. Of particular interest is the thermodynamic condition for solute "trapping" in which the crystal-melt interface sweeps through the solution without redistribution of solute in the interfacial region. The molar change in Gibbs free energy accompanying this process [Baker and Cahn's equation (23)] is

$$\Delta G = \chi_B^{\ell} \; \overline{\Delta\mu}_B + (1 - \chi_B^{\ell}) \; \Delta\mu_A \tag{1}$$

where χ_B^{ℓ} is the atom fraction of solute B in the liquid solution and to be trapped in the solid, $\Delta\mu_B$ and $\Delta\mu_A$ are the chemical potential changes upon crystallization of, respectively, solute B and solvent A. When the laws of the dilute solution apply:

$$\Delta\mu_A = RT \; \ln \; \frac{(1- \chi_B^{s}) \; (1 - \overline{\chi}_B^{\ell})}{(1- \chi_B^{\ell}) \; (1 - \overline{\chi}_B^{s})} \tag{2a}$$

and

$$\Delta\mu_B = RT \; \ln(\frac{\chi_B^{s} \; \overline{\chi}_B^{\ell}}{\chi_B^{\ell} \; \overline{\chi}_B^{s}}) \tag{2b}$$

where s and ℓ denote, respectively, the crystalline and liquid phase and $\overline{\chi}_B$ denotes the phase equilibrium atom fraction in the designated s or ℓ.

There may be one or more metastable states with free energy between the energized state reached in microprocessing and the most stable state of the mass. Which of the states the mass goes through or is quenched into will, of course, be determined by kinetic as well as thermodynamic considerations. According to an empirical rule, due to Ostwald, a metastable phase will tend to transform first into that phase which lies closest to it in free energy. There seems to be no microscopic foundation for this rule but it may reflect that the interfacial tension between phases, which in simple nucleation theory is a major determining factor in nucleation resistance, tends to scale as their standard free energy difference.

II. KINETICS AND MORPHOLOGY OF CRYSTALLIZATION AND MELTING

Many of the structural changes (e.g. epitaxial regrowth, glass formation) resulting from laser processing are determined by the kinetics and morphology of nucleation and growth of crystals from the energized state. These processes will be discussed within the framework of the simple nucleation theory, following the papers of Spaepen and the writer [5,6].

A. Nucleation of Crystals

In the simple theory the frequency/volume, I, of homophase (homogeneous) nucleation of crystals, at steady state, within a uniform phase is of the form:

$$I = Ak_i \exp[- b\ \sigma^3/(\Delta G_v)^2\ kT] \qquad (3)$$

where A is a constant which can be specified by kinetic analysis, k_i is the atomic exchange frequency/site across the interface formed by the nucleus, σ is the interfacial tension/area, here ass umed constant, between the crystal and the external phase, b is a geometric factor ($16\pi/3$ for a spherical nucleus) and ΔG_v is the change in Gibbs free energy/volume accompanying crystalliza- tion.

It is often convenient to scale σ in terms of the crystal- lization enthalpy/gm.atom, ΔH_c, thus:

$$\alpha = - (\bar{N}\ \bar{V}^2)^{1/3}\ \ \frac{\sigma}{\Delta H_c} \qquad (4)$$

where $\bar{V}$ is the volume/gr.atom of the crystal and $\bar{N}$ is Avogadro's number. Under conditions where the variation of the entropy of crystallization/volume, ΔS_v, with temperature is negligible the nucleation frequency can be expressed as :

$$I = A\ k_i\ \exp\left\{- \frac{b\ \alpha^3\ \beta}{(\Delta T_r)^2\ T_r}\right\} \qquad (5)$$

where $\beta = - \bar{V}\ \Delta S_v/\bar{N}\ k$, $T_r = T/\bar{T}$, $\Delta T_r = (\bar{T} - T)/\bar{T}$ and $\bar{T}$ is the thermodynamic equilibrium temperature.

Th e variation of the exponential factor in eq.(5) with reduced undercooling, ΔT_r for various choices of the parameter $\alpha\ \beta^{1/3}$ is shown in Fig. 2a [7,8]. The horizontal line rep- resents roughly the minimum value of the exponential, (corre- sponding to A k_i at an estimated maximum $\sim10^{35}$ cm^{-3}x sec^{-1}) at which nucleation would be detectable in a practicable experiment. We note that, according to the simple theory (1) the exponential rises very steeply with ΔT_r at small departures from equilibrium and (2) there is a range of undercooling over which nucleation would not be detectable, even at maximum A k_i, which is larger the larger is the parameter $\alpha\ \beta^{1/3}$.

a priori it is not clear how k_i should scale with the other

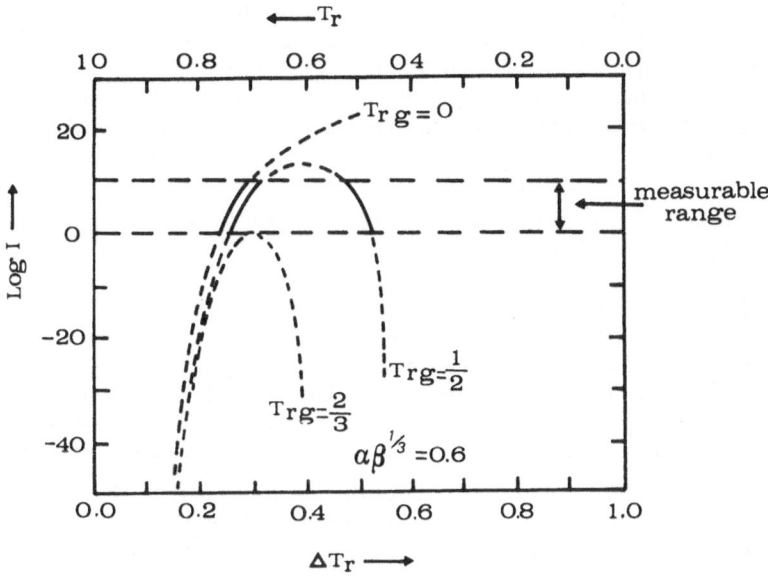

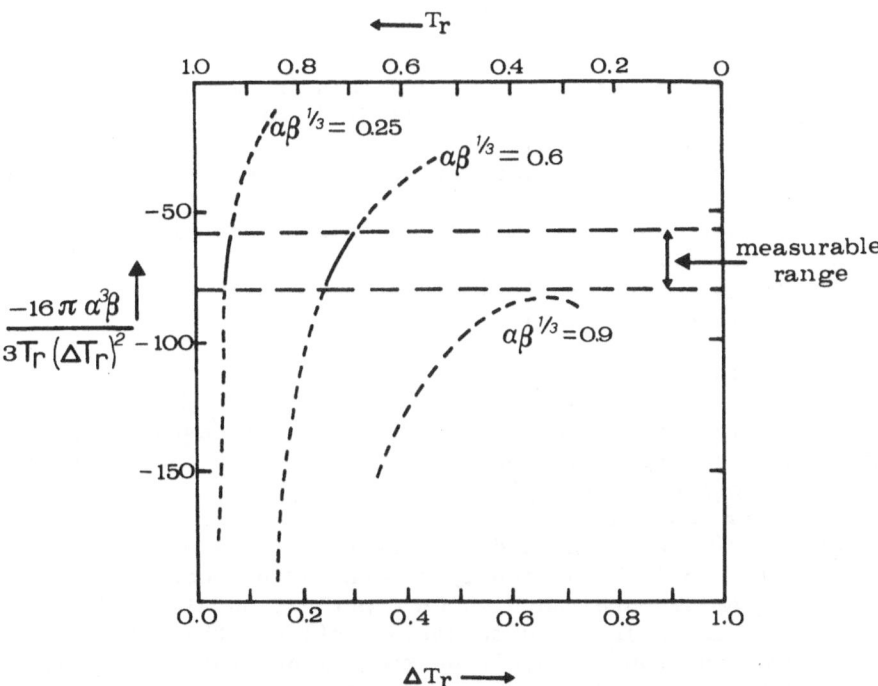

Fig. 2 Exponential factor of eq.(5) (Fig.2a) and nucleation
frequencies (Fig.2b) as a function of reduced undercooling ΔT_r

frequencies of atomic motion in the amorphous state, i.e., k_D for
diffusion, k_η for viscous flow and k_v for volume relaxation.
However, experience shows that for some systems k_i does scale
roughly with the other frequencies. Nucleation frequencies calculated with this scaling assumption and a particular choice of
$\alpha \beta^{1/3}$ and several choices for $T_{rg} = T_g/T_m$ are shown in Fig.2b. We
note that (1) I is negligible at $T_r \lesssim T_{rg}$ (2) as T_{rg} is increased the log I (T_r) curve is depressed and its maximum is
sharpened and displaced to smaller ΔT_r.

Under most conditions nucleation of crystals in amorphous
media occurs exclusively on heterophase impurities at relatively
small undercoolings. When the heterophase effects are eliminated
or minimized it is found that very large undercoolings, ΔT_r^0, are
required for the onset of measurable homophase nucleation in most
materials. For example, $\Delta T_r^0 \gtrsim 0.25$ to 0.3, corresponding to
$\alpha \beta^{1/3} \gtrsim 0.6$, for a number of pure metals. Also there are a considerable number of materials, fused SiO_2, B_2O_3 and o-terphenyl
are examples, in which the frequency of homophase nucleation is
negligible under all accessible experimental conditions. We see
that such a complete suppression of nucleation could occur, according to the simple theory, with $\alpha \beta^{1/3}$ and T_{rg} assignments
which would be quite plausible.

In summary, the nucleation of crystals in amorphous media by
the homophase fluctuation route is difficult and quite infrequent
excepting in certain materials at large undercooling. In the simple
nucleation theory this behavior reflects that the crystal-amorphous
phase interfacial tension is large relative to $|\Delta H_c|$. Consequen-
tly, crystallization in most materials is generally initiated by
heterophase impurities and the melts of many materials can be
quenched through their metastable ranges to configurationally
frozen states without crystallization.

B. Nucleation of Growth Steps

The kinetics of growth of crystals will, of course, depend
strongly upon the crystallographic orientation and roughness of
the crystal-amorphous interface.
Burton, Cabrera and Frank [9] showed that, at equilibrium, interfaces between dilute fluids and the crystal planes of densest
atomic packing should be virtually smooth and perfect on an atomic
scale over a temperature range which might extend almost to T_m.
Such resistance to interface disordering reflects that to form a
new configuration from a densely packed surface the atomic displacements and their attendant energy increases have to be very large.
Consequently, at small departures from equilibrium fluctuations
which would lead to growth steps on these planes should occur
only rarely. At some high temperature the entropic advantage of

disordering should become decisive enough to permit the interface to undergo a roughening transition.

Simple nucleation theory has been applied to the analysis of growth step formation on densely packed surfaces in the sub-roughening regime with results which are quite analogous to those for forming 3-dimensional crystal nuclei. In particular, disc shaped island monolayers formed by fluctuations on the crystal surface will tend to shrink or grow depending upon whether their radii are, respectively, less or greater than the critical value:

$$r^* = - \frac{\sigma_e}{\Delta G_v} \qquad (6)$$

where σ_e is the work/area of forming the vertical edge ("ledge") of the island.
The frequency/area, I_s, of forming these 2-dimensional growth nuclei may be expressed as :

$$I_s = A_s k_i \exp \{ \frac{\pi \lambda \sigma_e^2}{\Delta G_v kT} \} \qquad (7)$$

where A_s is a constant and λ is the ledge height.

Putting plausible estimates of σ_e into equation (7) it is found that the probability of nucleating island monolayers on the surfaces of perfect densely packed planes should be appreciable only when the departures from equilibrium are very large. A corollary to this result is that, excepting at such large departures from equilibrium, perfect crystals bounded by these planes should not grow measurably into dilute fluids. The usually observed growth of polyhedral crystals into dilute fluids under near equilibrium conditions generally reflects operation of the Frank screw dislocation mechanism [9] . In this mechanism growth occurs only by addition of material at the step edges formed by the winding up of a screw dislocation emergent in the crystal surface. The spacing, ΔR, of the growth steps so formed is proportional to r^* :

$$\Delta R \overset{\sim}{=} K_s r^* \qquad (8)$$

and the resulting fraction, f, of growth sites (i.e. ledge sites) in the surface is:

$$f \overset{\sim}{=} \frac{\lambda}{\Delta R} \overset{\sim}{=} \frac{\lambda}{K_s r^*} \qquad (9)$$

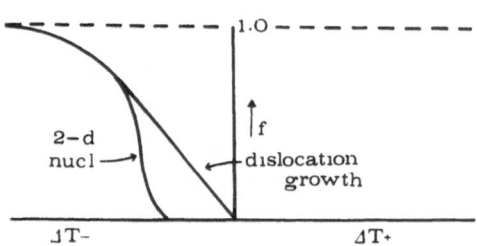

Fig. 3   Behaviour of f with ΔT

The variation of f with departure from equilibrium for dislocation and step nucleation limited growth is shown schematically in
figure 3. Presumably f is constant and near unity after the occurence
of the roughening transition.

    When the average crystallographic orientation of the inter-
face plane deviates a little from that of a closely packed one it
is thermodynamically advantageous [10], [11] for the interface
to break up into a series of terraces, which expose the close-
packed configuration, and monatomic ledges. These ledges would
provide sites for growth of the crystal into a dilute fluid.
When the orientation deviation can be described by a single angle,
Θ, of inclination of the average to the close packed orientation
the growth size fraction, f, would increase as sin Θ.

    There seems to be no generally accepted microscopic theory
for the equilibrium structure of the interface between the crystal
and its melt or a concentrated solution. Under these conditions
the interface roughness is often inferred from the morphological
stability of the crystal when it grows at small departures from
equilibrium. Thus it is supposed that a polyhedral growth form

reflects an atomically sharp and smooth interface while a dendritic form indicates a rough interface on which f $\sim$1. Jackson [12] noted that the growth form of crystals in their melts tended to be polyhedral when the entropy of fusion, ΔS_m, >2 $\bar{N}k$ and dendritic when ΔS_m < 2 $\bar{N}k$.

C. Kinetics and Morphology of Crystal Growth

1. Phenomenological Theory: Plane Interface

Usually crystal growth occurs in the two step sequence: (1) rearrangement of the interfacial configuration, with some frequency k_i per rearrangement (2) transport of the heat of crystallization away from the interface. Transport of matter to or from the interface also may be required but in the following formulation only heat transport is considered.

In steady state growth the interfacial temperature, T_i, assumes some level between the equilibrium temperature, T_m, and that, T, of a heat sink which may, e.g., be the ambient temperature of the system. The interfacial rearrangement is driven by the temperature difference $T_m - T_i$ and the heat transport by $T_i - T$.

The velocity, u, of the planar interface is related to the interfacial rearrangement frequency by the equation:

$$u = f \, k_i \, \lambda (1 - e^{\Delta G_c / RT_i}) \tag{10}$$

where λ is the interface displacement per rearrangement and ΔG_c is the free energy of crystallization/gm.atom. This rate is the product of a large forward rate and the thermodynamic factor $(1 - e^{\Delta G_c / RT_i})$ which reduces the forward by the reverse rate. When $RT_i \gg |\Delta G_c|$ the velocity reduces to :

$$u \cong - f \, k_i \, \lambda \frac{\Delta G_c}{RT_i} \tag{11}$$

and if the entropy of crystallization, $\Delta S_c (= - \Delta S_m)$, is constant, to :

$$u \cong - f \, k_i \, \lambda \frac{\Delta S_c (T_m - T_i)}{RT_i} \tag{12}$$

The flux $\dot{Q}$ of crystallization heat from the interface may be expressed as :

$$\dot{Q} = - \kappa \ (grad \ T)_i \ = u \ \frac{\Delta H_c}{\bar{V}} \tag{13}$$

where κ is the thermal conductivity of the heat transporting medium, $(grad \ T)_i$ is the thermal gradient at the interface and ΔH_c is the heat of crystallization/gm.atom.
In the steady state we have :

$$u = - \ \frac{\kappa \bar{V}(grad \ T)_i}{\Delta H_c} = f \ k_i \ \lambda(1-e^{\Delta G_c/RT_i}) \tag{14}$$

and in the linear regime:

$$T_m - T_i \overset{\sim}{=} \ (\frac{\kappa}{f \ k_i}) \ (\frac{\bar{V} \ RT_m T_i}{\lambda(\Delta H_c)^2}) \ (grad \ T)_i \tag{15}$$

Equation (14) or (15) can, in principle, be solved for the interface temperature, T_i.

When $f \ k_i << \kappa \ T_i$ will (See Fig.4) closely approach the ambient temperature for normal values of the thermal gradient. In this limit we say that the growth is "interface limited" although there must remain a small displacement of T_i from the ambient to drive the heat flow. In the alternative limit $f \ k_i >> \kappa$ and T_i becomes hardly distinguishable from the equilibrium temperature; in this event we say that growth is " heat flow limited" but we should realize that if the interface moves T_i must be somewhat displaced from T_m.

2. MORPHOLOGICAL STABILITY IN GROWTH

Consider the advance of a planar crystal-melt interface in the heat flow limited regime with the thermal gradient grad T imposed normal to the interface; grad T is taken to be positive when heat is flowing into the crystal. Suppose that a protuberance forms on the interface. With a positive gradient, $(grad \ T)_i$ at the protuberance tip must be less than on the flat so that the tip will slow down relative to the flat. Thus the protuberance decays and the interface maintains its planar form during growth.

In contrast, when the gradient is negative $|(grad \ T)_i|$ at a protuberance tip is increased relative to the flat and the tip

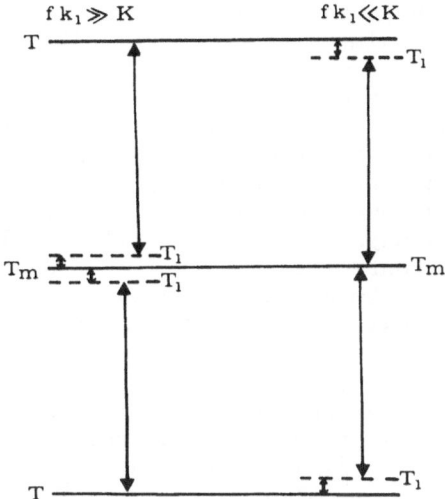

Fig. 4   Temperature diagram

then advances more rapidly than the flat. Thus the planar interface
is unstable to shape fluctuations and a dendritic morphology
develops. The smallness of the scale of this morphology will be
limited by capillarity, i.e. by the work required to create the
additional interfacial area. Taking account of capillarity Mullins
and Sekerka[13] showed that, in trasport limited growth driven by
negative gradients, shape instability may develop whenever the
interface curvature radii become roughly an order of magnitude
larger than the radius, r^* , of the critical nucleus.

Shape instability also develops when the growth of the crystal is limited by transport of rejected impurity away from the
interface and into the body of the external medium. The formal
theory for the onset and evolution of this instability is quite
analogous to that for heat flow limited growth.

Shape instability may occur even in the presence of a positive thermal gradient when a negative concentration gradient is super-posed. In this superposition the melt may become "constitutionally undercooled" provided the thermal gradient is not too large. The theory for constitutional undercooling and its morphological effects has been developed by Chalmers and coworkers [14].

We have noted that, according to Mullins and Sekerka's analysis, the onset of shape instability in diffusion controlled growth should occur when the crystal reaches a size only 10 times larger than that of the critical nucleus, i.e. $r \gtrsim 10 \ r^*$. The manifest stability, under certain conditions, of crystal growth forms at dimensions $\gg 10 \ r^*$ generally reflects interface rather than diffusion controlled growth [15] . Under these condi-tions the gradients, in temperature or composition, will be too small to overcome the restraints of capillarity on the onset of instability.

Crystals which have grown dendritically are likely to contain more dislocations,injected to accommodate the structural mismatch resulting from lateral impingement of dendrite arms, than those formed by planar growth [16]. Thermal stresses develop in crystals as they are cooled from their solidification temperature. These stresses can be accommodated elastically when the thermal gradients are uniform. However when the gradients are nonuniform, as in radial cooling, the stresses may become large enough to cause plastic deformation, with injection of dislocations, or fracture [16].

3. IMPURITY TRAPPING

Having noted the thermodynamic limits on impurity trapping by the moving crystalline front we now consider the kinetic possibi-lities [17]. An impurity atom will spend a time:

$$\tau_r \sim \lambda^2/D \tag{16}$$

where D is the diffusivity in the melt, diffusing one atom spa-cing, λ, in the direction of interface motion. The impurity should be trapped when this time exceeds that, λ/u, taken by the inter-face to move the distance, λ;that is, if

$$\frac{\lambda^2}{D} > \frac{\lambda}{u} \quad \text{or} \quad u > \frac{D}{\lambda} \tag{17}$$

For example, if $D \sim 10^{-5} \ cm^2 sec^{-1}$ and $\lambda \sim 3 \times 10^{-8} \ cm$ the impurities should be trapped at interface speeds exceeding ~ 3 meters/sec.

4. GRAIN COARSENING

Solids formed by condensation or rapid quenching often are composed of very small crystalline grains. Such a solid should, motivated by capillary forces, ultimately transform to a single crystal by grain boundary motions and interfacial diffusion. These capillary forces derive from the intercrystalline boundary, solid-external phase and solid-substrate tensions ($\sigma_{\beta\beta}$, $\sigma_{\beta\alpha}$ and $\sigma_{\beta s}$, respectively). The formal theory for the rate, u, of migration of a grain boundary is quite analogous to that for the crystal-melt interface. In particular u,

$$u = f \, k_i \, \lambda F_{cap} \qquad (18)$$

has the same form as equation (10) excepting for the thermodynamic factor F_{cap}, which will be determined by the interfacial tensions and by the geometrical configuration of the migrating boundary. Generally, F_{cap} will be an increasing function of $\sigma_{\beta\beta}/<r_g>$, where $<r_g>$ is the average grain size, but it may have important contributions from the other tensions, $\sigma_{\beta\alpha}$ and $\sigma_{\beta s}$, when $<r_g>$ approaches or exceeds the film thickness. Typically the magnitude of F_{cap} is one to several orders smaller than that of the thermodynamic factor in melt crystallization at moderate undercoolings.

D. Crystal Melting

In crystal melting or dissolution the dynamic balance of the interface exchange rates is shifted so that the interface moves into the crystal by the two step sequence, transport and inter-facial rearrangement, which is the reverse of that in crystal growth. The kinetic analysis is formally the same, see equations 10-15, as for crystal growth. Heat flow, e.g., is driven by $T - T_i$, where T ($>T_i> T_m$) is the temperature of the heat source , while interfacial rearrangement is driven by the interfacial super-heating, $T_i- T_m$ (See Fig.4). The criterion for interface control, $f \, k_i << \kappa$, is the same in melting as in crystal growth.

Consider interface controlled growth or melting. If, at equilibrium the interface is rough, with $f \sim 1$, the interface velocity, when plotted against $T_i - T_m$, goes through the origin with no slope discontinuity (see Fig. 5), as expected from the principle of microscopic reversibility.

Now suppose that the interface is atomically sharp and smooth at equilibrium so that the interfacial rearrangements are confined to the ledge sites. If, as is normally the case, the melt is the external phase the crystal will evolve into and grow in a poly-hedral form bounded by densely packed faces. However, in

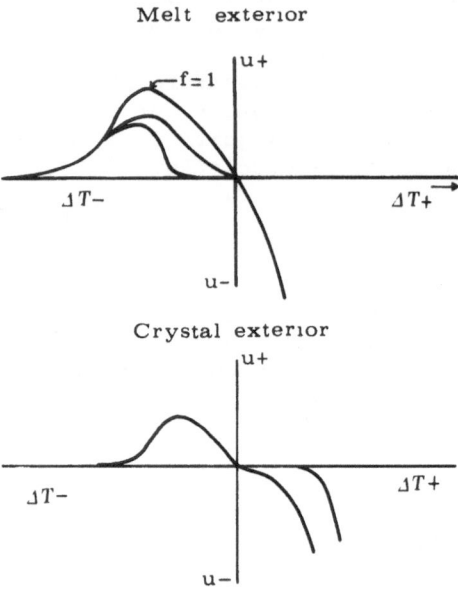

Fig.   5   Behaviour of u as a function of ΔT

melting the crystal will take on an ellipsoidal form since there
is no thermodynamic barrier to the initiation of layers at crystal
corners. Under these conditions u (T_i - T_m) will exhibit a slope
discontinuity at the origin, as shown schematically in Figure 5.
This discontinuity reflects the finiteness of the interface; the
microscopic reversibility principle applies only to interfaces
which are, effectively,infinite in area.

The morphological evolution in ledge limited interface motion
should be reversed when the melt is the interior phase. That is,
the melt should take a polyhedral form when it grows and an ellipsoidal form when it recedes. These predictions have not been
checked for the crystal-melt transition because of the difficulty

in forming an interior melt. However, they have been verified for
the growth and recession of droplets of concentrated solution in
the crystal interior.

III. EXPERIENCE ON PARTICULAR SYSTEMS

  A. Metals and other Monatomic Systems

    1. Nucleation and crystal growth

    When free of heterophase impurities pure liquid metals
exhibit quite a remarkable resistance to crystal nucleation. This
resistance is reflected by the very large undercoolings, often
0.25 to 0.30 on the reduced scale, required for measurable homophase nucleation. Also large masses, of order 1/2 kilogram, of
some pure liquid metals, - e.g. Fe, Co, Ni and Ag - have been
shown to sustain undercoolings. ΔT_r, of 0.15 to 0.20 for quite
long periods [18]. This behavior is consistent with the simple
nucleation theory if a value $\sim$ 0.6 is assigned to the scaled interfacial tension, α. Spaepen [19] has shown that this large value
of α can arise from the negentropy of localizing the atomic positions in the liquid on the interfacial layer. When such interface
localization exists little or no additional localization of positions is required to form a ledge on the interface.
Thus the ledge tension, σ_e and the thermodynamic barrier to the
nucleation of island monolayers on the interface should be negligible.

    Experience shows that crystals grow into pure liquid metals
very rapidly and with developing shape instability even at small
undercooling. Thus it appears that f k_i >> κ and growth is largely
controlled by the rate of heat transport. The onset of shape instability at small undercooling is in accord with the Jackson [12]
entropy correlation. It indicates that the crystal-melt interface
in metals is rough with f $\sim$ 1 at or very near equilibrium and
that σ_e is negligible in magnitude.

    With the normally available thermal gradients the interfacial
undercooling at a planar crystal-melt growth front in a metal is
so small that it is extremely difficult to measure or calculate.
We know that it is non-zero but almost negligible compared with
the undercooling, T_m - T, relative to the heat sink.

    Perhaps the most impressive evidence for the great rapidity
of the interface process in liquid metal crystallization comes
from the experiments of Walker [20,21] and Colligan and Bayles
[22] on the velocity of dendrite tips along columns of undercooled liquid nickel or cobalt. The velocity was determined from

the record at points along the column of thermal impulses generated by passage of the tip. In these experiments the thermal gradient driving growth must have been very large, perhaps $>10^5$ °/cm,
because of the high curvature at the dendrite tip.The velocities
in both metals increased roughly as the reduced ambient undercooling, ΔT_r, squared and were quite well characterized and
reproducible to $\Delta T_r = 0.1$ where $u \sim 50$ meters/sec. Considering
the thermal gradient and the depression of T_m due to tip curvature
it is evident thatthe interfacial undercooling, ΔT_{ri}, must have
been much less than the ambient ΔT_r. Thus the very high observed
velocities are certainly lower limits to the values at ΔT_r.

It is reasonable to suppose [5,23] that the upper limiting
velocity of the crystal-melt interface will be set by the impingement frequency of atoms from the liquid onto the interface.Thus

$$u \lesssim z_1 \lambda\beta\Delta T_{ri} \qquad\qquad (19)$$

where z_1 the impingement frequency/site replaces k_i in equation
12 and the other symbols are defined as before. It is difficult
to estimate z_1 but we expect that $z_1\lambda$ should be roughly equal to
the sound speed, u_s, in the liquid [1]. Thus :

$$u \leq u_s\beta\Delta T_{ri} \qquad\qquad (20)$$

Setting $\beta \sim 1$ and using the estimate $u_s \sim 2000$ meters/sec we
calculate for liquid Ni

$$u \gtrsim 200 \text{ meters/sec at } \Delta T_{ri} = 0.1$$

This limiting value is only a factor of 4 larger than the velocity
of a dendrite tip in liquid Ni or Co at ambient undercooling
$\Delta T_r = 0.1$. The calculation, if valid, implies that under these
conditions:

$$\Delta T_{ri} > \frac{\Delta T_r}{4} = 0.025$$

I estimate from equation (15) that for ΔT_r to reach 0.025, (grad
T)$_i$ must be of the order of 10^5 to 10^6 °/cm.

The collision limited interfacial process requires no thermal
activation. Also it needs much less correlation of atomic motions
than is necessary for diffusion within the liquid. Consequently

we expect that k_i will be much larger than, and not scale with, k_D.

2. GLASS FORMATION

According to the collision model the crystal growth rate in a pure metal, liquid or glassy, would not be fully suppressed by quenching. Therefore a metal glass would always be unstable against the growth of crystal nuclei. Actually, certain impurity admixtures greatly facilitate, and may be essential for, the condensation or melt quenching of metals to amorphous solid forms. If impurity redistribution, either local ordering or rejection by the crystal, is required for crystal growth, the growth rate will be controlled by the impurity transport rate and, under these conditions, we expect that k_i would scale with k_D rather than with Z_1. In this event the crystal-amorphous phase interface would be immobile at temperatures well below that of configurational freezing. Metal-metalloid glasses (e.g. Au_4Si) generally crystallize to an intermetallic phase with long range compositional order. Their crystallization may be controlled by the rate of the local compositional rearrangements required to achieve the long range order.

B. Covalent Systems

1. Nucleation and Growth

In covalently bound liquids and glasses the frequencies of the various atomic transport processes - viscous flow, crystal growth, diffusion and configurational relaxation -- are very low even at T_m. These frequencies generally scale together and their temperature dependences are described by activation energies with magnitudes which often approach the covalent bond energy. Thus, it is reasonable to suppose that the rate limiting step in the transport is disruption of the covalent bond.

The reduced temperature, T_{rg}, at the melt-glass transition in these materials is generally high, often $\geqslant 2/3$. In fused silica and some other covalent liquids the frequency of homophase crystal nucleation never reaches measurable levels so that their crystallization is always initiated at heterophase impurities.

Instead of meters/sec, as in pure liquid metals, the crystal growth rates in covalent amorphous phases tend to fall in the nanometer/sec range even at quite large departures from equilibrium [5, 6]. Under these conditions $\kappa \gg f k_i$ so that growth is interface controlled with the interfacial temperature closely approaching that of the ambient.

In the framework of the transition state theory k_i may be expressed as[5] :

$$k_i = n_r \nu_I \; e^{\Delta S'_1/R} \; e^{-\Delta H'_1/RT} \tag{19}$$

where ν_r is the normal frequency of the internal motion leading to rearrangement, $\Delta S'_1$ and $\Delta H'_1$ are, respectively, the entropy and enthalpy of activation and n_r is the number of rearrangements per activation. Generally, a pair of dangling half bonds are produced by the activation. These defects may, with little added activation, migrate over several interfacial sites before recombining, thus effecting the several, n_r, rearrangements. The process is analogous to that which occurs in the free radical mechanism of homogeneous chemical reactions.

2. Impurity and Photo Effects

The sequence, bond disruption, half bond migration and annihilation can be highly sensitive to traces of homophase impurities. Certain of these impurities may catalyze growth by reducing the energy for bond disruption while others may inhibit growth by annihilating the migrating defects. Both types of impurity effects have been observed in the growth of crystals in fused silica[24] and amorphous silicon[25].

According to the foregoing model irradiation by photons having energy enough to disrupt the covalent bonds ought also to catalyze crystal growth. Such irradiation enhancement of growth rate seems well established for the crystallization of amorphous selenium[26,27] but evidence for the effects in other materials is rather inconclusive.

3. Crystallization of Si and Ge

Crystalline Si and Ge melt to metallic liquids with substantial contractions in volume. In contrast, both elements form tetrahedrally coordinated amorphous semiconducting solids (a-sc) when condensed on substrates held at temperatures lower

than 400 to 500°K. The crystallization enthalpy of the a-sc form
of Ge was found to be only about 1/3 of that of the metallic
liquid (lm)[28]. These results suggested that there should be
a thermodynamic transition, a-sc$\rightleftharpoons$lm, in the temperature regime
$T < T_m$ where both phases are metastable relative to the crystal.
The independent thermodynamic analyses of Bagley an Chen[29] and
Spaepen and the writer[5] indicated that such a transition should,
indeed, occur at a reduced temperature of the order of 0.8 T_r. It
was presumed that a similar transition in Si would occur at
about the same T_r but the thermal data needed for the calculation
was lacking.

Mayer and associates[25] measured the epitaxial regrowth
rates, at $T_r < 0.8$, of crystalline Si and Ge into amorphous
overlays formed by self ion implantation.
At a given temperature the rate varied sharply with crystallographic orientation of the interface, being highest on interfaces
||to (100) and lowest on those || to (111) planes. However, the
temperature dependence of u for either element was described by a
single activation energy independent of interface orientation and
near to the energy of single bond formation. The orientation dependence of u indicated that growth was limited to ledge sites on
the (111) terraces. For both elements, prefactors, u_o, in the
Arrhenius temperature dependence fell well within the range of
those expected,

$$u_o = fn_r \nu_r e^{\Delta S_1' /R} \qquad (20)$$

from transition state theory[5].

We expect that the a-sc phase of Si or Ge would, if crystallization did not intervene, melt rapidly to the lm phase at
$T_r > 0.8$. Therefore, growth of these crystals in the temperature
range $T_r = 0.8$ to 1.0 should, unless f is orders of magnitude less
than unity, proceed at the rapid rate characteristic of the crystal-melt front in metals. Since there is a marked change in coor-
dination at the semiconductor-lm interface it is likely that σ_e
will be large enough to give rise to considerable resistance
to nucleation of growth layers on (111) planes. However for
large interfacial undercooling and most interface orientations we
expect that f will be within an order of magnitude of unity and
the crystal growth rates will be in the meters/sec range.

C. Interpretation of Laser Induced Structure Changes

1. Epitaxial Regrowth

Based on the foregoing analysis we expect two regimes divided

by the a-sc⇌lm transition temperature, T_{rt}, of epitaxial regrowth
of Si or Ge into amorphous overlays[5]. At $T_r < T_{rt}$ the growth
should proceed slowly, with thermal activation, in accord with the
measurements of Mayer and coworkers[25]. At $T_r > T_{rt}$ the overlay
should melt and regrowth would occur, with little   thermal
activation, at the rapid pace characteristic of that in liquid
metals. However, upon heating from low temperature the overlay will
persist, in substantial part, to T_{rt} only if the heating time is
too short for the a-sc →c-sc regrowth and for copious crystal nucleation within a-sc. Qualitative observations suggest that the
nucleation is quite infrequent and extrapolation of Mayer et al's
results give growth rates of 1 micron/sec at T_{rt}.
Thus in micron thick overlays the rapid growth regime should be
reached easily upon irradiation by short duration pulses (e.g. in
the micro - or nanosecond range) which provide sufficient thermal
energy.

In laser pulsing the heat of crystallization is normally
extracted through the crystal so that the crystallization front
stays planar. Also the thermal gradient is likely to be linear
through much of the central area irradiated. Thus, there is no
dislocation formation from dendritic growth or radial cooling and
the regrown layers may, despite their rapid formation, be relatively free of structural imperfections[17].

Impurities which are normally used to dope semiconductors,
e.g. P or Al, reside substitutionally in both the c-sc and a-sc
phases. They diffuse slowly, typically $D \sim 10^{-11} cm^2 sec^{-1}$ at
T_m, in c-sc and presumably, with no greater rapidity in a-sc.
Under these conditions, their residence time at one atom position
would exceed 10 microseconds and their spatial distributions would
not be appreciably altered by shorter duration energy pulses which
did not produce melting. In contrast, the impurity diffusivities
should be $10^{-5} cm^2 sec^{-1}$ or more in the molten metal, lm, phase
corresponding to diffusive displacements of $\gtrsim$ 300 A in 100 nanoseconds.

We have noted that impurities in the melt will be trapped in
the crystal, provided the thermodynamic conditions are fulfilled,
at interface speeds $\gtrsim$ 3 meters/sec. Such trapping can lead to
impurity concentrations which far exceed the equilibrium solubilities in the crystal. The excess impurity solubilities so far
reported seem to fall well within the realm of the thermodynamic
possibilities for epitaxial regrowth into the molten metal phase.

There are certain impurities which diffuse rapidly in crystalline semiconductor phases by the dissociative mechanism. In
this mechanism the transport occurs, predominantly, when the
impurity is in an interstitial state and in this state the dif-

fusivity often exceeds that in the liquid by considerable
factors. Such impurities, when in the overlay, can redistribute
themselves rapidly by crystalline state diffusion after the passage of the crystallization front.

J.van Vechten[30] has argued that pico - or nanosecond
pulsing periods in Si may be too short for thermal equilibration
of the charge carriers, produced by irradiation, with the Si ions.
He proposes that the irradiated state is one of zero shear modulus
consisting of cool Si ions in a hot plasma. With such properties
the van Vechten state, after a nanosecond, should be no less
liquid than that formed by near equilibrium melting[17]. Therefore it should crystallize by growth kinetics and morphology
similar to those in the crystallization of the normal metastable
liquid. Because of its lesser stability the van Vechten liquid
would provide greater thermodynamic margins than the normal one
for the trapping of impurities during regrowth. However, such
excess margins seem not to be demanded by the present results.

2. Graphoepitaxy

Geis, Flander and Smith[31] have developed a process for
transforming amorphous silicon films deposited on fused silica
sub-strates into singly oriented crystalline layers in which a
cube plane lies parallel to the substrate and a cube direction in
the plane lies parallel to a set of shallow grooves, with square
wave cross section, formed by etching the substrate. The monocrystalline layer was formed, by linear laser scanning, in two
stages : (1) crystallization of the amorphous film to a polygrain
structure in which a cube plane in each grain was parallel to the
substrate while the grains were misoriented by random rotations
about the direction (<100> fiber axis) perpendicular to this
plane. The fused silica substrate was treated so that this
texture formed whether or not the surface was grooved. (2) gradual
coarsening, by normal grain growth, of the grain structure to the
monocrystalline cube orientation. The grooves were essential for
the occurrence of this stage.

The structure at the end of stage one could have resulted
from growth of crystallites nucleated on the substrate preferentially in the fiber orientation ("oriented" nucleation) during
the Si deposition or, as Geis et al proposed, in the laser scanning. It could also have arisen from the more rapid growth of
crystallites in the fiber orientation ("oriented" growth) from a
randomly oriented population nucleated on the substrate. Indeed,
the results of Mayer et al showed that <100 > is a direction of
rapid growth of c-sc in a-sc.

The grain growth step is most readily explained by supposing, as Geis et al did, that the interfacial tension between fused silica, as conditioned, and c-sc is at a minimum when the interface is parallel to the cube planes of c-sc. Then the capillary forces along the walls of the groove will act to favor the growth of grains in the cube orientation.

In principle, the interfacial tension between an amorphous and crystalline phase can depend on the interfacial orientation relative to the crystal. As we implied, such a dependence is the basis for Geis et al's explanation for both stages of the crystallization. However, there is a possibility, perhaps remote, that a thin devitrified layer with a fiber texture had formed on the surface of the fused silica. If so, an interfacial tension minimum at the cube plane orientation of c-sc would reflect crystal-crystal rather than crystal-amorphous state interactions at the interface.

3. Glass Formation

In molten overlays formed on crystal surfaces by laser pulsing the thermal gradient through the melt may be so extreme that the crystal-melt interfacial temperature falls below T_g and the melt is quenched to a glass. Indeed in nano- , and especially in pico-, second laser pulsing the quench rates (10^9 to 10^{12} °/sec)[32] may far exceed those attainable in splat quenching or melt spinning. Thus, the range of glass forming compositions should be extended considerably by the heating and quenching in such pulses when they are sufficiently energetic.

Indeed, it has been shown that thin glassy overlays are produced by laser scanning of alloys known to form glasses in splat quenching[33]. However, the overlays formed on metals by very short laser pulses are so thin that they are difficult to isolate or characterize. Also such thin layers are often transformed, in large part, to oxide.

Liau et al[34] have reported the formation of thin layers of a-sc on areas of silicon crystals irradiated by high energy picosecond laser pulses. The a-sc structure generally appeared in the outermost regions of the irradiated areas while the central regions consisted of c-sc in good crystallographic registry with the base crystal. Liau et al supposed that a thin molten, lm overlay formed in the energization phase of the pulse. The depth of this overlay would decrease from the center to the edge of the irradiated region. Thus the quench rate in the cooling phase would have increased from center to edge and in the thinner regions it may have been great enough to reduce the interfacial temperature levels below T_{rt} before the completion of epitaxial regrowth. It was assumed that upon undercooling into this temperature range, $T_r < T_{rt}$, the liquid rapidly solidified to the a-sc phase identified at the end of the experiment. In contrast, T_{rt} in the central region did not fall below T_{rt} and the liquid there solidified by epitaxial regrowth at estimated rates of the order of 100 meters/sec.

REFERENCES

1. D.Turnbull, Contemporary Physics 10, 473 (1969).
2. W.Kauzman, Chem.Rev. 43, 219 (1948).
3. a) N.G.Ainslie, J.D. MacKenzie and D. Turnbull, J.Phys. Chem. 65, 1718 (1961);
    b) R.L. Cormia, J.D. MacKenzie and D. Turnbull, J. Appl.Phys. 34, 2245 (1963).
4. J.C. Baker and J.W. Cahn, "Solidification", pp.23-58, Am.Soc. Metals, Metals Park, Ohio (1971).
5. F. Spaepen and D. Turnbull, "Laser Solid Interactions and Laser Processing", AIP Conference Proceedings 50, pp. 73-83 (1979).
6. F. Spaepen and D. Turnbull, to be published in "Laser and Electron Beam Processing of Semiconductor Surface", J.M. Poate and J.W. Mayer, Academic Press.
7. D. Turnbull, "Physics of Non-Crystalline Solids", (ed. J.W. Prins) pp.41-56, North Holland, Amsterdam (1964).
8. F. Spaepen and D. Turnbull, "Rapidly Quenched Metals", (ed. N.J.Grant and B.C. Giessen), pp. 205-229, M.I.T. Press, Cambridge, MA (1976).
9. W.K. Burton, N.Cabrera and F.C. Frank, Phil.Trans.Roy.Soc. (London) 243, pp. 299-258 (1951).
10. W.C.Herring, Phys. Rev. 82, 87 (1951).
11. W.W. Mullins, "Metal Surfaces", pp. 17-66, Am.Soc.Metals, Metals Park, Ohio (1963).
12. K.A. Jackson, "Growth and Perfection of Crystals", (ed. R.H. Doremus, B.W. Roberts and D. Turnbull, pp. 319-324, Wiley, NY, (1958).
13. W.W.Mullins and R.F. Sekerka, J.Appl.Phys. 34, 323 (1963); 35, 444 (1964).
14. B. Chalmers, "Solidification", pp. 295-310, Am.Soc.Metals, Metals Park, Ohio, (1971).
15. J.W. Cahn, "Crystal Growth", (ed. H.S. Peiser), pp.681-690; Pergamon Press, "London, (1967).
16. K.A. Jackson, "Solidification", pp. 121-154, Am.Soc.Metals, Metals Park, Ohio (1971).
17. D. Turnbull, J. de Physique C4- 1980, 109.
18. D. Turnbull, Solid State Physics, 3, pp.225-306, Academic Press, N.Y. (1956).

19.   a) F. Spaepen, Acta Met. $\underline{23}$, 729, (1975); F. Spaepen and
        R.B. Meyer, Scripta Met. $\underline{10}$, 257 (1976).
20.   J.L. Walker, pp. 114-115, pp. 122-124, in Ref. 21.
21.   B. Chalmers, "Principles of Solidification", Wiley, N.Y.
        (1964).
22.   G.A. Colligan and B.S. Bayles, Acta Met. $\underline{10}$, 895 (1962).
23.   D. Turnbull and B.G. Bagley, "Treatise on Solid State
        Chemistry" (ed. N.B. Hannay) $\underline{5}$, pp. 513-554, Plenum Press,
        NY, (1975).
24.   V.J. Fratello, J.F. Hays and D. Turnbull, J.Appl.Phys.
        $\underline{51}$, 4718 (1980).
25.   L.Czepregi, E.F. Kennedy, J.W.Mayer and T.W. Sigmon, J.Appl.
        Phys. $\underline{49}$, 3906 (1978).
26.   J.Dresner and G.B. Stringfellow, J.Phys. Chem.Solids $\underline{29}$, 303
        (1968).
27.   a) K.S. Kim and D.Turnbull, J.Appl.Phys. $\underline{44}$, 5237 (1973); $\underline{45}$,
        3447 (1974);
        b) J.C. Carballes, R. Clement and B. de Cremoux, Rev. Tech.
        Thomson-CSF$\underline{5}$, 225 (1973);
        c) Geraldine Gross, R.B. Stephens and D. Turnbull, J. Appl.
        Phys. $\underline{48}$, 1139 (1977).
28.   H.S. Chen and D. Turnbull, J. Appl.Phys. $\underline{40}$, 4214 (1969).
29.   B.G. Bagley and H.S. Chen, "Laser Solid Interactions and
        Laser Processing", AIP Conference Proceedings $\underline{50}$, 97 (1979).
30.   J. van Vechten, in press, J.de Physique C4-1980 , 15
31.   M.W. Geis, D.C. Flanders and Henry I. Smith, Appl.Phys.Lett.
        $\underline{35}$, 71 (1979).
32.   N. Bloembergen, "Laser Solid Interactions and Laser Proces-
        sing", AIP Conf. Proc. $\underline{50}$, 1 (1979).
33.   E.M. Breinan, B.H. Kear, C.M. Banas and L.E. Greenwald,
        "Proc. 3rd. Int.Symp. on Super Alloys", p. 435, Claitors
        Pub.Div., Baton Rouge, LA (1976).
34.   P. Liu, R. Yen, N. Bloembergen and R.T. Hodgson, Appl.Phys.
        Lett. $\underline{35}$, 433 (1979).

CO_2 LASER MATERIALS PROCESSING

Conrad M. Banas

United Technologies Research Center
East Hartford, Connecticut   06108

INTRODUCTION

Since the first successful operation of a laser in 1960, the laser's potential for materials processing has been clearly recognized and actively pursued. Initial applications included drilling, microwelding and jet-assisted cutting. In these tasks the laser was demonstrated to offer unique processing capabilities as well as reliable performance.

With the advent of high-power, continuously-operating, carbon-dioxide lasers in the late 1960s, the laser's capability for materials processing was significantly enhanced. In 1971, for example,[1] it was reported that a high-power laser had been used to form deep-penetration welds similar to those previously obtained only with electron beams.

Within the last decade, three main areas of laser processing have received considerable attention. These are: welding,[2] jet-assisted cutting[3] and surface treating.[4] The latter has included transformation hardening of ferrous alloys, surface alloying and rapid surface heating and cooling[5] in order to achieve desired metallurgical surface characteristics.

Within the next decade, it is anticipated that production utilization of high-power laser systems will increase dramatically. The following brief status review of high-power laser materials processing is offered to assist the potential user in evaluating

prospective applications. This review is limited to CO_2 laser
applications because this laser is currently the only unit suitable
for continuous industrial use at multikilowatt power levels.

Fundamental Concepts

From an engineering standpoint, the laser can be considered
simply as an energy conversion device. Energy from a primary
source (electrical, thermal, nuclear, chemical, optical, etc.) is
delivered to the unit in controlled fashion in order to induce
laser operation. A fraction of the delivered energy is converted
into coherent light (the laser beam) and the residual energy is
discarded as waste heat. Electrical excitation is used for the
high-power industrial CO_2 laser. Such systems typically exhibit
conversion efficiencies of the order of 10-15%; ideal, or Carnot,
efficiency for the system is 42%.

The individual photons which comprise the laser output beam
exhibit an energy corresponding to the laser transition. Photon
energy is related to the frequency by

$$E = h\nu$$

in which E is the photon energy, h is Planck's constant and ν is
the frequency of radiation. Since the product of frequency and
wavelength equals the velocity of light,

$$\lambda\nu = c \qquad \text{then} \qquad \lambda = hc/E$$

i.e., the wavelength, λ, is inversely dependent on photon energy.
For the CO_2 laser, the photon energy corresponding to the 10.6
micron wavelength is 0.12 e.V. Since the x-ray threshold is of
the order of 10^4 volts, this explains why x-rays are not generated
on interaction of the beam with a workpiece. On the other hand,
it should be noted that extremely-high-power lasers are capable of
generating high-temperature plasmas from which x-rays can be
radiated. Under conditions normally encountered in materials
processing, however, x-rays should not be anticipated.

Because the energy in a laser beam is highly ordered, the
beam can be focused to provide very high power densities. From
basic principles, the minimum spot diameter to which the beam can
be focused is of the order of the beam wavelength. For the CO_2
laser beam, this corresponds to a diameter of about .01 mm. Since

the beam must pass through a finite aperture, the minimum spot
diameter is further constrained by diffraction in accordance with
the relationship

$$d_s = K\lambda f$$

in which f is the ratio of focal length to aperture of the focusing
optics and K is a constant which depends on the mode characteristics of the laser beam. For example, for a plane wave, the value
of K is 2.44.

Practical considerations dictate the mode which is utilized
for high-power laser operation. A Gaussian mode output provides
good focusing, but does not yield the highest operating efficiency.
In addition, partially-reflecting mirror materials for high power
levels are not available. For heat treating, a uniform-intensity
energy distribution is often advantageous; such beams are, however,
not sufficiently focusable for welding and cutting applications.
The smallest attainable spot diameters are usually desirable for
cutting. For welding, larger spot diameters are often dictated by
the characteristics of the weld joint. Further, spot diameters
promoting power densities in excess of about 5×10^7 W/cm^2 are
undesirable for welding due to the tendency for excessive
vaporization.

From the above comments, it should be evident that the nature
of the laser beam material interaction can be conveniently referenced in terms of the power density incident on the workpiece
surface. Accordingly, important laser materials processing areas
have been identified against coordinates of power density and
interaction time in Fig. 1. For a continuous process, the interaction time can be defined as the time that it takes the incident
laser spot to move one diameter relative to the workpiece surface.
It is further noted that the product of the power density and
interaction time is the specific energy delivery to the material.

To obtain a relevant physical interpretation of the intensity
of the laser source, it is useful to associate an "equivalent"
temperature with power density. It is recalled that the Stefan-
Boltzmann relationship for the total energy radiated per unit area
by a perfect thermal source is

$$E_T = \sigma T^4$$

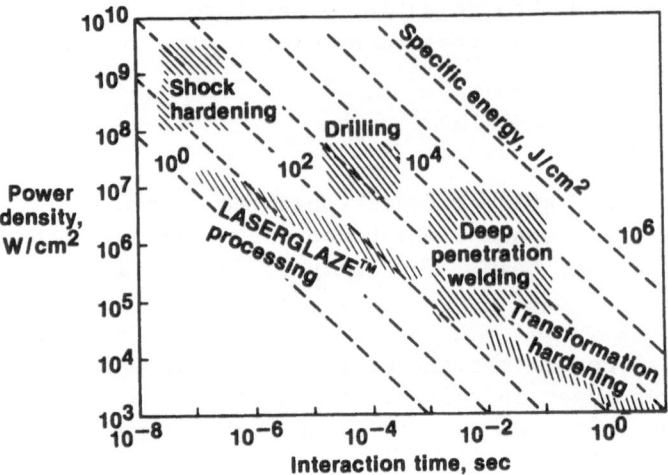

Figure 1. Laser beam-material interaction spectrum.

in which σ is the Stefan-Boltzmann radiation constant and T is the
absolute temperature. For reference, it is noted that a power
density of 10^6 W/cm^2 corresponds to a thermal source operating at
20,500 K. Compared to conventional heating devices, therefore, the
laser provides much higher energy concentrations which facilitate
melting and vaporization of any known material.

As shown in Fig. 1, the laser processing spectrum spans many
orders of power density. At the relatively low power density
required for transformation hardening, the laser serves primarily
as a surface-heating device. As power density is increased to the
10^6 W/cm^2 level, local vaporization occurs at the interaction point
and deep-penetration welding is facilitated. LASERGLAZE^{TM} processing occurs as a limiting case of welding in which extremely high
processing speeds (short interaction times) are used and shallow
melt depths are obtained. Addition of jet-assist converts the
laser welder into a cutter.

At higher power densities, the fraction of the affected
material at the interaction point which is vaporized increases and
material expulsion takes place. Under these conditions, normally
attained with pulsed lasers, drilling is obtained. At still higher
power densities, rapid vaporization takes place which can induce
acoustic waves in the material and promote shock-hardening.
Shorter interaction times and higher power densities than shown in
Fig. 1 are being explored for plasma production and other
applications.

Having identified the processing regimes for high-power laser applications, we shall consider the nature of the individual processes separately.

Transformation Hardening

It has been well-known for centuries that heating steel or cast iron to red heat and cooling rapidly results in hardening. The elements of the transformation process are illustrated in the diagram presented as Fig. 2. Steel in its nonmagnetic form

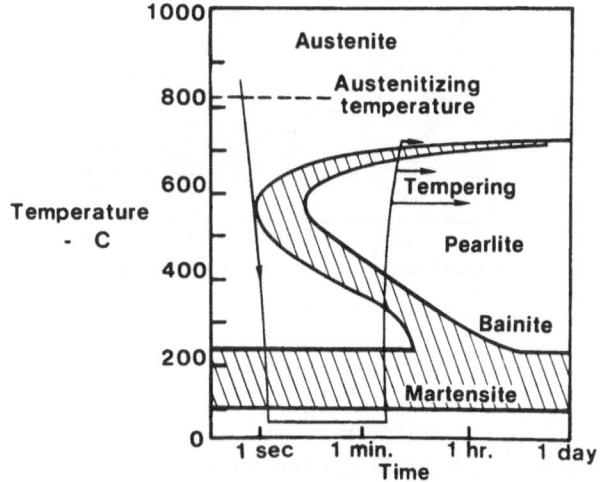

Figure 2. Typical isothermal transformation diagram for a ferrous alloy.

(austenite) at high temperature transforms to a hard, martensitic structure on cooling at a rate sufficiently rapid to pass the knee of the Bain "S" curve shown.

Transformation hardening is extensively used in industry to improve wear characteristics, erosion resistance and durability of parts. The laser brings a new dimension to the process because it permits flexible, convenient local heating to facilitate spot heat-treating. It is in this mode, in particular, that the laser offers advantages as a processing tool.

Another advantage of laser utilization for transformation hardening is the potential for close process control. As shown in Fig. 3, the laser surface-heating process can be referenced to

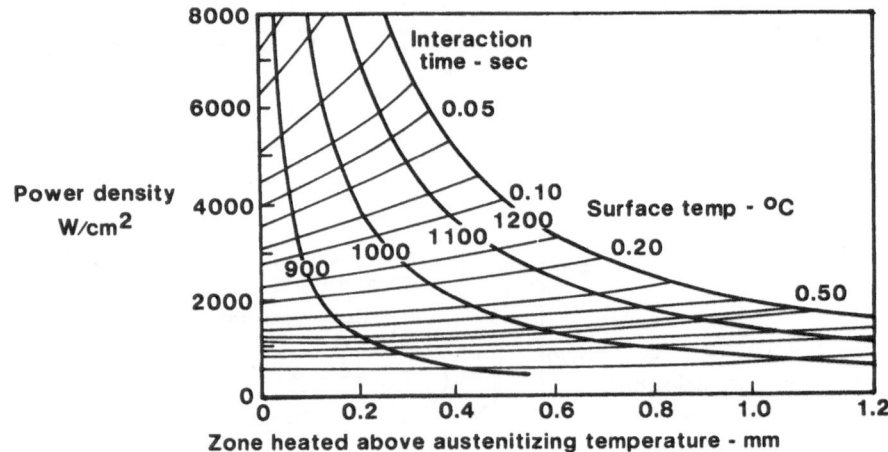

Figure 3. Cast iron heat-treating parameters.

power intensity and interaction time. The curve shown in Fig. 3 was constructed for a 3 1/2% carbon cast iron which exhibits a melting point of the order of 1200 C. The abscissa of the curve indicates the depth to which the material is heated to at least the transformation (austenitizing or Curie) temperature. It is noted that the depth to which this can be accomplished decreases as power intensity increases since more rapid surface-heating occurs.

At the other extreme, low power densities and long interaction times promote substantial thermal diffusion into the base material. Under these conditions, when the laser beam passes, the thermal gradient is not sufficient to provide the rapid cooling of the surface layer that is required to produce a hard structure.

A practical limit on case depth is approximately 3 mm, with the maximum decreasing as the carbon content of the ferrous material increases and the melting point, correspondingly, decreases.

In practice, the process is not quite as simple as the curves might imply. A principal difficulty is interposed by the high reflectivity of metals at the laser wavelength. For that reason, the power absorbed by the material surface, represented by the ordinate in Fig. 3, is some fraction of that incident. An "as received" steel surface may reflect 50% of the incident energy; machined and/or polished surfaces are even more reflective. It is, therefore, essential to improve beam absorption for laser transformation hardening by use of an absorptive coating. Recalling that insulators are, generally, poor reflectors (good absorbers) at long wavelengths, coatings used for CO_2 processing are usually insulative.

For steels, manganese phosphate is often used as a rust preventive during processing; it is a good absorber at 10.6 microns. Similarly, paints containing oxide pigments form good absorbers. A flat black paint, for example, is an excellent absorber. To underscore the fact that visual appearance does not necessarily identify the infrared properties, it is noted that a flat white paint is also a highly effective absorber.

When a coating is used, it is well to take note that the material forms an effective thermal barrier which must be considered relative to the heat flow to the underlying material. Thermal flux to the base material is inhibited by the coating and is dependent on coating thickness. Process repeatability is therefore dependent upon the generation of uniformly reproducible coatings. In this respect, the laser stands at a disadvantage to the electron beam, which is not reflected by metallic surfaces.

A typical laser transformation hardened zone in cast iron is shown in Fig. 4. We note that there is a sharp line of demarcation between the zones. Also to be noted is that the depth of the hardened zone tapers slightly at the outer edges of the pass due to the influence of lateral heat conduction.

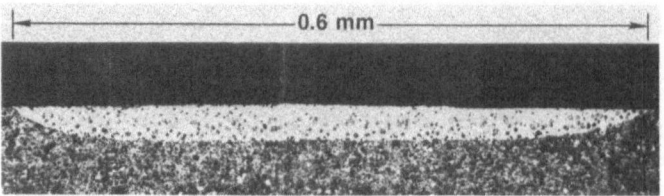

Figure 4. Laser transformation-hardened cast iron. (Power: 2.5 kW,
           Speed: 106 mm/sec, Spot diameter: 0.62 mm,
           Depth: 0.35 mm).

    The use of multiple overlapping passes to harden an extended
surface is indicated. In this case, special consideration should
be given to the overlap zone. Within this zone, some previously-
hardened material will be heated to a temperature below the austenitizing point. Since the material remains at an elevated temperature for a finite time, tempering occurs (note Fig. 2). It is
therefore possible that soft regions will occur in the overlap
zone. If these cannot be tolerated, then processing techniques
which will permit entire surface coverage in one pass must be
sought. For example, a cylindrical part can be rapidly rotated
under the laser beam to produce circumferential heating. Axial
motion will then draw the hot ring over the material, resulting in
uniform hardening over the entire surface without overlap.

    A final factor to be considered in laser hardening is distor-
tion. Although laser processing is accomplished with low energy
input relative to conventional techniques, thermal gradients
established are large and some thermal distortion may occur. Further, there is a change in the volume of the processed material
since there is a change in the lattice structure of the material
when it is transformed from ferrite to martensite. With localized
heat-treating, local stresses may be high.

    Processes related to transformation hardening are hardfacing
and surface alloying.[4] In the former, a wear- or corrosion-
resistant coating material is melted onto the workpiece surface.

Conditions are sought which will provide a tight bond but which
will inhibit dilution of the coating material with the substrate.
In contrast, the latter requires uniform mixing of an additive into
the layer melted on the workpiece. In this manner, the chemical
composition of the surface of the material can be modified. A common procedure, for example, would be the addition of chromium to
the surface layers of a steel workpiece to improve wear and/or
corrosion characteristics. In situ generation of a stainless steel
surface is, in principle, possible.

Welding

     At low power intensities, the laser can be utilized to perform
fusion welding similar to that obtained with oxyacetylene. Under
such conditions, the reflectivity of the surface must be dealt
with. Further, the process is relatively slow, inefficient, and
provides limited penetration capability. In most instances, this
mode of operation does not produce results justifying the cost of
a laser system.

     As power intensity is increased to approximately 10^6 W/cm^2,
however, the nature of the beam interaction with material changes
dramatically. Energy input at the workpiece surface is locally
too intense to be removed by conduction, convection and radiation
from the incident spot. Local vaporization occurs and a "cavity"
(Fig. 5) is established in the material. Metal vapor pressure

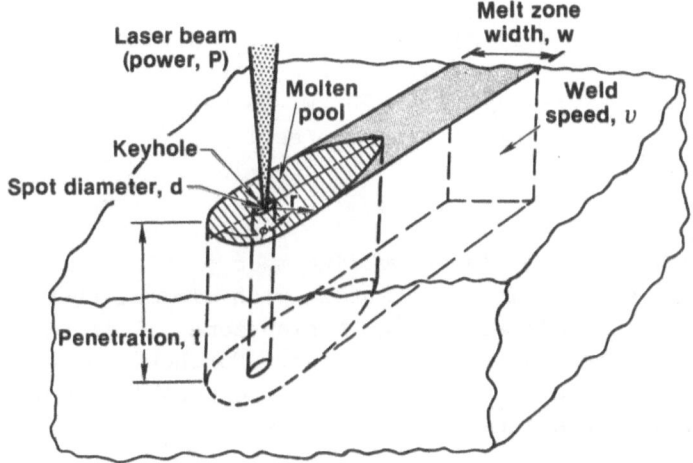

Figure 5. Deep-penetration weld characteristics.

forces support this cavity against the hydrostatic forces of the
liquid metal surrounding it. With appropriate relative motion
between the beam and the workpiece, the cavity becomes dynamically
stable and moves through the material with melting preceding and
solidification following it. The presence of the cavity permits
laser energy to be deposited at depth within the material and not
just at the surface. Thus a narrow, high depth-to-width ratio weld
(Fig. 6) is formed in contrast to the roughly hemispherical fusion
zone obtained with conventional welding processes. The deep-
penetration process may be modeled by an unsteady line heat source.[6]

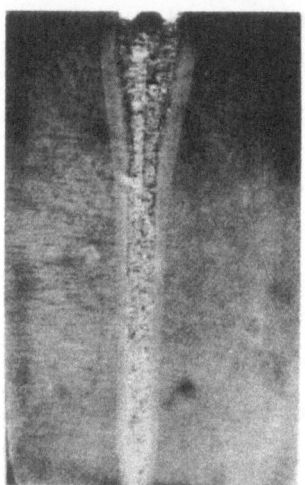

Figure 6. Deep-penetration laser weld
profile in alloy steel.
(Power: 90 kW, Thickness: 38 mm,
Speed: 50 mm/sec).

     The deep-penetration laser welding mode provides the capa-
bility for high-speed joining of metals with minimum thermal
effects on material adjacent to the fusion zone. This, in some
cases, leads to enhanced weld properties. Further, in contrast to
the electron beam, the laser provides this capability in the atmosphere and without the generation of x-rays.

     Away from limiting conditions, weld penetration at a given
laser power level is inversely proportional to the welding speed.

Correspondingly, at fixed speed, the penetration is directly
related to the power. This latter characteristic is illustrated
in Fig. 7, which represents laser performance in terms of speed,
power and penetration. As welding speed is decreased at constant

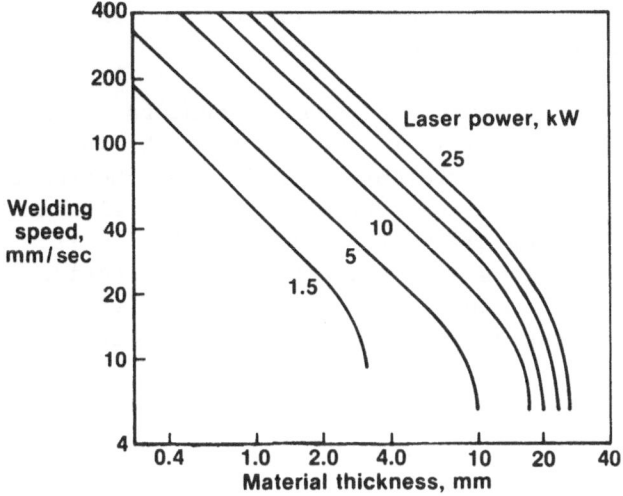

Figure 7. Laser Welding Performance

power level, the extent of the molten metal zone surrounding the
cavity is increased in proportion to the increased energy deposition per unit length. Eventually the metal vapor pressure is
insufficient to counter the fluid dynamic forces of the liquid
metal and the deep-penetration cavity collapses. Under these conditions, an abrupt decrease in penetration occurs and a roughly
hemispherical fusion zone characteristic of surface energy input
is obtained. The conditions just prior to collapse of the cavity
therefore serve to define a maximum weld penetration, as evidenced
in Fig. 7. It has been shown[2] that the maximum penetration is
proportional to the 0.7th exponent of laser power.

Fusion Zone Purification

One of the unique characteristics which has been observed in
laser welding has been the process of fusion zone purification.
This process has led to the attainment of exceptional weld properties in some high-strength low-alloy steels and is due, in large
measure, to the optical characteristics of the laser beam.

As noted previously, the carbon-dioxide laser system is currently the only suitable unit for industrial applications at the continuous, multikilowatt level. The output wavelength of this system is at 106,000 Å in the infrared. It is a general characteristic of this wavelength that it is highly reflected by metals and absorbed by insulators. Thus, in passing into the deep-penetration cavity, the beam is substantially reflected on interaction with a clean molten metal surface.

If, on the other hand, the beam encounters an impurity (typically an oxide or a silicate), preferential absorption occurs. Therefore, the impurity element is selectively heated and vaporized out of the weld zone. Direct evidence of this cleaning action has been obtained by chemical analysis (Table I) as well as by direct count of the inclusions within the weld cross sections. As shown in Table I, reduction in oxygen content within the weld zone relative to parent material by as much as 50% has been attained.

Table 1:   Chemical Analysis of Alloy Steel Laser Welds

Element	Weld Speed: 8.5 mm/sec		12.7 mm/sec	
	Base Metal	Weld Metal	Base Metal	Weld Metal
Carbon %	0.127	0.120	0.122	0.117
Nitrogen (ppm)	125	103	130	115
Oxygen (ppm)	44	21	78	51
Sulfur (ppm)	69	74	73	70
Hydrogen (ppm)	10	10	10	10

Indirect evidence of purification has been obtained from weld mechanical properties. Typically, a laser weld in a high-strength, low-alloy steel exhibits a tensile strength equivalent to or greater than that of the parent material and a corresponding increase in hardness which is related to the carbon content. Generally speaking, an increase in hardness is associated with a corresponding increase in brittleness. Impact tests of some laser welds in high-strength, low-alloy steels have, however, shown surprisingly high values of toughness. The simultaneous attainment of good tensile characteristics, increased weld zone strength and retention of good toughness characteristics is attributed to the

reduction of impurities within the fusion zone during the welding
process.

Weld Configurations and Filler Metal Addition

Although the majority of reported laser welding information
has been obtained for the "downhand" welding position, the laser's
capability for out-of-position welding has been clearly demonstrated. In particular, comparison welds have been formed at the
same power and speed in the downhand, horizontal and overhead positions. It has been found that there is little difference in the
weld fusion zones provided that welding proceeds well within the
deep-penetration mode. Further, vertical-up and vertical-down welding has been attained in the generation of tee joints.

One of the beneficial characteristics observed in the generation of tee joints has been the tendency for the beam to follow the
seam. As shown in Fig. 8, formation of a tee section with the laser

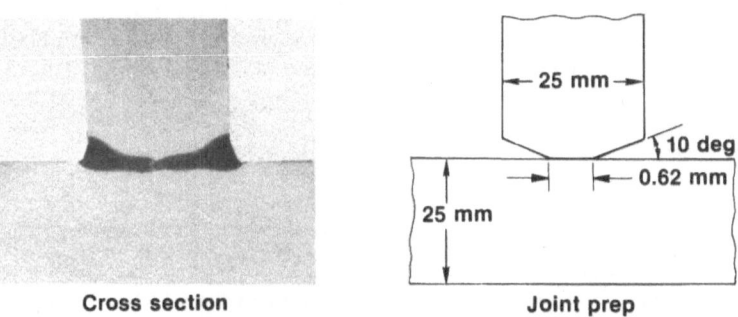

Cross section   **Joint prep**

Figure 8.   Dual-pass laser tee weld with filler addi
            (Power: 13 kW, Weld speed: 12.6 mm/sec,
            Wire (.9 mm) speed: 126 mm/sec,
            Edge prep: 10 deg vee)

beam incident at a shallow angle to the tee promotes fusion of the
entire seam. This behavior yields a strong joint even though the
extent of the reinforcement at the corner is substantially smaller
than that in a conventional fusion-welded tee section. The bead
reinforcement which occurs in a direct fusion laser weld is due to

a slight contraction of the material normal to the weld. Such
contraction also accounts for the bead reinforcement in butt welds.

Another practical consideration in applications of laser welding is the ability to add filler material to the fusion zone. This
requirement stems both from nonperfect joint fitup and from the
need for modification of fusion zone chemistry to attain desired
weld properties. In spite of the relatively narrow molten zone
associated with laser welding, continuous addition of filler material in wire form has been demonstrated, as evidenced also by Fig. 8.
Using filler material, sound weldments have been formed between
1-cm-thick alloy steel plates separated by a 1.5 mm gap.

An interesting feature of laser welding with filler addition
is that the general characteristics of the deep-penetration weld
are retained, provided that filler metal is not added so rapidly
that it results in collapse of the deep-penetration cavity. This
behavior is quite significant in that it indicates that effective
control of weld chemistry and limited compensation for imperfect
fitup may be attained without significant loss of welding performance. That is, the laser weld profile is essentially the same at
constant specific energy input regardless of whether a bead-on-
plate penetration, a tight butt weld or a gapped butt weld with
filler addition is formed.

Nonferrous Materials

In general, the weldable titanium and nickel-base alloys
respond well to laser welding. The latter, which exhibit high
resistance to oxidation, can often be welded with only minimal
inert gas-shielding provisions. With appropriate shielding, high-
quality laser welds can be formed in titanium alloy, as shown in
Fig. 9. Sound, full penetration can often be generated over a
broad speed range. Such welds exhibit no visible defects under
radiographic inspection. Microstructure, however, varies with
specific energy input. At low speeds (high specific energy input),
a relatively coarse-grained structure reminiscent of plasma arc
welding is attained while at higher speeds the weld zone is very
fine-grained. Tests have shown that the latter structure provides
excellent fatigue characteristics while the former exhibits improved
fracture toughness in the as-welded condition. These differences
underscore the laser's potential for the attainment of specific
weld properties by appropriate selection of welding process
parameters.

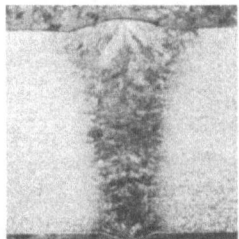

Figure 9. Laser weld in titanium alloy. (Thickness: 0.6 cm,
          Power: 5.5 kW, Weld speed: 24 mm/sec)

The response of aluminum alloys to laser welding has not been
as favorable as with other metal alloys. In large measure, this
has been due to the initially high reflectivity of clean aluminum
surfaces to the 106,000 Å laser welding beam. Typically, the
reflectivity of a mechanically-scraped (standard weld prep) aluminum surface is greater than 90%. The deep-penetration process
must therefore be initiated with less than 10% of the incident
power. This leads to the requirement for very high incident power
densities in order to initiate deep-penetration welding.

Once the deep-penetration cavity has been formed, however,
beam absorption due to the geometric blackbody characteristics of
the cavity is greater than 90%. It appears that the intensity
required to establish penetration is greater than that required for
smooth fusion. Thus, once the cavity is formed, overheating of the
molten zone occurs. Since hydrogen contributes significantly to
weld defects in aluminum alloys, and since hydrogen solubility in
aluminum increases drastically with temperature, this overheating
has obvious deleterious results. In addition, excessive vaporization occurs which triggers sporadic formation of a beam-absorbing
plasma at the surface which interrupts penetration and produces
surface craters.

In spite of the inherent difficulties experienced with high-
power laser welding of aluminum, some acceptable joints have been
formed in the 5000-series (marine corrosion-resistant) alloys. In
particular, lap fillet and tee joint welds have been formed with
promising characteristics. The general applicability of laser

welding to the aluminum alloys, however, requires further process
development.

Laser Welding Efficiency

One of the disadvantages cited for laser welding is that of
the relatively low conversion efficiency of electrical to optical
power, typically of the order of 15%. The overall welding efficiency, however, can be placed in proper perspective only by consideration of the effectiveness with which this optical energy is
utilized to form a fused joint. This effectiveness is defined in
terms of the melting efficiency,

$$\eta_m = \frac{mC_pT_m}{P}$$

in which m is the total mass of fused material, C_p is the specific
heat, T_m is the melting temperature and P is the laser power. This
efficiency represents the ratio of the energy required to melt the
material within the weld zone to the absorbed laser power. Absorption efficiency, η_A, depends on material and surface characteristics. The overall welding process efficiency is then given by

$$\eta_a = \eta_A \eta_m.$$

Direct calorimetric measurements of absorption in deep-penetration
welding show values in excess of 90%. Melting efficiency has been
determined by a series of controlled weld penetrations in a range
of pure metals, as shown in Fig. 10.

Melting efficiency was computed from material thermal properties and direct measurements of the weld zone cross-sectional area
obtained in the controlled penetrations noted above.

With reference to Fig. 10, it is seen that melting efficiencies exceeding 70% were demonstrated in the lower thermal diffusivity materials. This means that more than 70% of the absorbed
energy participates in the melting process and that less than 30%
is lost due to conduction from the fusion zone.

Due to the high melting efficiency attained, the specific
energy input to laser welds is typically about an order of magnitude lower than for conventional fusion welding processes. Thus,
the electrical energy required to form a unit length of joint is

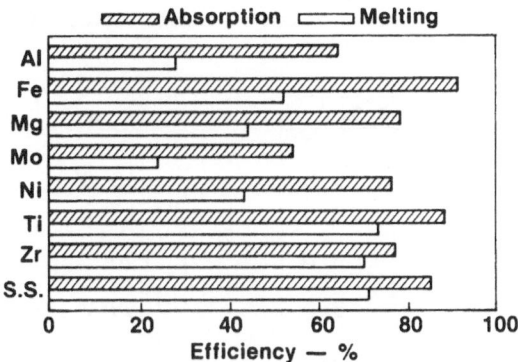

Figure 10.   Laser welding efficiency.

comparable to that required for arc processes despite the penalty
paid for conversion to optical energy. This performance, coupled
with enhanced weld characteristics and high welding speeds, renders
the laser a serious candidate for many significant welding tasks.

CUTTING

     As shown in the interaction spectrum in Fig. 1, laser cutting
is performed at essentially the same power intensities as welding.
Although a higher intensity is desirable, it is not currently
available with continuously-operating, industrially-suited, CO_2
systems. In general, a gas jet-assist must be used to facilitate
laser cutting of metals. If jet-assist is not utilized, the material melted by the beam remains in place and a bead-on-plate weld
penetration is obtained. Both inert gases, which provide clean cut
surfaces, and reactive gases, which tend to speed the cutting
process, may be used to assist laser cutting.

     Representative high-power laser cutting performance is shown
in Table II. It may be noted that thin material can be cut at
 relatively high speeds and that 2-inch-thick steel has been cut
with a 6 kW laser system. It has been found, however, that the
laser is not currently a cost-competitive cutter for thick-section
material. One of the reasons for this is that jet-assist parameters cannot be scaled with laser power. For example, a 0.3 kW,

Table II. Typical Jet-Assisted Laser Cutting Performance

	Thickness, mm	Speed mm/sec	Power, kW
Aluminum	3.0	50.8	3.0
	12.5	12.7	3.0
Boron/Aluminum	3.0	190	3.0
Nickel Alloy	3.0	59.2	3.0
Stainless Steel	3.0	42.3	3.0
Steel	3.0	67.7	4.0
	54.0	5.5	6.0
Titanium	3.0	21.2	4.3
	50.6	8.5	3.8

CO_2 laser is a highly effective cutting tool when utilized with a
sonic assist gas jet. If a 3 kW laser is utilized, direct scaling
of the fluid dynamic characteristics of the jet would require a
flow velocity ten times that used at the 0.3 kW level. On the
basis of fundamental fluid dynamics, it is obvious that such scaling is not possible.

Another factor which limits high-power-laser cutting performance in thick metals is a consequence of the narrow cut width
achieved. Although narrow cuts are desirable from the materials
standpoint, the narrow zones introduce two primary problems. The
first stems from the fact that a small-diameter assist jet must be
utilized to match the width of the cut. Since the coherence length
for an overexpanded free jet is typically of the order of a few jet
orifice diameters, this leads to jet expansion in a short distance
with resultant flaring of the cut zone. Secondly, the thin layer
of liquid metal in the narrow cut zone is strongly bound to the
solid by surface tension forces. Clean removal of the liquid
material with the aid of the limited flow possible through a small-
diameter assist jet is found to be difficult. Laser-cut edges in
thick metals are therefore generally rougher than plasma arc or
oxyacetylene cuts, which benefit from large volume flows due to
the wider cut zones. It is to be anticipated that improvements in
thick metal laser cutting performance will accrue primarily from
improvements in matching jet-assist and laser parameters.

In thin materials, on the other hand, the laser has demonstrated impressive capability in many applications. Fully automated laser sheet metal cutting systems are available from at least a half-dozen manufacturers. Such units normally utilize laser systems with up to about 1 kW of power, which is adequate for steel. Higher power is indicated for generation of smooth cuts in aluminum alloys with minimal thermal damage to the base material.

An example of such performance is illustrated in the cross sections of a CO_2 jet-assisted laser cut in 1.5-mm-thick aluminum alloy shown in Fig. 11. This cut was formed with 3 kW of laser power at a

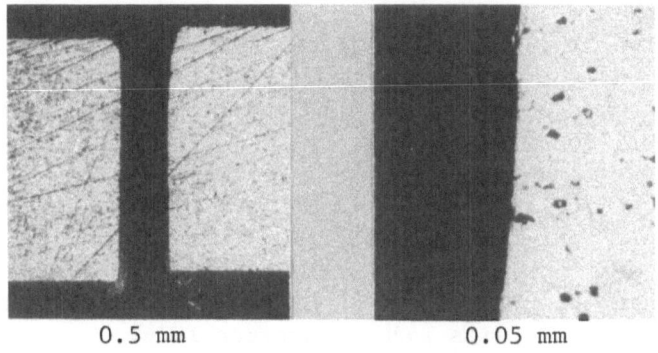

      0.5 mm   0.05 mm

Figure 11. CO_2 jet-assisted laser cut in aluminum alloy
(Power: 3 kW, Speed: 127 mm/sec, Thickness: 1.5 mm)

speed of 127 mm/sec. Because of the high cutting speed, evidence of thermal damage to the cut edge extends only about 1/20 of a mm into the base material. This is, perhaps, surprising in view of the high thermal diffusivity of aluminum and the fact that this is a thermal cutting process. Fatigue tests of aluminum alloy specimens cut under such conditions have indicated endurance levels comparable to those obtained on mechanically-blanked parts.

CONCLUDING REMARKS

High-power CO_2 laser materials processing is currently undergoing intense development. Thus, even as this is being written, new processing capabilities are being evolved which may render some of the cited performance data obsolete. On the other hand, the fundamental principles of physics, thermodynamics and heat transfer which govern the nature of the beam-material interaction will continue to form a sound basis for a priori evaluation of potential

applications. The laser's unique and exciting capabilities for
materials processing should not serve to obscure the usefulness of
these principles.

REFERENCES

1. C. O. Brown and C. M. Banas, Deep-Penetration Laser Welding.
    Paper presented at the AWS 52nd Annual Meeting, San
    Francisco, California, April 26-29, 1971.

2. C. M. Banas, High Power Laser Welding - 1978. Optical
    Engineering 17:3 (May-June 1978).

3. S. L. Engel, Laser Cutting of Thin Materials. SME Technical
    Paper MR74-960 (1974).

4. D. A. Belforte, High Power Laser Surface Treatment. SME
    Technical Paper 1077-373 (1977).

5. E. M. Breinan, B. H. Kear and C. M. Banas, Processing Materials
    with Lasers. Physics Today (November 1976).

6. H. Tong and W. H. Giedt, Depth of Penetration During Electron
    Beam Welding. ASME Paper No. 70-WA/HT-2.

7. E. M. Breinan and C. M. Banas, Fusion Zone Purification During
    Welding with High Power CO_2 Lasers. Presented at the Second
    International Symposium of the Japan Welding Society, Osaka,
    Japan (August 25-29, 1975).

HARDENING AND ALLOYING OF STEEL SURFACES

BY HIGH-POWER LASER BEAMS

Gerd Sepold

Bremer Institut für angewandte Strahltechnik

Ermlandstr. 59, 2820 Bremen 71, Germany

ABSTRACT

Heat fluxes near the surface of steel are calculated as a function of laser parameters such as power, density distribution and absorbed power. Additionally, experiments are carried out to improve the melting and alloying of defined areas in steel and to compare the experimental results with the calculated ones. It is shown that hardened zones and fusion lines in laser treated steel can be calculated whereas the geometry of alloyed areas cannot be fully derived from heat equations.

INTRODUCTION

CO_2-lasers are modern heat sources for material processing. One of their main applications is hardening and alloying of steel leading to partial or even large-area improvement of such surface properties as erosion or corrosion restistance. Those properties are drastically influenced by impinging high power laser beams and the resulting heat flux into the material leading to transformation hardening and melting of the material; sometimes the fusion zone is alloyed, adding different elements to the melt. In this field first communications on experimental results were given from [1] and [2], including the hardening of gears, crankshafts and cyclinder-bores as well as the alloying of automotive valves and valve seats. These applications may be extended to surface alloying of piston rings, vanes and the hardening of saw blades. The knowledge of the dimensions of hardened zones is one of the conditions for the economic use of high-power CO_2-lasers for prolonging the life of a construction. The size of hardened zones depends mainly on the speed of the heat source, the laser parameters and the amount of alloying elements.

It is therefore the aim of this work to calculate heat fluxes near the surface of steel as a function of the laser parameters such as power, density distribution and absorbed power. Further, experiments are carried out to improve the melting and alloying of defined areas in steel and to compare the experimental results with the calculated ones.

CALCULATIONS OF THE SIZE OF LASERTREATED ZONES

A focussed laser beam impinging on a metal surface generates a heat flux with resulting local and time-dependent temperature fields which are described by the differential equation of heat diffusion (1)

$$\frac{\partial T}{\partial t} - a \cdot \Delta T = I\ (r,\ t) \qquad (1)$$

t = Temperature
t = time
a = heat diffusion constant
Δ = Laplace operator
I (r,t) = generated power density at a place r, during t

There are many solutions of this equation available to calculate e.g. the size of a transformed zone, represented by the 850° C isotherm or to calculate the size of the melt represented by the 1600° C isotherm. For heat treatment of steel some interesting results are given by [3], which partially agree with experimental results. The discrepancy possibly depends on the fact that during heating the boundary conditions have to be clearly defined and compared with the real situation for each thermal problem.

For the calculations reported here the following assumptions where made to solve eq. (1).:

1. The two-dimensional heat pole at the point of impingement consists of an infinitesimally thin layer on the surface and passes over it at a constant speed.
2. The absorbed power density distribution of the beam has one of two forms:

$$I\ (r) \ = \ I_{max} \cdot \exp\left(-\frac{r^2}{2\delta^2}\right) \qquad (2)$$

or after [4]

$$I\ (r) \ = \ I_{1max} \cdot \exp\left(-\frac{r^2}{2\delta_1^2} - I_{2max} \exp\left(\frac{r^2}{2\delta_2^2}\right)\right) \qquad (3)$$

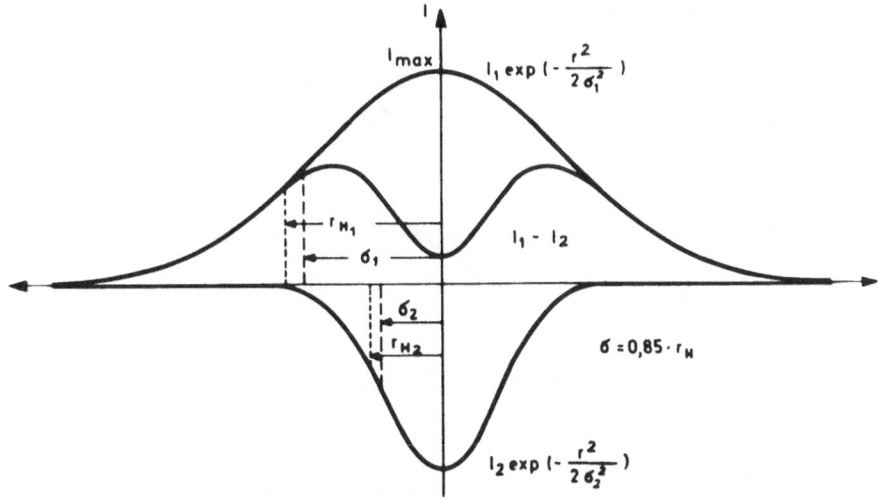

Fig. 1.   Power density distributions of the laser
beam leading to uniform heat flux $(I_1 - I_2)$
in the material.

The second power density distribution, see eq. (3) and figure 1,
was chosen to gernerate a uniform temperature field in the material,
thus producing uniform properties of the material in the laser
affected zones.

3. The size of the workpiece to be heated at its surface should be
large in all directions compared to the heat pole so that the
material may be treated as a semi indefinite body.

4. Losses of heat and the latent heat for the solid-liquid transformation should be negligeable.

The distribution of the resulting quasi stationary heat flux
due to the afore mentioned heat source moving over the metal surface
with velocity v can be described as follows:

$$T(r,z,t) = \frac{8 \cdot a^2 \cdot t_o \cdot I(r)}{\lambda \, (4 \pi \, a)^{3/2}} \cdot \exp\left(-\frac{v \cdot x}{2a}\right) \cdot \tag{4}$$

$$\cdot \int_0^t \frac{dt'}{\sqrt{t'} \, (t+t')} \cdot \exp\left[-\frac{t^2}{4at'} - \frac{r^2}{4a \, (t_o+t')} - \frac{v^2 \, (t_o-t')}{4a}\right]$$

with $t_o = \dfrac{\delta^2}{2a}$, $r^2 = x^2 + y^2$ and $I(r)$ see eq. (2)

The temperature at a given time t is equal to the sum of the
temperature rises of all elementary heat sources (principe of
superposition [5]), which occur within the duration of heating
effect of the real sources during the time $t' = 0$ to $t' = t$.

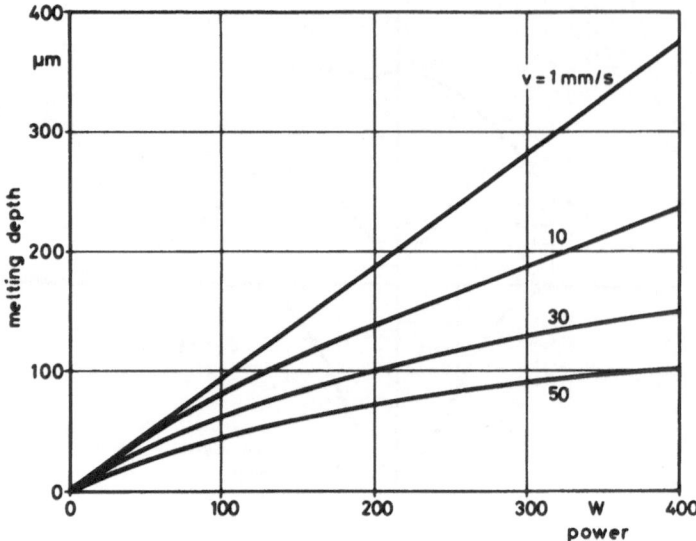

Fig.2.   Maximum melting depth as function of
         absorbed laser power and speed.

        Equation (4) can be numerically solved leading to the results
shown in figure 2. For a   and λ the mean values were taken, which
is permissible according to [6] . Further, it is permissible to use
the 90 % reflectivity of steel [7] for CO_2 -laser which means that
only 10 % of the laser power is transferred to heat by interactions.
From figure 2 it follows that

1. for a given absorbed constant power the melt depth is not
   directly proportional to the speed,

2. working speeds seem to be advantageous and economical at
   10 to 50 mm/s, leading to melting depths in the range of
   0,1 to 0,2 mm at power levels of more than 4 kW. It still
   has to be proved that this depth is good enough for alloying and

3. using the same laser power higher speeds and depths can
   be achieved by an increase of absorption by "blackening"
   the surface or by the formation of a small capillary at
   higher focussing degrees. The capillary acts like a black
   body.

EXPERIMENTS WITH A 5 kW CO_2 -LASER

Experimental Set-up

A gas transport laser of continuous 5 kW output had been used
to perform some experiments.

The active region of the laser head, where a glow discharge is
generated for the activation of the gas mixture, consists of a
cathode and seperated anodes. The gas mixture, directed by vanes,
is moved by an axial blower through the glow discharger region and
cooled by a heat exchanger. Then it is recirculated back to the
optical cavity through the laser head chamber. A principal cross
section of the laser head is given in figure 3.

The optical cavity consists of an output mirror, a rear mirror
and a pair of folding mirrors, causing photons from the excited gas
to make five passes through the gas. The output mirror is partially
transparent for extraction of the CO_2 -laser beam which is guided
towards a moving workpiece by an additional deflecting mirror and
focussing lens of 5" focal length.

The power density distribution of the beam was measured with
a calorimetric method using a watercooled copper pinhole being moved
stepwise through the beam in different planes. Each step leads to
heating the unterlying workpiece which was registered by thermocoup-
les, see figure 4. At each step the temperature rise of the thermally insulated steel block is a measure of the absorbed energy and
gives a value for the power density distribution curve.

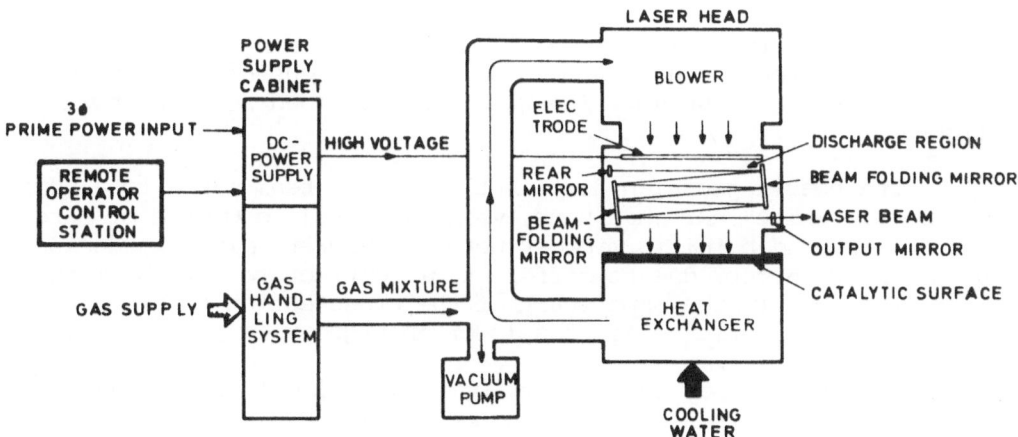

Fig. 3.   Schematic view of the laser assembly

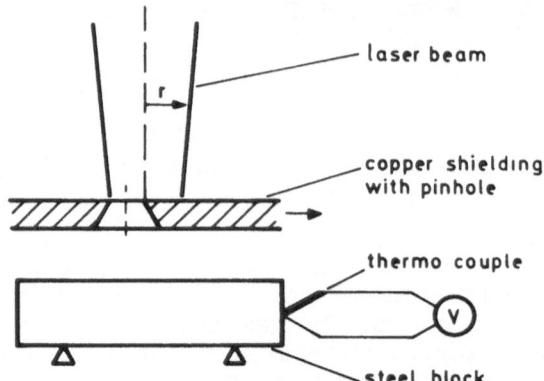

Fig. 4.   Set-up for measuring beam
           intensities.

Laser Affected Zones in Steel

A first series of experiments was carried out to heat steel
at its surface to elevated temperatures of 1000°C in order to study
the hardening effect. A second series was performed to heat steel
to approx. 2000°C at its surface. Sections were taken from heat
affected zones and after etching in 2 % HNO_3, micrographs were prepared to obtain a view of the size of the heated zones. Furthermore,
microhardness curves give informations on the hardenability of the
steels and microprobe analysis on the distribution of the alloying
elements in the melt after rapid cooling.

For transformation hardening a steel containing 1 % carbon
and 1 % tungsten was heated by a slightly defocussed laser beam
(L_a = 500 W; r_H = 400 μm). The resulting hardened zone is shown in
figure 5.

The nearly gaussian distribution of the beam leads to a hemispherical "white" hardened zone, which is mainly composed of hard
martensitic and bainitic microstructures. Vickers micro hardness
values HV 0,05 vary due to different microstructures of the matrix
from HV = 1000 in the upper part of the crosssection to HV = 400 in
the starting material - see the lower enlarged view of figure 5.

In the same way experiments were carried out with unalloyed
steel of 0,2 % carbon taking into account a gaussian and a double
gaussian distribution as has been demonstrated in figure 1. The
resulting heated zones are shown in figure 6.

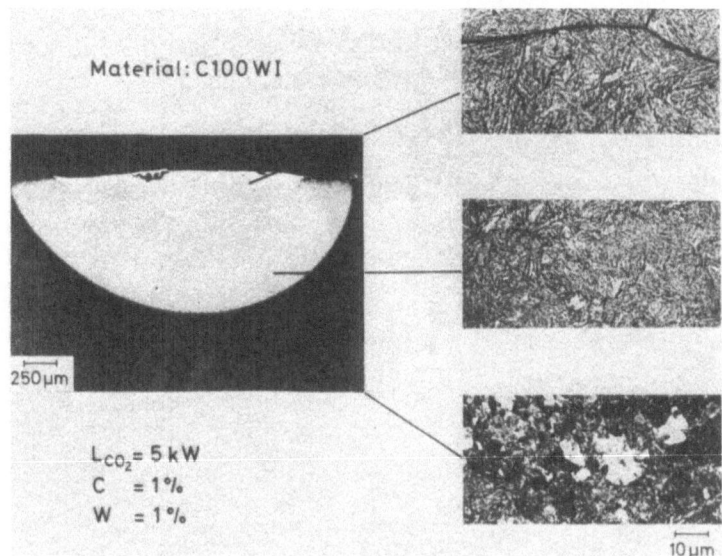

Fig. 5.   Hardening of steel grade C 100 W 1 by 5 kW
          CO_2-laser beam.

The flat melt isotherm (b) depends on the power density distribution $I_1 - I_2$ and correlates quite well with the predicted heat
flux into the material. The dotted lines in figures a and b were
derived from calculations. They do not fit very well with the experimental results due to possible changes in energy absorption at
elevated temperatures and due to exothermic reactions with air.
However, it may be concluded that intensity distributions affect
the hardened zones in such a way that it is possible to attain a
uniform geometry of the hardened zones, avoiding overheating and
high temperature gradients.

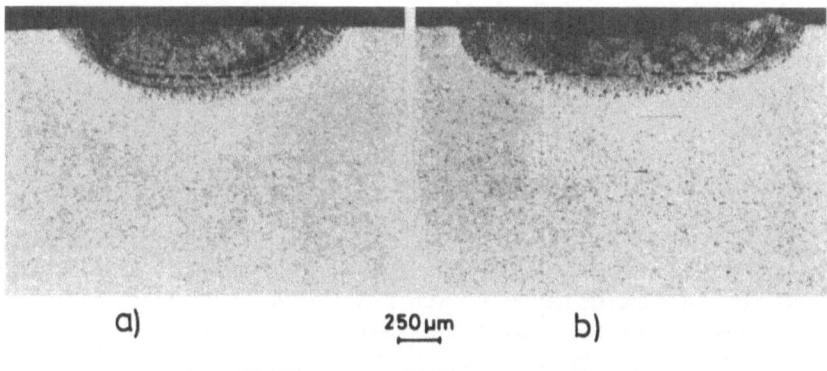

Fig. 6.   Fusion zones in unalloyed steel (c = 0,2 %)
          L_A= 450 W; r_H= 1000 µm; v = 5 mm/s; $I_{(r=o)}$= 10^4-$5 \cdot 10^3$ W/cm²

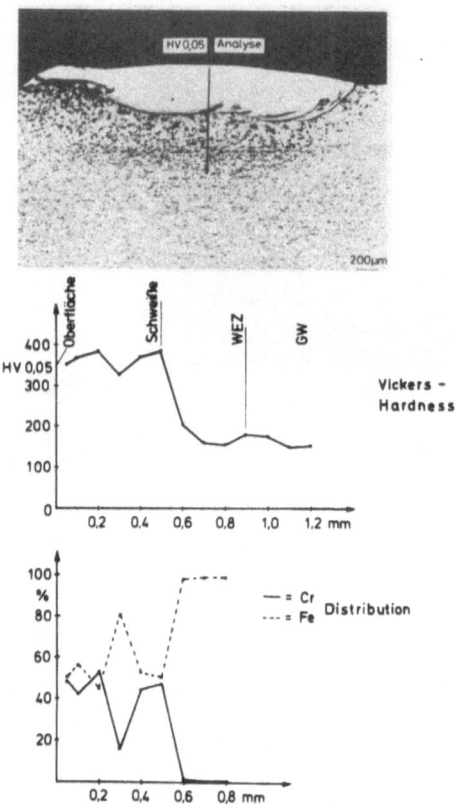

Fig. 7.   Surface alloying of steel with chromium;
          Vickers hardness curve and Cr-Fe-distribution versus depth.

        For surface alloying, chromium and tungsten carbide with 10 %
cobalt powder had been added in the laser-melted layer of unalloyed
steel. In the case of chromuim alloying a ferritic and heat
resistant layer should be generated whereas tungsten carbide was
used for hardfacing the surface. In both cases a thin layer of the
mentioned elements was placed onto the steel surface. Iron and
chromium melted together forming a thin and smooth surface layer
of increased hardness, see figure 7.

        The distribution of chromium in the melt is not uniform which
has no large influence on hardness values. The size of the fusion
zone can be roughly calculated by eqs. 3 and 4 if one assumes that
chromium and iron form a homogeneous layer with thermophysical
properties, simular to those of the elements themselves. Contrary
to these first results, melting of tungsten carbide with 10 % cobalt
leads to difficulties, as the powder agglomerates forming a rough
surface, figure 8. It seems to be clear that the former equations
are not valid in this case for the determination of the size of the
melt-isotherm.

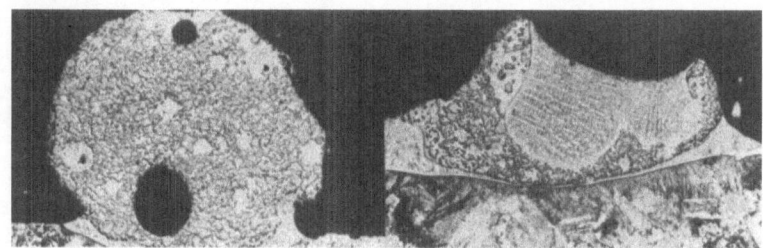

100 µm

$r_H = 500 \mu m$; $L_{CO_2} = 5\,kW$; $v = 30\,mm/s$

Fig. 8   Surface of steel with agglomerated melted Co with
         WC-particles.

PRODUCTION OF AMORPHOUS LAYERS

     The results of theoretical calculations show that cooling rates
of up to 10^7 K/s to 10^8 K/s may be attained under realistic assumptions with regard to the experimental conditions. This cooling rate
is sufficient to produce amorphous layers of many technically interesting alloys, and not only 'model alloys' such as PdCuSi, which
require only a few hundred K/s.

     An unknown factor in this process is a possible nucleation
effect of the underlying, crystalline material, which may inhibit
the amorphous solidification of surface liquid layers, even at
cooling rates above the limit for homogeneous crystallization. The
method of laser glazing was therefore tested experimentally on the
alloy $Fe_{40}Ni_{40}P_{14}B_6$, which has a critical cooling rate of 10^5 K/s
to 10^6 K/s, in the homogeneous state. As to the theoretical calculations two simplifying assumptions were made: Firstly, the
thermodynamic properties were assumed to be temperature independent.
(In particular, phase transformations were ignored.) Secondly,
energy losses due to radiation at the specimen surface were neglected.

     Following experiments were performed with a 5 kW CO_2 laser.
A crystalline specimen was attached to a rotating disc and was
moved with a velocity of 5 m/s relative to the laser beam. The
beam radius was 0,2 mm and the power density $2 \cdot 10^6$ W cm $^{-2}$.

     The surface material on basis of $Fe_{40}Ni_{40}P_{14}B_6$ in the path of
the laser beam was subsequently investigated by transmission electron microscopy. Fig. 9a is a micrograph showing amorphous and
crystalline regions in the melted zone. The diffraction pattern
(Fig. 9b) shows both diffuse rings from the amorphous material and

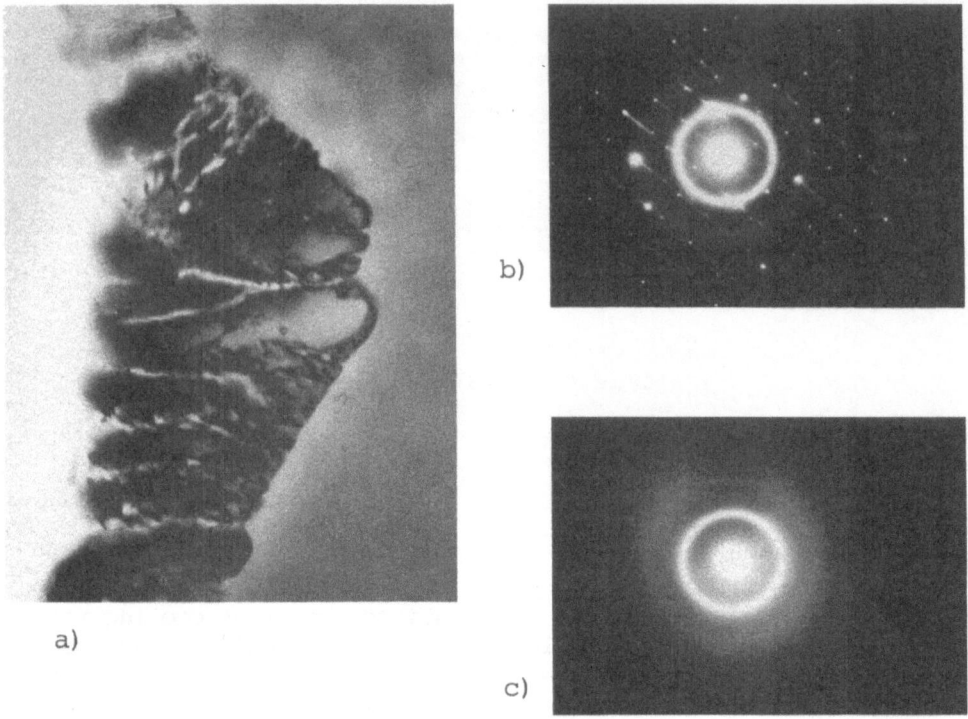

a)

b)

c)

Fig. 9. Transmission electron micrograph and diffraction patterns
from $Fe_{40}Ni_{40}P_{14}B_6$.

a) Amorphous and crystalline region in the laser melting
   zone
b) Electron diffraction pattern of the region shown in
   fig. 9a, showing the diffuse ring of the amorphous
   structure together with spots from the crystalline area
c) Electron diffraction pattern of a completely amorphous
   region.

spots from the crystalline regions. The diffraction pattern (Fig. 9c)
was taken from a purely amorphous region of the laser beam trace.

Electron microscopy investigations of various regions of the
melted zone suggest that the crystallization fronts grow from the
crystalline substrate. Teh thickness of the amorphous lyer was estimated to be of the order of 10 μm.

The results of this investigation show that laser glazing of
the used power is possible if the critical cooling rates are in the
range of 10^6 K/s. In the near future it is hoped that a amorphous.

layer may be produced by scanning the laser beam over the surface. The aim is ultimately to coat cheap base materials with coatings of composition suitable for wear and corrosion protection and transform these layers by laser treatment into the glassy state.

CONCLUSIONS

Solutions are given to calculate the transformation hardened zones and fusion lines in laser treated steel. The calculations are in quite good agreement with experimental results. The geometry of alloyed areas cannot be derived from the mentioned solution of the equation of heat diffusion, due to different and unknown absorption coefficients, thermophysical data, surface tensions and distributions of the alloying elements in the melt. Future work should be directed towards overcoming these difficulties, using special laser techniques such as circular sweeping of the beam in order to achieve better mixing of the metals and a possible correspondence between theory and practice.

6. REFERENCES

1   R. A. Hella and D. S. Gnanamuthu, High-Power Lasers in Materials Processing, Research Rep. 412, 1975, AVCO Everett Res. Labs. USA

2   P. Dumonte, D. Maurin and G. Sayegh, Possibilités D'utilasation d'un Laser de 10 kW dans le Travail de Métaux, I.I.W.-Dokument IV - 179-75

3   H. E. Cline and T. R. Anthony, Heat Treating and Melting Material with a Scanning Laser or Electron Beam, J. Appl. Phys. 48(1977)9, pp. 3895-3900.

4   K. H. Drake Kontaktieren von Halbleiterbauelementen mit Laserstrahlen nichtgaußförmiger Intensitätsverteilung, DVS-Berichte, Schweißen in der Elektronik und Feinwerktechnik S. 105 bis 109, DVS-Verlag 1977

5   H. S. Carlslaw and J. C. Jaeger, Conduction of Heat in Solids Oxford Univ. Press., N.Y. 1959

6   G. Sepold, Habilitationsschrift, bisher unveröffentlicht.

7   T. J. Wieting and J. T. Schief, Infrared Absorptances of Partially Ordered Alloys at Elevated Temperatures. J. Appl. Phys. 47 (1976) 9, pp. 4009-4011.

LASER ANNEALING OF SEMICONDUCTORS

Mario Bertolotti

Istituto di Fisica,Fac.Ingegneria,Università Roma

and GREQP of CNR, Italy

LASER DAMAGE ON SEMICONDUCTORS AND FIRST APPLICATIONS FOR ALLOYING

a - effects on reflectivity

The study of laser effects on semiconductor surfaces started shortly after the invention of the laser itself. The first measurements can be found in 1964 [1,2] when it was observed that a polished semiconductor surface used as a reflector in a ruby laser, can function as a passive Q-switch. The increase in reflectance was attributed to the high density of generated electron-hole pairs [1]. Sooy et al. [2] repeated the experiment and tested Si, Ge, InP, InSb, Ga ($As_x P_{1-x}$) finding in all the materials an increase in reflectivity. Damaging of most materials was also observed. The time development of the reflectivity during illumination with a Q-switched ruby laser, was studied with the disposition shown in Fig. 1, obtaining the results shown in Fig. 2a (Ge) and 2b (InSb). At low power irradiation the reflectivity followed the development of the exciting pulse.
At higher excitations, the reflectivity saturated and stayed on longer. This effect was attributed to melting, and the decay of reflectivity was stated to follow the cooling and solidification of the surface.
The following year M. Birnbaum [3,4] studied the large increase in the reflectivity of GaAs, InSb, Ge, and Si, using a ruby laser, and confirmed these first results.
He observed also a large mechanical damage at the surface, consisting of regular patterns of cracks which were correctly attributed to the cleavage habits of the irradiated semiconductors [4].

175

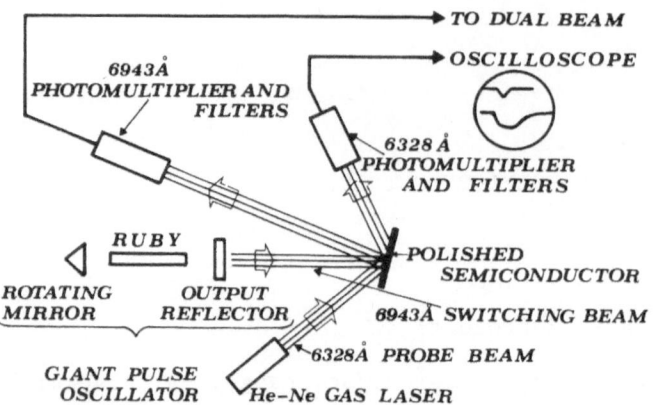

Fig. 1 Set-up for reflectivity measurement used by Sooy et
al.ref. (2).

Somehow later he and Stocker [5] performed a detailed study of
reflectivity enhancement of semiconductors by a Q-switched ruby
laser pulse, and interpreted the results as related to the
occurrence of a molten surface layer of the semiconductor. They
wrote "The temporal dependence of the enhanced reflectivity, in
particular, the persistence of the enhanced reflectivity and its
continued increase after termination of the ruby-laser giant pulse,
can be readily explained with the aid of a model with a liquid
semiconductor of higher reflectivity than the solid semiconductor".
Nearly in the same period several other researchers started to
study different effects of laser radiation on semiconductors[6-15].
In the Soviet Union great attention was given to the effects of
laser irradiation on properties involving changes in the large
amount of injected excess carriers, such as f.e. reflectivity[8-13]
Grinberg et al.[8] were among the first to correctly understand the
transfer mechanism of light energy to the crystal via the
recombination of the light generated electron-hole pairs. They
considered the following cases

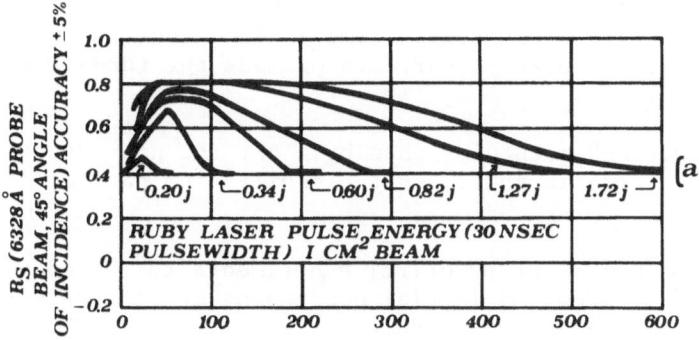

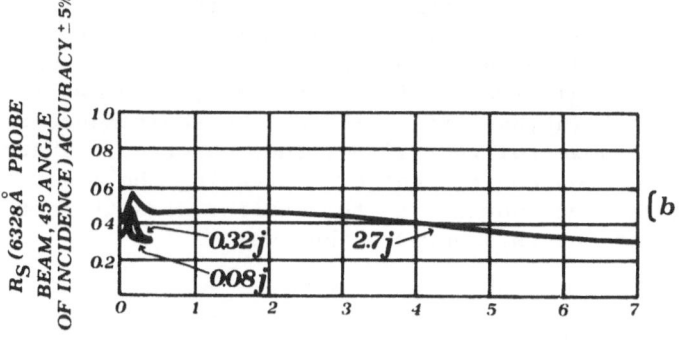

Fig.   2   Reflectivity behavior versus time a) germanium,
             b) InSb. From ref. (2).

1. "Metallic" mechanism when $h\nu <$ ΔE (bandgap) but there are free
   carriers in the permitted band. Absorption is possible by
   interaction with these free-carriers.

2. "Induced metallic" mechanism, where $h\nu > \Delta E$.
   The non radiative transition time is long and the absorption
   by non equilibrium free-carriers exceeds the fundamental
   absorption.

3. "Semiconducting" mechanism when $h\nu > \Delta E$. The nonradiative
   transition time is short and the equilibrium carrier density
   is low.

    Through a number of beautiful experiments the Lebedev group [9]
[10,11,13] studied the initial decreasing of reflectivity due
to the non-equilibrium injected carriers which shows a minimum
as a function of the exciting power.
Such a minimum is characteristic of plasma resonance, and they
measured it in several semiconductors using a CO_2 laser beam
as a probe (s.Fig.3). From this minimum they derived the values
of the density Δn of non-equilibrium carriers and the relaxation (scattering) time τ of the carriers, as shown in table I.
In the case of intense excitation of the semiconductors by
light from a powerful ruby laser, they observed a sharp increase
in the reflection coefficient by increasing the laser power that,
as they noted, was not due to the carrier density increase
(insufficient for this purpose) but was

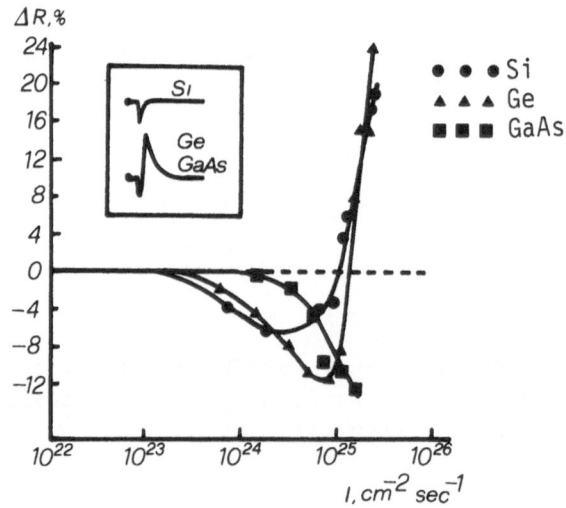

Fig. 3 CO_2 laser reflectivity changes as a function of ruby
         laser power. From ref.(13).

TABLE   I

Semic.	Δn	τ
Ge	$2.10^{19} cm^{-3}$	$10^{-14} sec$
GaAs	$0.9.10^{19}$	2.10^{-14}
Si	$< 2.10^{19}$	$> 2.10^{-13}$

possible only as a result of the sharp increase (by two orders of magnitude) of the density of the free equilibrium carriers when the surface layer of the crystal was molten.

So already in the 60's, people had found convincing evidence of melting of the surface. Problems relative to the various recombination mechanisms, change in recombination time and maximum value of the non-equilibrium carrier concentration were treated in the works ref. [9,10,11,12,13].

b- mechanical damage

In Italy great attention was given to the mechanical damage produced at the surface [7,15,16,17] (see also ref.14 for a similar study in USSR).

These studies on the mechanical damage produced at the surface were performed with both an optical and an electron microscope and revealed the existence of a large mechanical damage when some level of irradiation was surpassed.

A comparative study among several different semiconductors (Ge,Si,GaAs,InSb,CdSe) was done in ref.(7). These first experiments were done by using ruby lasers, and were confirmed later by other researchers [18].

The general results can be summarized as follows treating separately the case of free-generation and Q-switched lasers.

1. Free-generation

     A free generation ruby laser (λ = 6943 Å) usually emits a
train of spikes lasting a time of the order of a millisecond or
less.

     At low energy density (5-10 J/cm^2) for Ge (about 17 J/cm^2 for
Si) the exposed surface, observed with an optical microscope after
irradiation, shows area which have lost their flatness and appear
sligthly moved. Increasing the light energy density in some
semiconductors (Ge,Si,GaAs, GaSb) sharp crack lines and thermally
produced etch pits appear. The crack-lines are formed along
cleavage directions. The threshold energy density at which the
crack-lines are formed depends on the pulse duration (s.fig. 4
ref. [18]). In Fig. 5 crack lines forming 60° angles among them are
shown formed onto a 111 Ge surface. Fig. 6 shows the effect onto
a 100 Ge surface. The angles among lines are now 90°. The mechanism of cracking is to be attributed to thermal shock [14,18],
i .e. the temperature gradient produced by the absorbed laser
energy, puts stress on the crystal. For simplicity we may assume
that the laser is flat in time, and uniform over the x-y plane,
and also that the sample is isotropic and described by temperature
independent parameters. Since the dimensions of the laser beam are
large compared to the depth to which heat is conducted during the
time of the laser pulse, the one-dimensional temperature distri-
buti on T(z,t) and the non-vanishing stress component at the

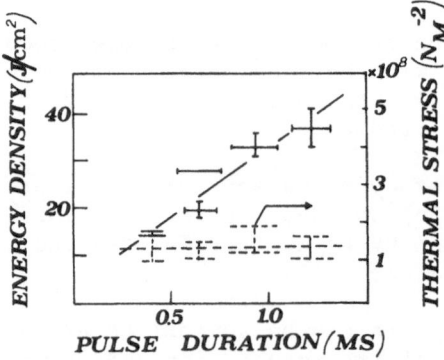

Fig. 4   Threshold energy density at which cracks are formed
          onto a silicon surface as a function of ruby laser
          pulse duration. From ref.(18).

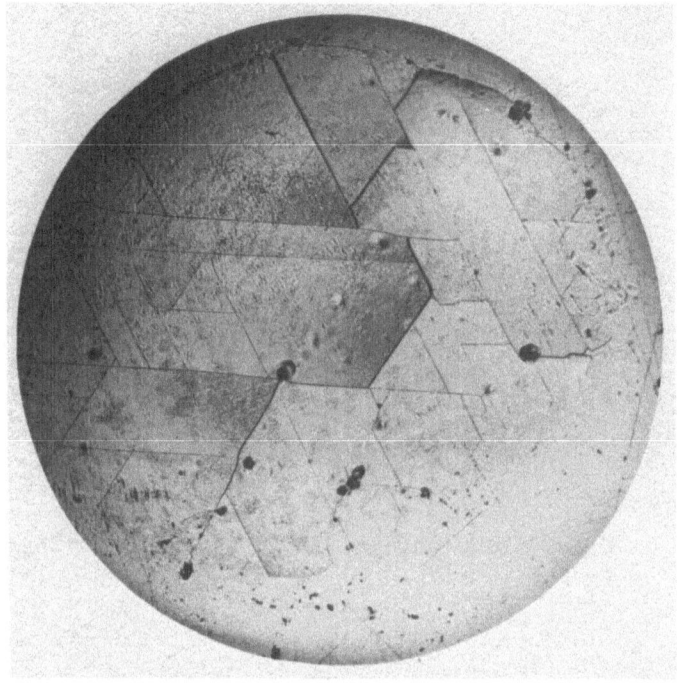

Fig. 5 Crack lines onto a [111]   germanium surface
          from ref.(7).

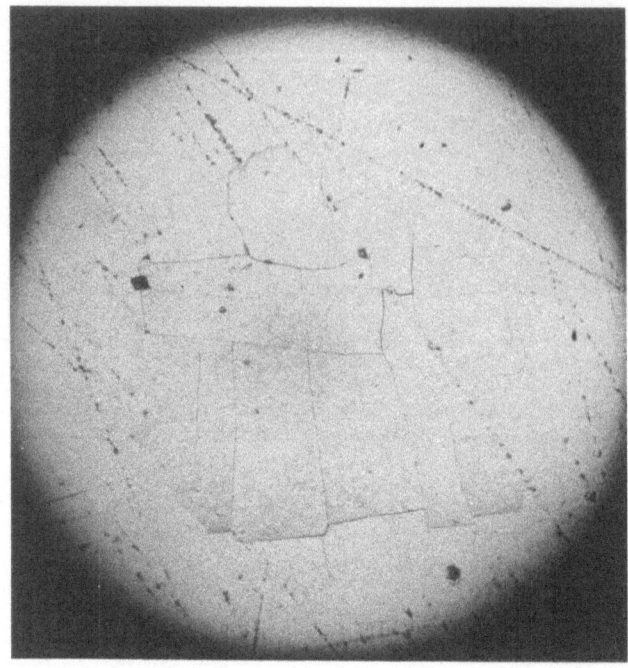

Fig. 6 Crack lines onto a ⌈100⌉ germanium surface
         (unpublished)

surface $\sigma(\sigma = \sigma_{xx} = \sigma_{yy})$ are considered [18]

$$T(z,t) = \frac{2(1-R)Q \sqrt{\lambda t_o}}{\chi} \quad erfc \left(\frac{z}{2 \sqrt{\lambda t_o}} \right) , \qquad (1)$$

$$\sigma = \frac{2 \cap E}{1- \nu} \quad \frac{(1-R)Q\sqrt{\lambda t_o}}{\chi} \left[- \frac{1}{\sqrt{\pi}} + \frac{D_t}{L} - \frac{1}{\sqrt{\pi}} \left(\frac{D_t}{L} \right)^2 \right]$$

$$(2)$$

where

R is the surface reflectivity
Q is the input laser flux (W/cm^2)
λ is the thermal diffusivity
t_0 is the pulse duration
χ is the thermal conductivity
β is the thermal expansion coefficient
E is the Young modulus
ν is the Poisson ratio
L is the thickness of the sample
$D_t = 2\sqrt{\lambda t_0}$ is the diffusion length of heat

For a free-generation laser Eq.(2) can be simply written as

$$\hat{\sigma} = - \frac{\beta \; E \; T_m}{1 - \nu} \tag{3}$$

where T_m is the surface temperature.

At larger energy densities (20-30 J/cm^2) a heavy damage appears concentrated along a circular line inside which a deep crater is eventually produced. As an example Fig. 7 shows the

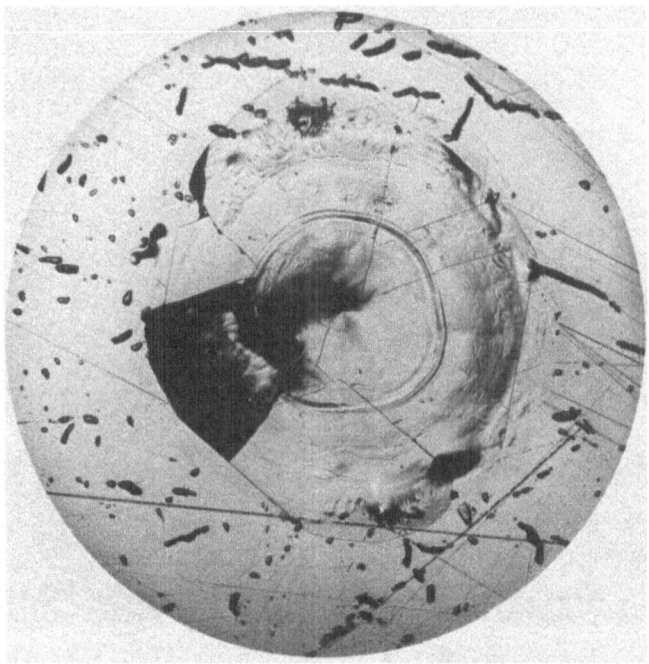

Fig. 7 Start of a developing crater onto a germanium surface
        (unpublished)

surface aspect of a laser bombarded surface of Ge as seen with an
optical microscope. Figs.8 and 9 show the craters obtained at

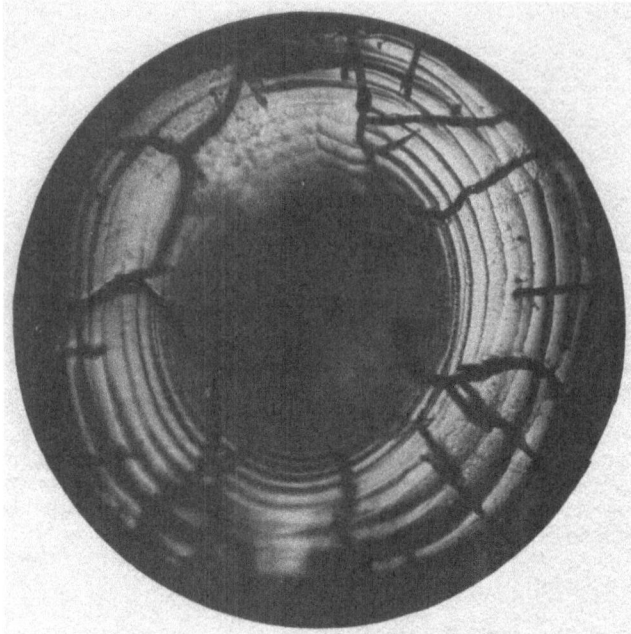

Fig.8 A top sight of a
well developed
crater in ger-
manium . From
ref. (7).

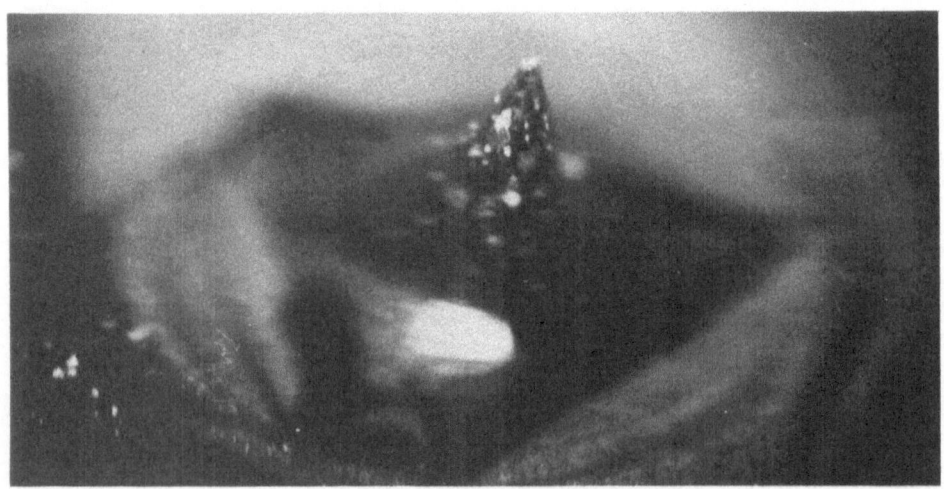

Fig. 9 Side view of a crater in germanium with a protuberance
         at its center due to the protrusion of resolidified
         material. From ref.(7) modified.

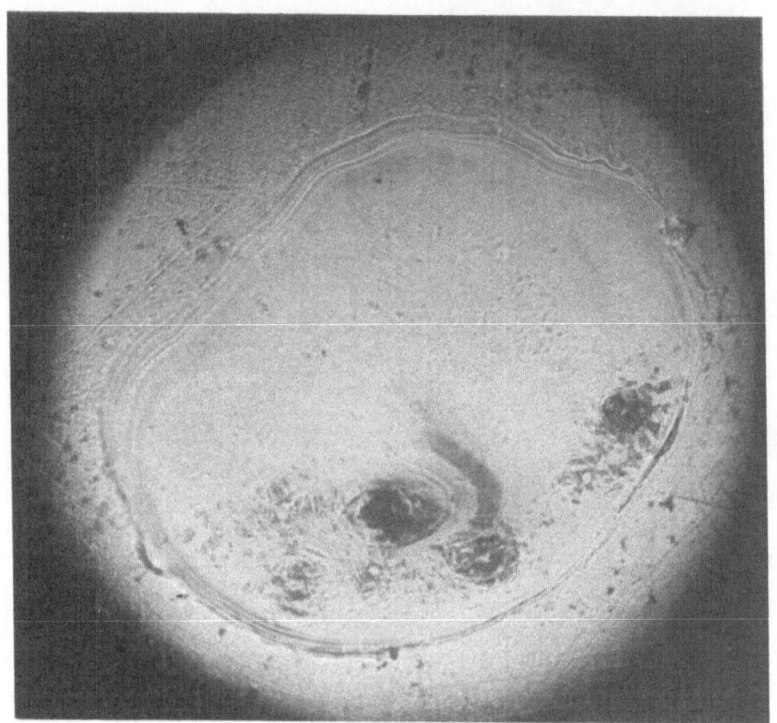

Fig. 10   Surface anspect of InSb from ref. (7).

high er energy. Fig.10 shows a sample of InSb at 17.5 J/cm^2.
Fig. 11 (ref.17) refers to CdSe at the same energy and Fig. 12
shows a crater in GaAs.

    The details of the observed effects depend upon

a – the crystal type
b – the electrical conductivity of the crystal
c – the crystallographic orientation of the exposed surface
d – the surface state of the crystal.

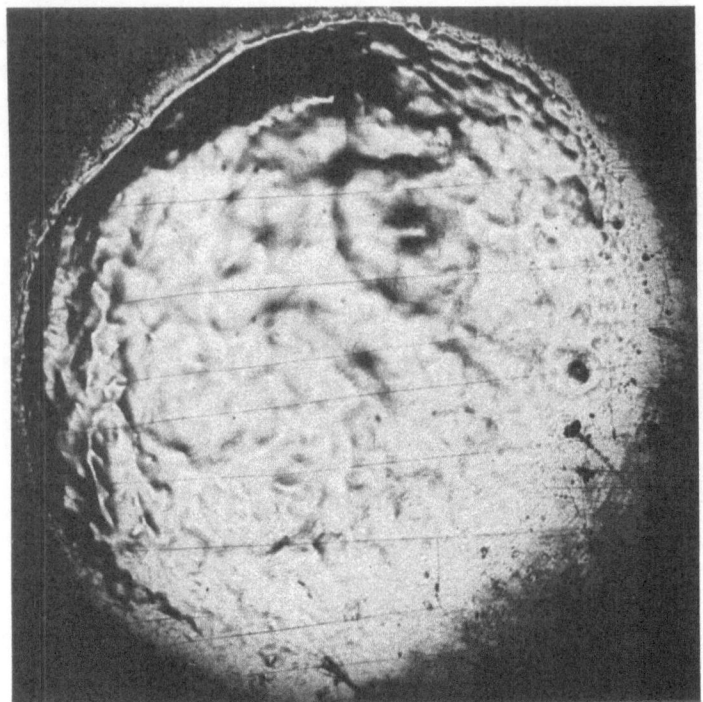

Fig. 11 Surface
anspect
of CdSe
from
ref.(7).

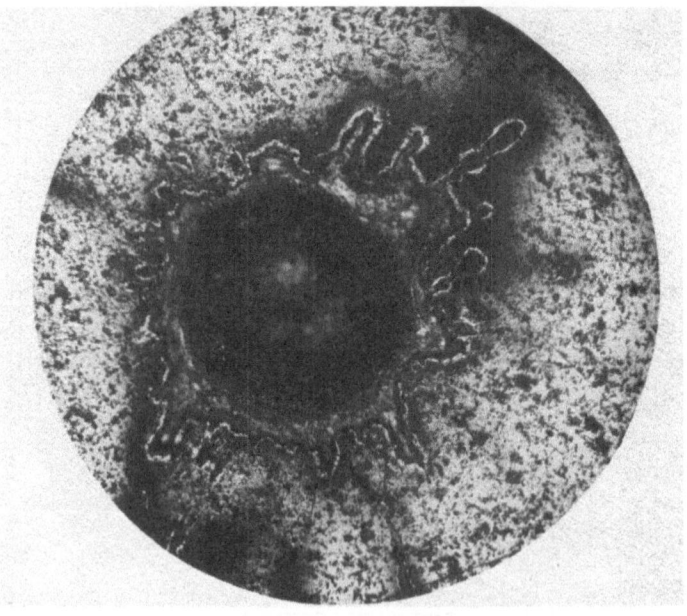

Fig. 12 Surface
anspect
of GaAs
(unpublished).

The damage is more clearly evidentiated with an electron microscope [15,16]. Three kinds of structures were observed in the central part of the irradiated zone in Ge (s.Fig. 13).

a - pseudo-hexagonal structures with a relief formation at their center (Fig. 14).

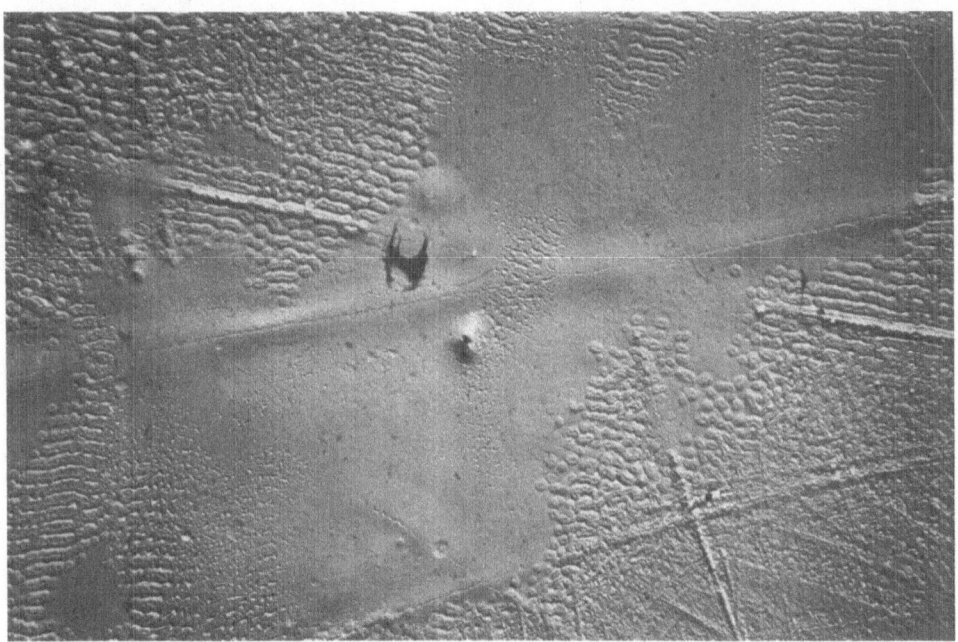

Fig. 13 General view of a laser irradiated Ge surface at the electron microscope (from ref.15).

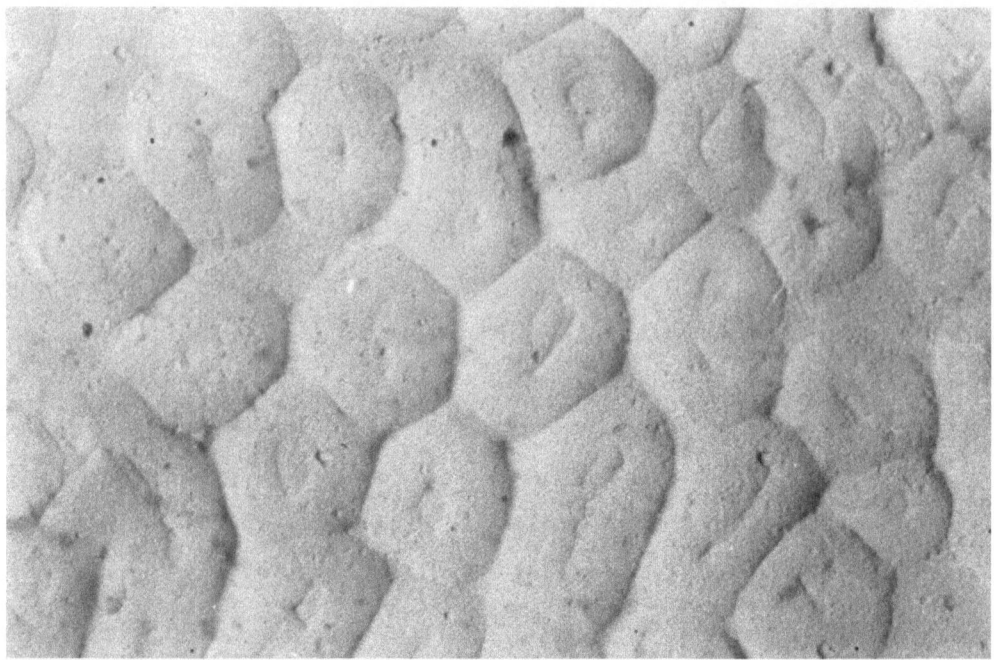

Fig. 14   Pseudo hexagonal structures with a relief formation at
          their center. From ref.(15).

b - depressions with triangular symmetry on (111) surfaces and
    rectangular symmetry on (100) surface with a hillock at
    their center (fig. 15 a and b).

c - clusters of conical hillocks without surrounding structures

In subsequent studies [19] it was shown that the triangular depressions
are associated with the formation of an oxide layer when irradiation
of Ge is made in air.

In compounds the effects were studied in ref.(16). GaAs was
found to behave in a rather similar way as Ge. Absence of
triangular or square (according to the surface orientation)
depressions was observed and attributed to the very high temperature of formation of the oxide in GaAs which prevents its

Fig. 15 Depressions. At left with a triangular symmetry produced onto [111] surfaces, at right with a rectangular symmetry produced onto [100] surfaces. From ref. (15).

formation. A study of the anspect of the irradiated surface showed
the production of a phase of metallic Ga which was formed due to
the high volatility of As.
GaAs and InSb behave in a completely different way from GaAs,
showing a plastic behavior.

The produced effects in Ge, Si, and partly in GaAs were explained as due to simple thermal effects produced by conversion of radiant
energy into heat, which gives rise to high temperatures with melting
of a surface layer followed by epitaxial growth.

Q - switching

In this case the crack-lines are formed at almost the same
power density as the molten zone is produced. The calculated
critical stress value is about 5.10^8 N/cm^2 for Si, higher than the
value of $1 \div 2.10^8$ N/cm^2 in the free-generation case, and this is
attributed to the difference in pulse duration, i.e. the difference
in loading time.
In Si at 50 - 55 Mw/cm^2 (~ 1.5 j/cm^2) a few discrete microcraters
are observed.
At about 60 Mw/cm^2, damage visible to the naked eye is produced.
At high power greater than about 350 Mw/cm^2 cracks extending into
the volume of the crystal are present.

c - mechanisms of heating

The mechanism through which light is absorbed in semiconductors is now to be considered. The light impinging on a
semiconductor surface is absorbed through two different
processes (s.lecture here by M.von Allmen),

a - interaction with free carriers in the bands. The energy so
    absorbed is then given as heat to the lattice in a very
    short time.

b - interaction with electrons in the valence band, thus
    creating electron-hole pairs. The energy absorbed in this
    case is then given to the lattice as heat, via different
    mechanisms. The excess energy absorbed by the electron
    raised in the conduction band is given as heat in very
    short times as it thermalizes in the band. Then the pair-
    recombination energy is restitued with times which can be
    much larger depending on the lifetime of the produced
    excess carriers. In this case it must be remembered that
    1) at high excitation levels the lifetime is a decreasing
    function of the number of produced carriers, and 2) that a
    source of heat is present at the surface due to recombina-
    tion of carriers at surface impurities. This is the reason

of the strong dependence of damage on the surface state of the semiconductor [17].

The more used semiconductors like Ge, Si, GaAs, GaSb, InAs, InSb, etc. have band gaps as shown in table II.

TABLE   II

Semicond.	band gap at 300°K (eV)
Si	1.12
Ge	0.66
GaAs	1.43
GaSb	0.67
InAs	0.33
InSb	0.16
CdSe	1.7

Therefore the absorption of light is very sensitive to wavelength. Figs.16 [20] and 17 [21] show the absorption coefficient of Si and GaAs as a function of wavelength.

Several calculations of the temperature raising in a semiconductor during optical excitation were performed in the past [8] [22 - 27].

More recently J.R. Meyer and collab. [28,29] have performed more refined calculations which are given here. By assuming uniform irradiation of a semi-infinite sample, the temperature increase is written as

$$\Delta T = \frac{Q \, t_o \, |1 - R(T_o)|}{\rho \, c(T_o) \, L_H}$$

where T_o is the initial temperature and ρ is the mass density of the material, c is the specific heat; the quantity L_H, which may be roughly interpreted as the depth of material heated, is given

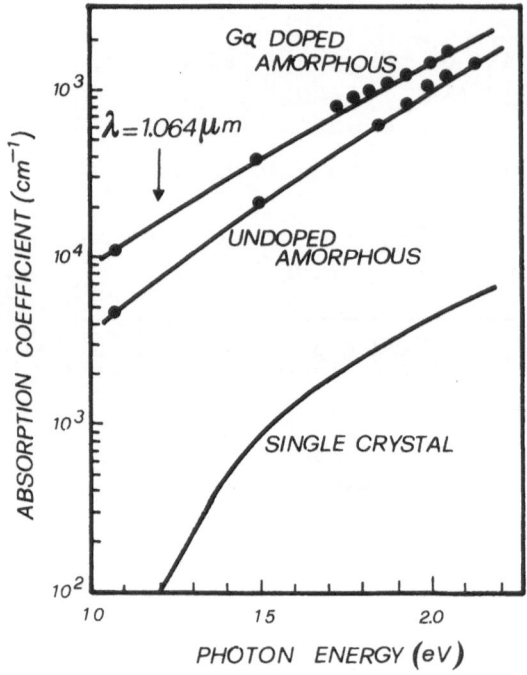

Fig. 16 Absorption
coefficient
of Si.

Fig. 17 Absorption coefficient
of GaAs.

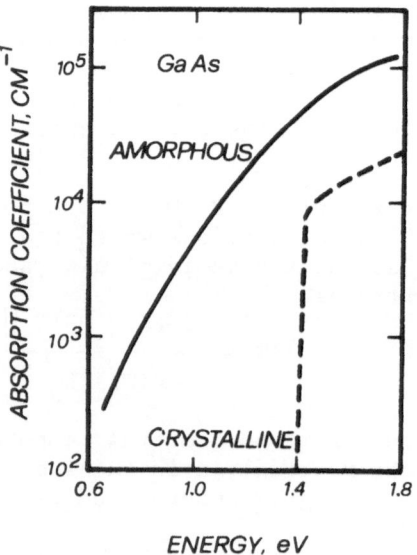

$$L_H = \frac{[1-R(T_0)]}{c(T_0)\ \Delta T} \int_{T_0}^{T_f} \frac{d\ T_C}{(1-R)\alpha} \left[\chi_T / (L_T\ \alpha + 1) \right.$$

$$(4)$$

$$\left. + \chi_B^{NR} / (L_T\ \alpha + L_D\ \alpha + 1) + \chi_S^{NR} / L_T\ \alpha \right]^{-1}$$

where $\alpha = \alpha_1 + \alpha_2 + \alpha_{FC}$ is the total optical absorption coefficient, α_1 and α_2 are the one-and two-photon band-to-band absorption coefficients, α_{FC} is the free carrier absorption, $L_D = (D\tau_B)^{\frac{1}{2}}$ is the free carrier diffusion depth, D is the ambipolar carrier diffusion coefficient, and τ_B is the bulk carrier recombination lifetime. The thermal diffusion depth, L_T, is given by

$$L_T(T) \approx \pi^{\frac{1}{2}} \left[k(T)t_p \right]^{\frac{1}{2}} (T-T_0) / \ \Delta T, \qquad (5)$$

where k is the thermal diffusion coefficient. The parameters χ_T, χ_B^{NR} and χ_S^{NR} are the fractions of the laser energy entering the sample which go respectively into the thermalization of excess carriers immediately after excitation, into nonradiative bulk recombination, and into nonradiative surface recombination, and are given by

$$\chi_T = \left[(1- \eta_Q E_R/h\nu)\alpha_1 + (1-E_R/2h\nu)\alpha_2 + \alpha_{FC} \right] / \ \alpha, \qquad (6)$$

$$\chi_B^{NR} = (\tau/ \tau_B^{NR}) (1 - \chi_T), \qquad (7)$$

and

$$\chi_S^{NR} = (\tau/\tau_S^{NR}) (1- \chi_T) L_D\ \alpha / (L_D \alpha + 1) , \qquad (8)$$

where η_Q is the quantum efficienty, $h\nu$ is the photon energy, E_R is the recombination energy $\tau = (\tau_B^{-1} + \tau_S^{-1})^{-1}$ is the effective carrier recombination lifetime $\tau_S = L_D/S$ is the effective surface lifetime, S is the surface recombination velocity, and

τ_B^{NR} and τ_S^{NR} are the nonradiative components of the two lifetimes.

Many of the parameters in Eqs. 4 to 8 depend on the optically excited carrier concentration. The rate of carrier generation, evaluated at the surface of the material (z = 0), is given by

$$g(z = 0) = g_1(z = 0) + g_2(z = 0)$$

$$= \eta_Q(1-R) \ P_o \ \alpha_1/h \ \nu + (1-R)^2 P_o^2 \ \beta/2h\nu \ , \qquad (9)$$

where g_1 and g_2 are the generation rates due to one-photon and two-photon absorption respectively and β is the two-photon absorption coefficient. It is shown in ref.(26) that the temperature-dependent carrier density at the surface is then given to a good approximation by

$$n(z=0,T) = n_i(T)+g(z=0) \ \tau \ /(L_D\alpha + 1) \ \big|_{T=T(z=0)} \cdot \quad (10)$$

Equation (10) approximates to within 20% the exact solution to the carrier diffusion equation for a broad range of experimental conditions. It should be noted that n appears on both sides of Eq.(10) because g, τ, L_D and α can all depend on carrier density.

Calculations performed with a computer which are shown in Figs. 18 to 23.

d – first experiments of laser annealing

Some technological applications of lasers in making junctions in semiconductors were considered starting from 1968.

Fairfield and Schwuttke [31] made solid-state diodes by focusing radiation from a pulsed ruby laser onto polished Si surfaces on which an appropriate impurity had been applied,and the same did Harper and Cohen with a Nd: YAG laser [32].

Also Marquardt and Giuliani [33,34] did similar experiments and in addition observed redistribution of dopants. Changes in concentration profiles were explained by the action of

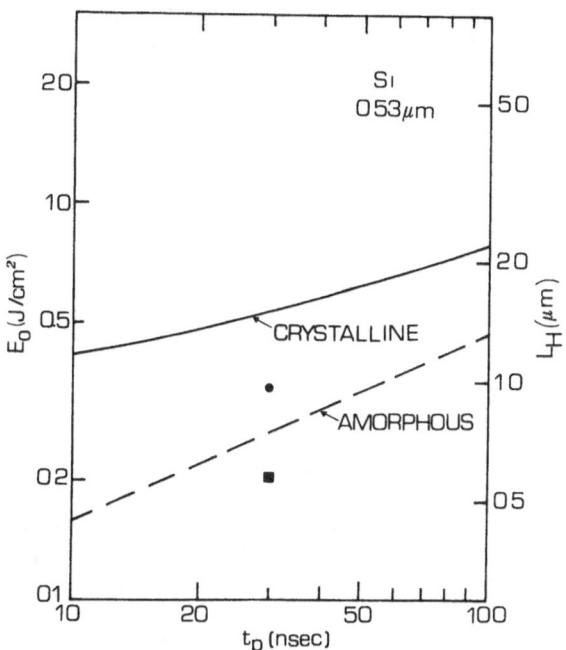

Fig.   18   Heating depth L_H and energy density E_0, which is required to raise the surface temperature from 300°K to T_m in Si under irradiation with λ = 0.53 μm as a function of the laser pulse duration t_p. From ref.(30).

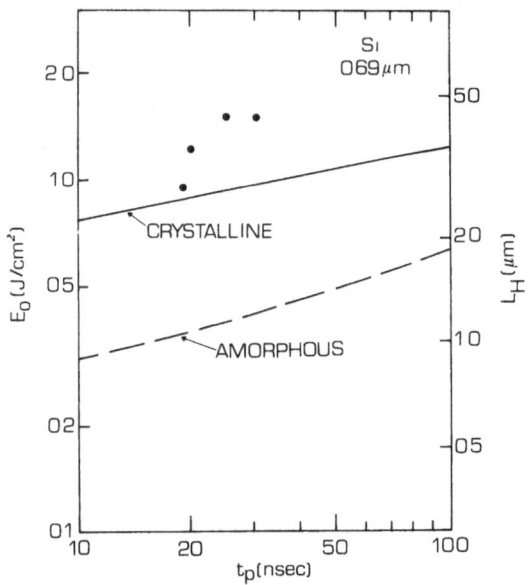

Fig. 19   Same as Fig.18 for ruby laser irradiation
            (λ = 0.69 μm). From ref.(30).

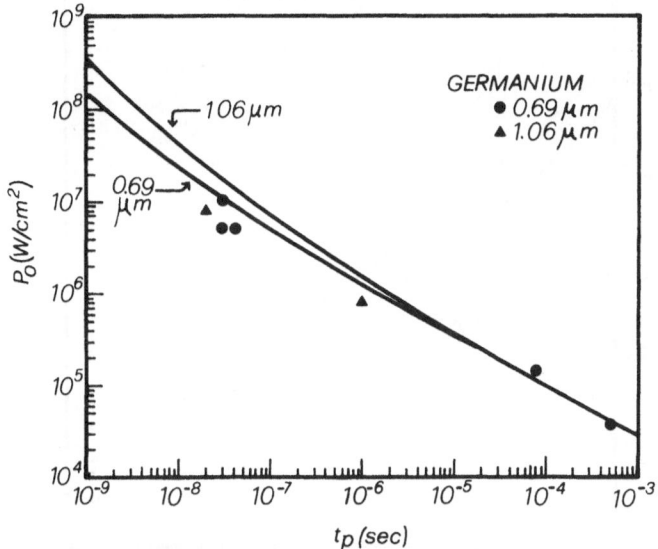

Fig. 20   Theoretical and experimental thresholds for melting
          vs   pulse duration in Ge at 0.69 and 1.06 μm   From
          ref.(29).

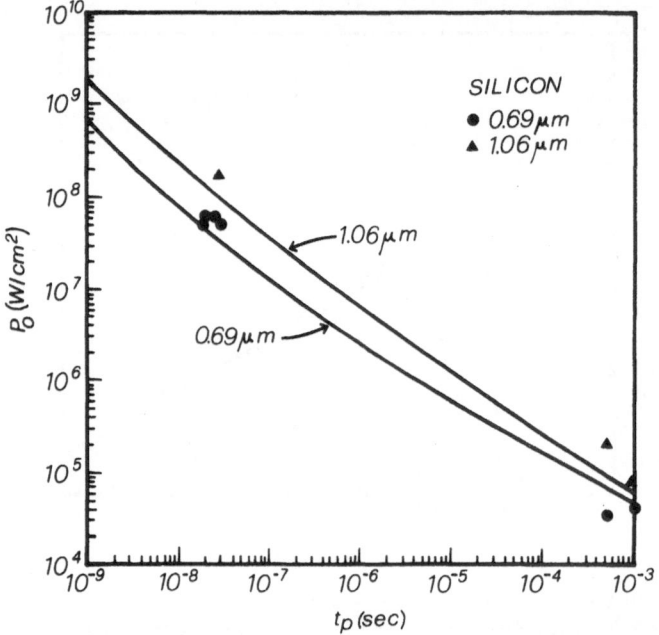

Fig. 21 Same as Fig. 20 for Si. (From ref.29).

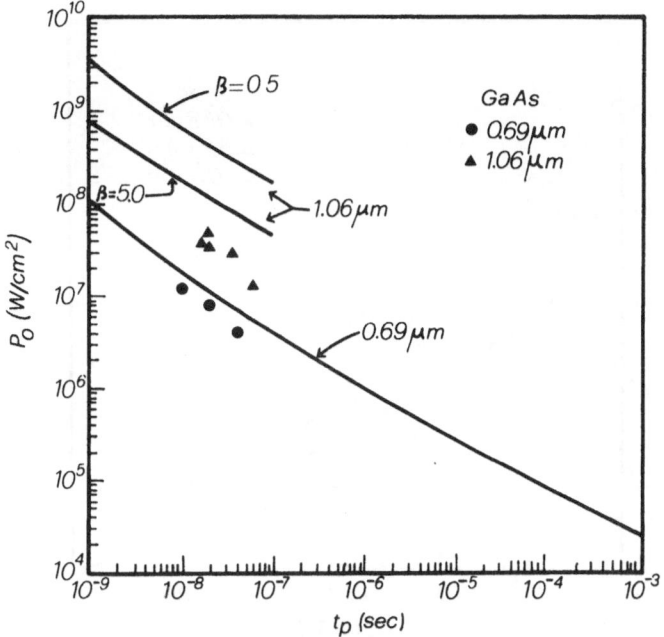

Fig. 22   Same as Fig.20 for GaAs. (From ref.29).

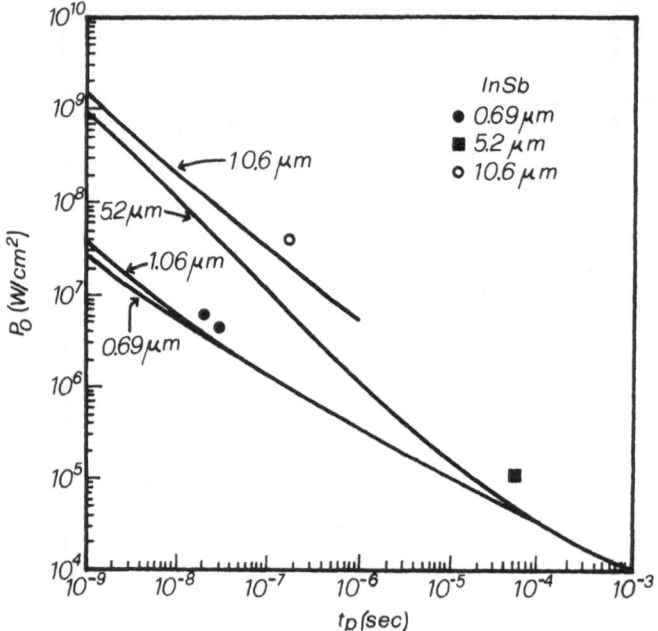

Fig. 23 Same as Fig.20 for InSb. (From ref.29).

diffusion and segregation. A simplified model was used consisting
of diffusion in the melt, followed by the segregration upon
refreezing.
The use of an Ar laser was considered by Laff and Hutchins [35].

It is the great merit of soviet scientists to have continued
the research and found finally that laser irradiation could make
to recover electrical properties of heavily ion implanted
(amorphous) semiconductors, and eventually even change their
desordered structure to a more ordered one [36-42].

In 1977 the matter exploded when the italian group repeated
the experiments finding clear evidence of the structural transitions [43-45]. The italian and russian results were contrasted at
the first USA-USSR seminar on ion implantation in Albany, N.Y.
(USA) July, 1977 and from that moment on the rush to research
started.

MAIN CHARACTERISTICS OF LASER ANNEALING OF SEMICONDUCTORS

Laser annealing is obtained by irradiating with one from a
large variety of lasers the surface of a disordered semiconductor
sample, as f.e. is the one implanted with ions.

Ion implantation, as it is well known creates a massive
crystal damage when ions dissipate their initial kinetic energy
in coming to rest in the crystal. A major part of this energy is
generally delivered to recoiling crystal atoms. The defects
resulting from the subsequent large number of displaced atoms are
usually extended and complex, and for high implantation doses the
region where ions come to rest is devoided of any-long range order
in a state that is usually referred to as "amorphous". The
electrical activity of such material is very poor.

Laser annealing has been found effective in recovering good
electrical activity of the implanted material and in producing a
reordering of the disordered region. It has been proved effective
on both elemental semiconductors, like silicon and germanium,and
compound semiconductors like GaAs,InSb. It is also effective with
other kinds of "disordered materials" like for example, deposited
amorphous or polycrystalline materials.

We can make a classification of the obtained effects according to which melting of the laser irradiated surface has
occurred or not [46].

a - Liquid phase epitaxy

In this case the laser energy is able to melt the semiconductor
surface for a depth which depends essentially on the power density
and the wavelength of the radiation and on the characteristics of
the semiconductor. This model can be used to explain most of the
pulsed ruby or Nd laser effects in the nanosecond region, and
focused Ar laser annealing, in the quasi-continuous regime in which
the surface is scanned with a continuous laser beam.

   The model is probably not valid in the picosecond pulse regime
in which other explanations may be valid [47]. Two cases can occur.

a_1) If the molten layer thickness is larger than the thickness of
     the amorphous layer, epitaxial regrown from the single-crystal
     sub-strate occurs, giving rise to a perfect crystal and it is
     possible to see that regrowth occurs in exactly the same
     crystallographic orientation of the substrate (s.Fig. 24). This

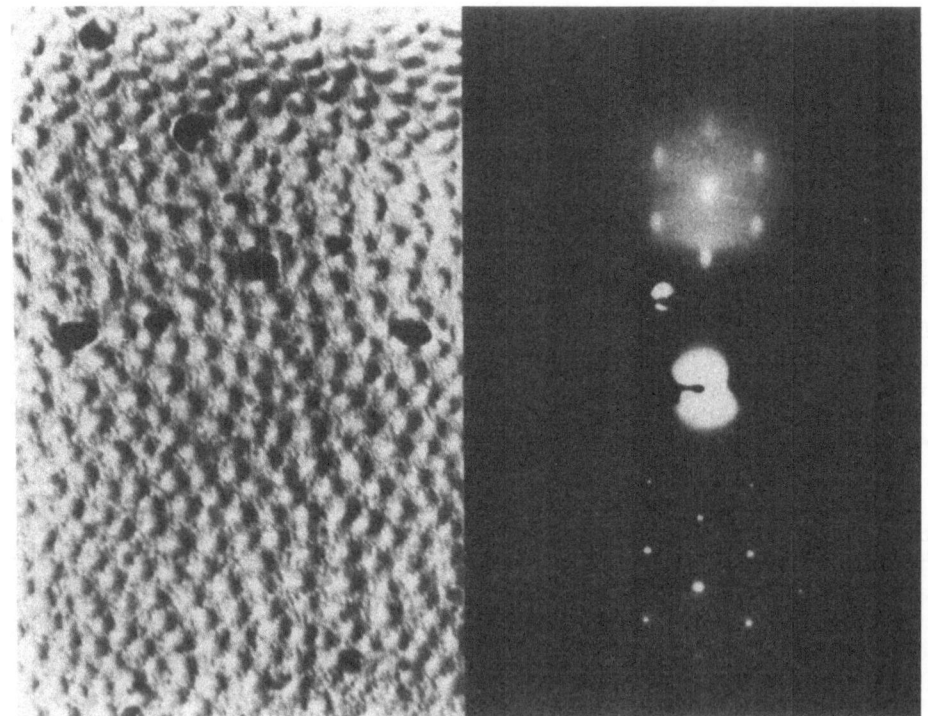

Fig. 24 RHEED double-pattern of the <100 > Si substrate (upper
          side) and of its surface irradiated with a laser pulse
          (lower side),taken along the <110> azimuth direction.At
          right electron micrograph showing a Ni—Cr shaded carbon
          replica of the laser irradiated surface.

mechanism is active both when the initial surface is
amorphous or when it is polycrystalline [48]. It works also for
amorphous materials deposited from a vapor or with some other
technique onto some material [49]. In this case it is necessary
that the base material is the same semiconductor of the
deposited amorphous layer or it has a higher melting temperature than the amorphous one, otherwise compounds can be
produced. Of course, except the case in which the base material
is the same of the amorphous one, heterogeneous nucleation
occurs giving rise to a polycrystalline material. Fig. 25 shows

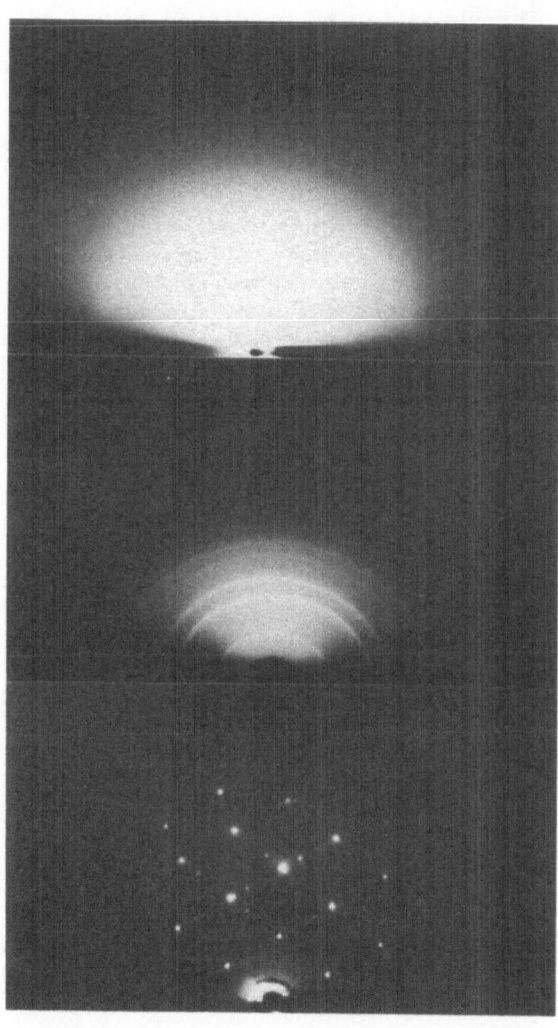

Fig. 25 RHEED patterns
        of (from top to
        bottom):amor-
        phous Si layers
        on Si < 100>
        substrate before
        laser irradia-
        tion, after a
        25 MW/cm^2 laser
        pulse, and after
        a 200 MW/cm^2
        laser pulse,
        (along the
        <100> azimuth
        direction).

some of these structure transformations.

a_2) If th e molten layer thickness is smaller than the thickness of
the amorphous,heterogeneous nucleation occurs and the material
is turned on to the polycrystalline state.
Again the mechanism works also in the case of deposited
amorphous semiconductors. Side inspection of the annealed
specimen shows that the grain size is decreasing going inward
into the material [50].

The two cases a_1 and a_2 are illustrated in Fig. 26.

In case a_1 of course a threshold exists tor the laser power
density needed to produce the amorphous to single crystal transition, and it depends, among others from the thickness ot the
amorphous layer (s.Fig. 27) [51]. Threshold can be predicted once
the thickness of the melted layer is calculated. A simple
analitical calculation is given in ref.(46) .

Some computer calculation s have been performed in the case
of a ruby laser [52] (s.Fig. 28 and 29).

At power densities near the threshold value a high residual
disorder is present in the the crystallized region (s.Fig. 30). This
can be interpreted by observing that just under the amorphous
layer in the case of ion-implantation, a disordered region exists
due to the tail of the distribution of ranges of the implanted
ions. Therefore the epitaxially regrown material starts from a
disordered substrate and mantain memory of this.

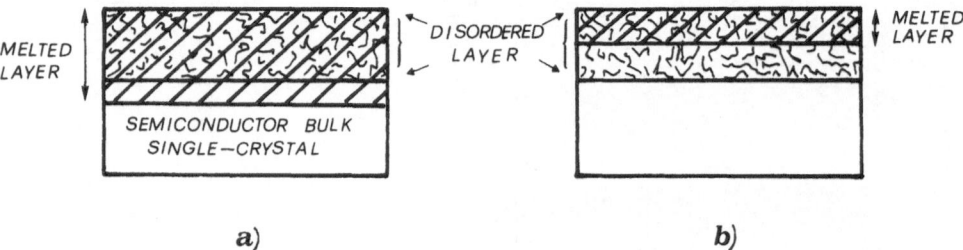

a)   b)

Fig. 26 Showing the conditions for producing a single-crystal
(a) or a polycrys tal (b) upon laser irradiation.

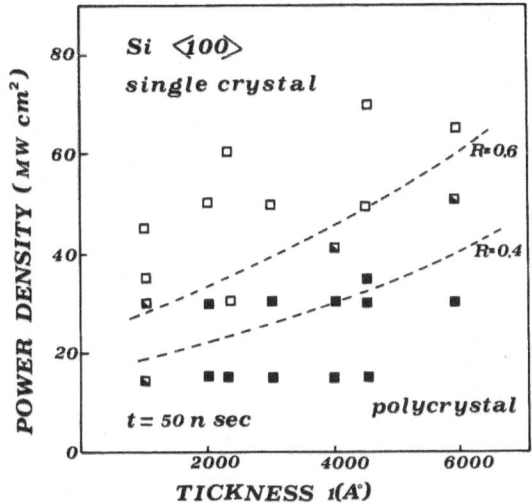

Fig. 27 Threshold for the single crystal transition.
Open, full, and partially full squares indicate
single-crystal, poly, and highly disordered single
crystal respectively. Dashed curves are calculations
(from ref.51).

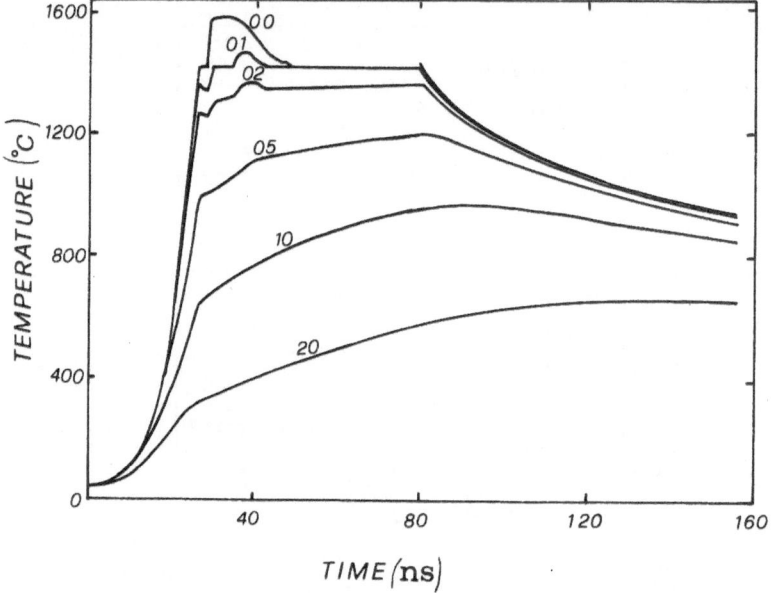

Fig. 28 The predicted time dependence of temperature in
          different depths of a virgin crystal of Si irradiated
          with a ruby laser pulse of 25ns and 1 J/cm^2 (from R.O.
          Bell et al. in ref.52).

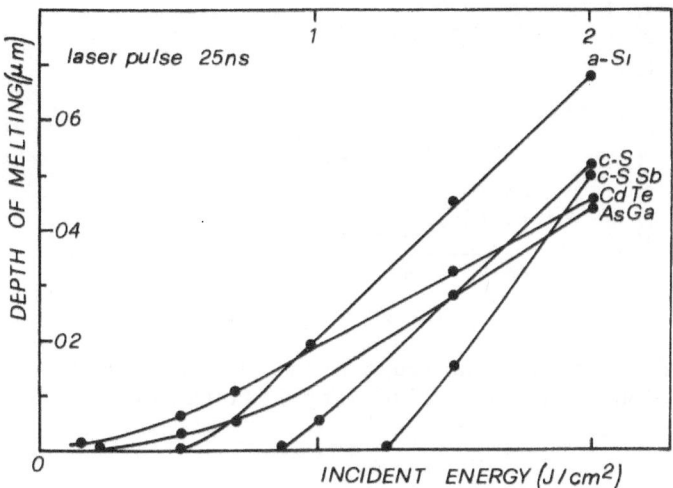

Fig. 29 Depth of melting as a function of incident energy with
        a pulse width of 25 ns for a single-crystal Si sample
        (c-Si) amorphous Si (a-Si), for a single-crystal Si
        covered by a 150 Å thick layer of Sb (c-Si) (Sb), for
        CdTe and for GaAs. Evaporation is not taken into account
        (from O.R.Bell et al. in ref.52).

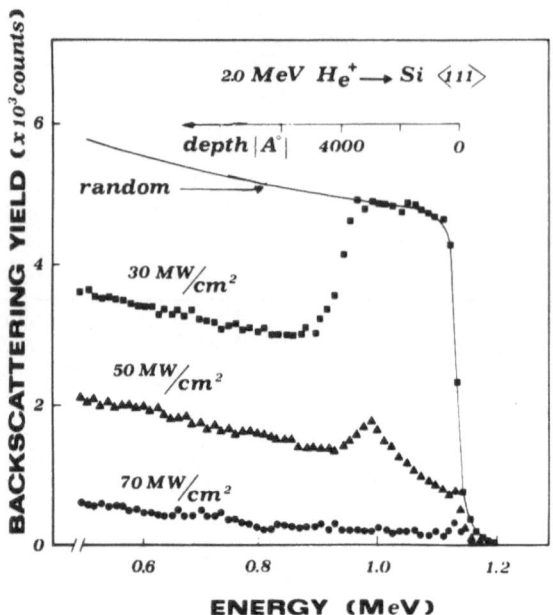

Fig. 30 Backscattering analysis of a 4000 Å thick a-Si for
          various laser power density irradiations (from ref.51).

        Substantial redistribution of the implanted dopants occurs in
case a_1 [s.Fig. 31][53]   .
The redistribution is substantially greater than can be expected
by diffusion in the solid. Measured profiles are instead in good
agreement with profiles calculated using diffusion coefficients
in the liquid state and this is one of the more convincing
prooves, together with the change in reflectivity, that melting is
the real process occurring.
Usual values of the diffusion constant in the solid state even near
the melting temperature are 10^{-11} cm^2/sec which give a diffusion
length $L = \sqrt{Dt_o} = 3.10^{-9}$ cm, for a time of 1 µsec which is
approximatively the time duration of the high temperature produced
by the laser pulse.
Motion of 1000 Å in 1 µsec requires a diffusivity of about
10^{-4} cm^2/sec which instead is consistent for diffusion in liquid
silicon. Pulsed laser annealing has also shown to cause significant
segregation to the surface of certain impurities in silicon,which

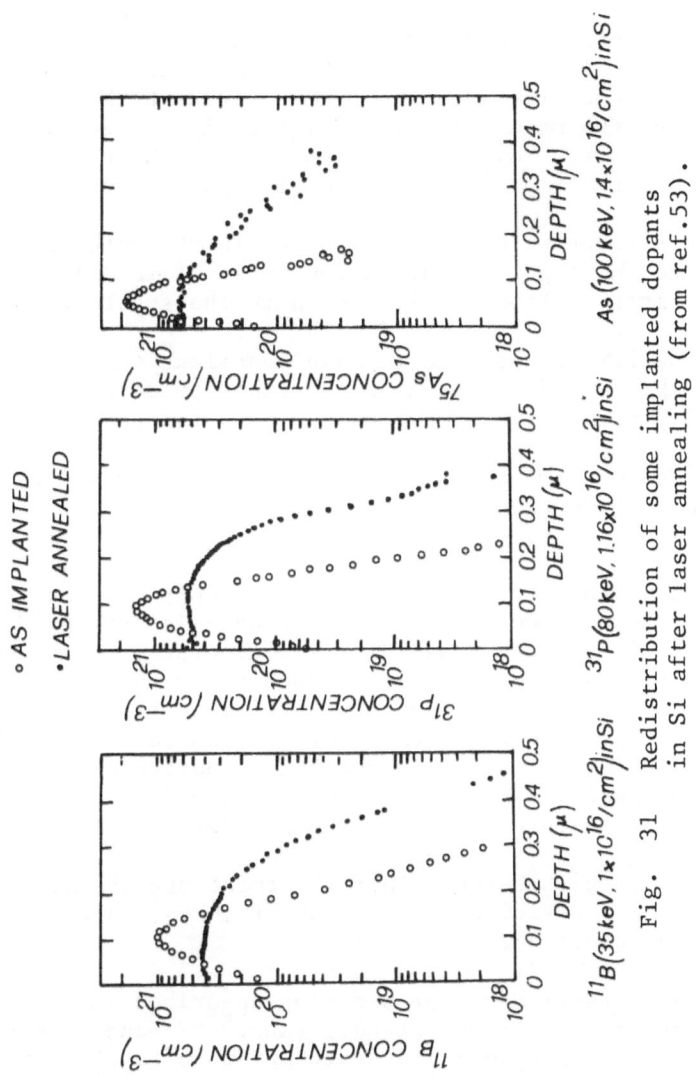

Fig. 31   Redistribution of some implanted dopants
in Si after laser annealing (from ref.53).

has to be interpreted in terms of a non equilibrium segregation at
the moving solid-liquid interface during the ultrarapid recrystallization[54,66].

The accepted model is therefore that under laser irradiation
the temperature near the semiconductor surface rises very rapidly
and melting occurs. The melt front propagates rapidly into the
crystal and reaches its maximum penetration depending on the laser
pulse characteristics and semiconductor constants a short while
after the pulse end. The melt front then sweeps back to the surface,
recrystallizing the material epitaxially as it proceeds. While the
crystal is molten, the dopant atoms have a very high diffusion
coefficient and the implanted profile can change markedly.

The redistribution of dopant species in the material may
largely exceed the depth of the amorphous layer in all the cases
in which the melting front arrives deep in the substrate material.

This behavior can be used to make junctions whose position
and type can be varied by changing the laser pulse characteristics
and eventually using time shaped pulses or a train of pulses [55].

The effect of time shaping a train of laser pulses is to allow
a better temperature distribution inside the material.

The explanation given above is not the only one and is
questioned by some authors. According to calculations by Yoffa[56]
the diffusion of the hot, dense carriers generated in pulsed laser
annealing of Si, can substantially reduce the rate at which energy
is transferred to the semiconductors lattice near the surface. The
extent of the region in which this energy transfer occurs is
consequently increased, and the expected temperature rise is
decreased. The observed structural changes would therefore be
produced through an interaction of the hot carrier plasma produced
[47].
Crucial to distinguish between this treatment and the melting
theory should be a definite measurement of the surface temperature
reached during annealing.

Reflection measurements of the kind described in refs.[2,5]
have been repeated by a group at Bell Labs.[57-59] and have been
interpreted as giving evidence of melting. E.Yoffa also studied
the wavelength dependence of reflection, finding no evidence of
plasma effects [60].
On the other hand N.W.Lo and A.Compaan are claiming to have
measured the temperature during pulsed laser heating of Si determining the ratio of Stokes to anti-Stokes phonon Raman scattering
finding a temperature rise of only 300°C [61].

This point needs therefore more attention and probably in the range of very-short pulse irradiation (psecs) the plasma model will have importance.

b) Solid State Epitaxy

It is possible to have the crystalline recover even without melting if the disordered material is taken for a sufficient long time to a temperature high enough[63,63]. This regime occurs in the milli-second region. Solid-state epitaxial recrystallization from the amorphous single-crystal interface can take place in this case. The regrowth phenomena closely follow those observed in furnace annealing except in time scale.

Williams et al.[64] showed that by extrapolating into the high-temperature regime furnace epitaxial regrowth rates at temperature of the order of 850-950°C the time needed to regrow a 500 Å amorphous layer silicon is just between 10^{-2} and 10^{-3}sec. Cw laser induced recrystallization can thus occur via simple thermal solid phase epitaxial regrowth. Recent calculations[62] show that the experimental data are consistent with the extrapolation of solid-phase epitaxial regrowth rates to the calculated laser-induced temperatures.

An implant dose dependence in which high doses result in poor regrowth, a substrate orientation dependence indicated by better recovery of (100) oriented wafers than of (111)[65], and other effects which have been identified in furnace annealed silicon, finally support this point of view.

There are also important differences between cw laser annealing obtained by swelling the laser onto the surface and furnace annealing. The msec regime of the cw laser annealing precludes significant implant diffusion in the solid state even at temperatures just below the melting point.

LOW POWER EFFECTS

a - multi-pulse irradiation

Experimental evidence for this was obtained first by using a free-generation laser[45]. In this case polycrystalline material was obtained from amorphous silicon with a single shot from a ruby laser with a power p $\sim$ 25 Kw/cm^2 (pulse duration 0.3 msec). The maximum temperature increase at the end of the laser pulse in this conditions is of the order of $\Delta T \sim 150°C$, by far too small to account for the transformation.

Moreover if a Q-switched laser is used at very low power so to have only a very small increase in temperature, but shot more than a single pulse on the examined area, even if the pulses are well separated in time one from the other (f.e. minutes), we find that gradually by increasing the number of shots the rings of polycrystalline material appear as shown in Fig. 32a and b [67]. Finally if the number of pulses is increased the single-crystal can be obtained [68] [69] (s.Fig.32c).

In this cases the size of the produced grains is found to depend on the laser power while the fraction of the crystallized material is a function of the total delivered laser dose (s. Fig. 33)[67,70].

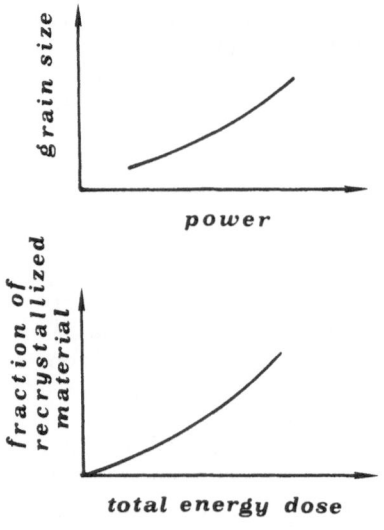

Fig. 33   Grain-size a function of laser power (upper curve) and fraction of crystallized material as a function of the total energy dose (lower curve) in the case of multi-pulse irradiation.

Fig. 32   a) Glow discharge a-Si: sample B-12 after 10 MW/cm^2; sample 4-13 after 2 MW/cm^2; sample 4-20 after 10 superposed pulses and sample 4-22 after 30 superposed pulses of 4,5 MW/cm^2 each.

b) 1000 Å -Si implanted amorphous layer onto a Si crystal-sample VO as irradiated. Samples V-2.3 and V-2.11 after a single pulse and three superposed pulses each one of 7 MW/cm^2 respectively.

c) Arrows point to small dots on the photograph which show the appearance of the single-crystal. (From ref. 68).

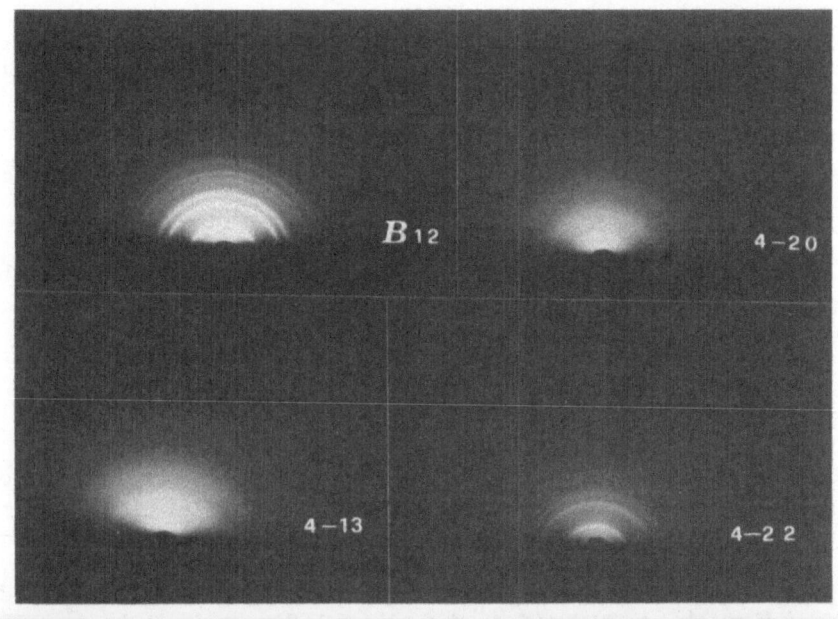

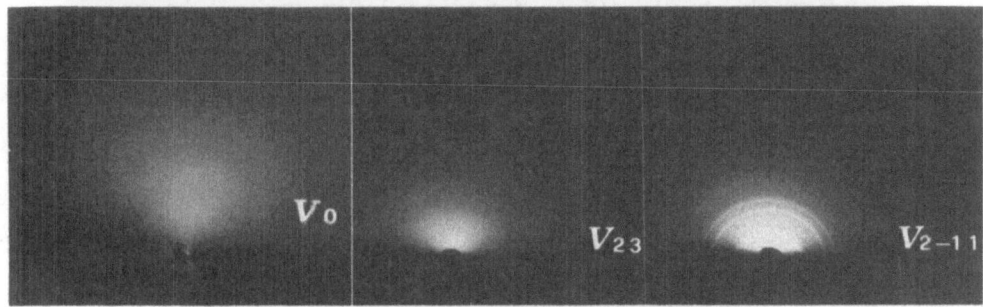

This behaviour could be explained by assuming that at each laser
shot a number of small crystallites are generated at random in the
amorphous matrix. The size of each crystallite is somehow connected to the value of the used energy density. When a second shot is
sent over the surface, other crystallites of the same size (or with
the same size distribution) are created in other points of the
matrix, due to the randomness of the creation process. Under
repeated laser shots eventually all the amorphous matrix is
filled with the crystallites.
The randomness of the creation process may be connected to several
causes. Among them we may quote a non-perfect homogeneity in the
space distribution of the laser light in the spot, which may
produce "hot" points over the irradiated area.

b - non thermal effects

    Examples of amorphous-polycrystalline transitions at
practically room temperature are reported in the letterature for
elemental semiconductors like Si and Ge.
We can here recall the results by Laude et al.[71] who found crystallization of amorphous Ge films irradiated with a dye-laser,
with pulses of 10^{-6}sec, repetition rate 16.7 Hz, h = 2.08 eV and
power density about 30 Kw/cm^2, irradiated for several hours. The
temperature increase was calculated as negligible.

    These results demand for some explanation which cannot be
found in a simple thermal model. They can be compared with other
well known results on the effects of ionization (low-energy
electron beams or ionizing visible radiation) on annealing of
defects in semiconductors [72,73]. More recently attention has been
given on the effect of laser on annealing or on the introduction
of defects. A new field is therefore opening which uses the laser
light not for producing structural changes but for obtaining a
rearrangement of defects in crystals.

CONCLUSIONS

The following exposition should allow to understand the basic
mechanisms involved in laser annealing. An interesting new domain
opens with this technique bearing on the thermodynamics of fast
processes.

    The first applications of this new powerful technique are
here treated by C.Hill.
    Finally I wish to thank Dr.G.Vitali for discussion and
criticism.

REFERENCES

1. C.H.Carmichael and G.N.Simpson, Nature 202, 787 (1964)

2. W.R.Sooy,M.Geller, D.P.Bortfeld - Appl.Phys.Lett. 5, 54 (1964)

3. M.Birnbaum - J.Appl.Phys. 36, 657 (1965)

4. M.Birnbaum - J.Appl.Phys. 36, 3688 (1965)

5. M.Birnbaum and T.L.Stocker, J.Appl.Phys. 39, 6032 (1968)

6. N.G.Basov, A.Z.Grasyuk, V.F.Efinkov, U.G.Zubarev,V.A.Katalin,
    Yu.M.Popov - J.Phys.Soc.Japan 21, suppl.277 (1966)

7. M.Bertolotti,F.de Pasquale,P.Marietti,D.Sette and G.Vitali-
    J.Appl.Phys. 38, 4088 (1967)

8. A.A.Grinberg, R.F.Mekhtiev, S.M.Ryvkin, V.M.Salmanov, I.A.
    Yaroshetskii, Sov.Phys.Solid State 9, 1085 (1967)

9. L.M.Blinov,V.S.Vavilov,G.N.Galkin,Sov.Phys.Semic. 1,1124 (1967)

10. L.M.Blinov,V.S.Vavilov, G.N.Galkin, Sov.Phys.Solid State 9, 666
    (1967)

11. L.M.Blinov,E.A.Bobrova, V.S.Vavilov,G.N.Galkin, Sov.Phys.Solid
    State 9, 2537 (1968)

12. A.M.Bonch - Bruevich, V.P.Kovalev, G.S.Romanov, Ya.A.Imas,
    M.N.Libenson - Sov.Phys.Tech.Phys. 13 507 (1968)

13. G.N.Galkin,L.M.Blinov, V.S.Vavilov, A.G.Golovashkin - Sov.Phys.
    JETP Lett. 7, 69 (1968)

14. A.Z.Grasyuk, J.G.Zubarev - Sov.Phys.Semic. 3, 576 (1969)

15. M.Bertolotti,P.Marietti,D.Sette,L.Stagni,G.Vitali - Rad.Eff.1,
    161 (1969)

16. M.Bertolotti,D.Sette,L.Stagni, Rad.Eff.16, 197 (1972)

17. M.Bertolotti,L.Stagni,.G.Vitali - J.Appl.Phys. 42, 5893 (1971)

18. Y.Matsuoka - J.Phys. D(Appl.Phys.)9, 215 (1976)
    Y.Matsuoka and A.Usami-Appl.Phys.Lett. 25, 574 (1974)
19. M.Bertolotti,D.Sette,L.Stagni,G.Vitali - J.Appl.Phys. 41, 818
    (1970)

20.   J.C.Bean,H.J.Leamy,J.M.Poate,G.A.Rozgonyi,J.P.van Ziel;
      J.S.Williams, G.K.Celler - J.Appl.Phys. 50, 881 (1979)

21.   R.A.Mc Forlane and L.D.Hess - Appl.Phys.Lett. 36, 137 (1980)

22.   F.Bartoli,L.Esterowitz, M.Kruer and R.Allen, J.Appl.Phys. 46,
      4519 (1975)

23.   L.Stagni - Appl.Phys. 12, 31 (1977)

24.   M.Lax - J.Appl.Phys. 48, 3919 (1977)

25.   M.Lax - J.Appl.Phys.Lett. 33, 786 (1978)

26.   K.T.Yang, J.Appl.Mech. 25, 146 (1958)
      K.T.Yang, and A.Szevczky - ASMD Trans. C81, 251 (1959)

27.   I.P.Dobrovol'skii, A.A.Vlov - Sov.J.Quant.Electr. 4, 788
      (1974)

28.   J.R.Meyer, F.J.Bartoli,M.R.Kruer, Phys.Rev. 21B, 1559 (1980)

29.   J.R.Meyer,M.R.Kruer, F.Bartoli, J.Appl.Phys. 51, 5513 (1980)

30.   J.R.Meyer, F.J.Bartoli,M.R.Kruer, internal report

31.   J.M.Fairfield and G.H.Schwuttke - Solid State Electr. 11, 1175
      (1968)

32.   F.E.Harper and M.I.Cohen - Solid State Electr. 13, 1103 (1970)

33.   C.L.Marquardt, J.F.Giuliani, F.W.Fraser - Rad.Eff. 23, 135
      (1974)

34.   J.F.Giuliani and C.L.Marquardt - J.Appl.Phys. 45, 4993 (1974)

35.   R.A.Laff and G.L.Hutchins - IEEE Trans.Electron Devices ED.
      21,743 (1974)

36.   E.I.Shtyrkov, I.B.Khaibullin, M.M.Zaripov, M.F.Galyatudinov,
      R.M.Bayazitov - Sov.Phys.Semicond. 9, 1309 (1976)

37.   O.G.Kutukova,L.N.Strel'tsov - Sov.Phys.Semicond. 10, 265
      (1976)

38.   G.A.Kachurin, E.V.Nidaev, A.V.Khodyachikh, L.A.Kovaleva -
      Sov.Phys.Semicond. 10, 1128 (1976)

39. I.B.Khaibullin, E.I.Shtyrkov, M.M.Zaripov, M.F.Galyantdinov, G.G.Zakirov - Sov.Phys.Semicond. 10, 81 (1976)

40. G.A.Kachurin, N.B.Pridachin,L.S.Smirnov, Sov.Phys.Semicond.9, 946 (1975)

41. V.V.Bolotov, N.R.Pridachin, L.S.Smirnov - Sov.Phys.Semicond. 10, 338 (1976)

42. I.B.Khaibullin, E.I.Shtyrkov,M.M.Zaripov, R.M.Bayzitov, M.F. Galjontdinov - Rad.Eff. 36, 225 (1978)

43. G.Foti,G.Vitali,M.Bertolotti,E.Rimini - Appl.Phys. 14, 189 (1977)

44. G.Vitali,M.Bertolotti,G.Foti,E.Rimini - Phys.Lett. 63A, 351 (1977)

45. G.Vitali,M.Bertolotti,G.Foti,E.Rimini - Proc.7th Int.Conf.on Amorphous and Liquid Semicond., Edinburgh June 27-July 1, 1977 ed.W.E.Spear - centre for Industr.Consultancy and Liaison, Univ. Edinburgh 1977 p.24

46. M.Bertolotti,G.Vitali,E.Rimini,G.Foti - J.Appl.Phys.50, 259 (1979)

47. J.A.Van Vechten - J.de Phys. C4 15 (1980)

48. G.Vitali,M.Bertolotti,G.Foti,E.Rimini - Appl.Phys. 17, 111 (1978)

49. M.Bertolotti,G.Vitali,W.E.Spear - in Laser Solid Interactions and Laser Processing - 1978 edited by S.D.Ferris, H.J.Leamy, J.M.Poate, American Institute of Physics, New York 1979 pag.492 for the transition of amorphous silicon deposited on a single Si crystal see also S.S.Lou, W.F.Tseng,M.A.Nicolet, J.W.Mayer, R.C.Eckardt, R.J.Wagner - Appl.Phys.Lett. 33, 130 (1978)
D.Hoonhout, CB.Kerdijk, F.Sans - Phys.Lett. 66A, 145 (1978)
P.Revesz, G.Farkas, G.Mezey, J.Gyulai - Appl.Phys.Lett. 33, 431 (1978)

50. A.G.Cullis, H.C.Webber, D.C.McCaughan, N.G.Chew, Academic Press, 1980 pag.183

51. G.Foti,E.Rimini,M.Bertolotti,G.Vitali - Phys.Lett. 65A, 430 (1978)
G.Vitali,M.Bertolotti,L.Stagni in Laser Interactions and Laser Processing - 1978 edited by S.D.Ferris, H.J.Leamy,

J.M.Poate - American Inst.Phys.New York 1979 pag.111

52. J.C.Wang, R.F.Wood, P.P.Pronko - Appl.Phys.Lett. 33, 455
    (1978)
    P.Baeri, S.U.Campisano,G.Foti,E.Rimini - Appl.Phys.Lett. 33,
    137 (1978)
    J.C.Schultz, R.J.Collins - Appl.Phys.Lett. 34, 84 (1979)
    R.O.Bell, M.Toulemonde, P.Siffert - Appl.Phys. 19, 313 (1979)

53. C.W.White, W.H.Christie, B.R.Appleton, S.R.Wilson, P.P.Pronko,
    C.A.Magee - Appl.Phys.Lett. 33, 662 (1978)

54. See f.e. R.F.Wood in Laser and Electron Beam Processing of
    Materials - Material Res.Soc.Annual Meeting, Cambridge, Mass.
    Nov. 26-30, 1979 Academic Press, 1980 pag.37

55. G.Vitali,M.Bertolotti,G.Foti - Appl.Phys.Lett. 33, 1018 (1978)
    M.Bertolotti and G.Vitali in Defects and Radiation Effects in
    Semiconductors, 1978, edited by J.H.Albany, The Institute of
    Physics, London 1979 pag.454

56. E.J.Yoffa - Appl.Phys.Lett. 36, 37 (1980)

57. D.H.Auston, C.M.Surko, T.N.C.Venkatesan, R.E.Slusher and
    Golovchenko - Appl.Phys.Lett. 33, 437 (1978)

    D.H.Auston, J.A.Golovchenko, P.R.Smith, C.M.Surko and T.N.C.
    Venkatesan - Appl.Phys.Lett. 33, 539 (1978)

58. D.H.Auston, J.A.Golovchenko, T.N.C.Venkatesan - Appl.Phys.
    Lett. 34, 558 (1979)

59. D.H.Auston, J.A.Golovchenko, A.L.Simons, C.M.Surko and T.N.C.
    Venkatesan - Appl.Phys.Lett. 34, 777 (1979)

60. E.Yoffa - AIP Conf.Proc.Academic Press (1980) pag.59

61. H.W.Lo and A.Compaan - Phys.Rev.Lett. in press

62. A.G.Klimenko, E.A.Klimenko, V.I.Donin - Sov.J.Quantum Electr.
    5, 1289 (1976)

    G.A.Kachurin, E.V.Nidaev, A.V.Khodyachikin, L.A.Kovaleva -
    Sov.Phys.Semicond. 10, 1128 (1976)

    A.Gat, J.F.Gibbon, T.J.Magee, J.Peng, V.R.Deline, P.Williams,
    C.A.Evans jr. - Appl.Phys.Lett. 32, 276 (1978)

63. R.B.Gold and J.F.Gibbons - J.Appl.Phys. 51, 1256 (1980)

64. S.J.Williams, W.L.Brown, H.J.Leamy, J.M.Poate,J.W.Rodgers,
     D.Rousseau, G.A.Rozgonyi, J.A.Shelnutt, T.T.Sheng - Appl.Phys.
     Lett. 33, 542 (1978)

65. Y.I.Nissim, A.Lietoila, R.B.Gold, J.F.Gibbons - J.Appl.Phys.
     51, 274 (1980) s.also H.E.Cline, and T.R.Anthony, J.Appl.Phys.
     48, 3985 (1977)

66. C.W.White, S.R.Wilson, B.R.Appleton, F.W.Young jr. - J.Appl.
     Phys. 51, 738 (1980)

67. D.H.Auston, J.A.Golovchenko, P.R.Smith, C.M.Surko, T.N.C.
     Venkatesan - Appl.Phys. Lett. 33, 539 (1978)

     M.Bertolotti,G.Vitali - AIP Conf.Proc.Academic Press (1980)
     pag.189

68. G.Vitali, Phys.Lett. to be published

69. J.L.Regolini, T.W.Sigmon, J.F.Gibbons, T.J.Magee and J.Peny
     in Laser Solid Interactions and Laser Processing 1978 ed.S.D.
     Ferris, H.J.Leamy, J.M.Poate, AIP, New York 1979 pag.393

70. G.Vitali and M.Bertolotti - J.de Phys. C4-37 (1980)

71. L.D.Laude, M.Lovato, M.C.Martin, M.Wautelet - Phys.Rev.Lett.
     39, 1565 (1977)

72. s. J.C.Bourgoin and J.W.Corbett, Rad.Eff. 36, 157 (1978)

73. J.Suski and J.Rzewuski - Rad.Eff. 40, 81 (1979)
     A.Kraitchinskii, H.Rzewuski, Z.Werner - Rad.Eff. 29, 137
     (1976)
     J.Suski, L.Gyulai, H.Rzewuski, Z.Werner - Rad.Eff. 29, 137
     (1976)
     D.V.Land and L.C.Kimerling - Phys.Rev.Lett. 33, 489 (1974)

LASER ANNEALING OF SEMICONDUCTOR DEVICES

Chris Hill

Plessey Research (Caswell) Limited
Allen Clark Research Centre
Caswell, Towcester, Northants, England

INTRODUCTION

Conventional isothermal heat treatment techniques have served
for semiconductor fabrication very well up until now. Most of the
physical processes involved in fabrication are rate-determined by a
solid state diffusion process (e.g. oxidation, dopant redistribution,
anneal of implantation damage) and such processes have activation
energies typically in the range 2-5 eV. The resulting sensitivity
to temperature has required temperature control to $\pm 1^oC$ on modern
semiconductor furnaces, and this is routinely met. What, then, is
the place of radiant beam processing? This technique opens up the
possibility of non-isothermal heat treatment, in which we have control over the spatial and temporal extent of the temperature profile
inside the silicon over a very wide range of times (100 picosecs to
100 secs) and distances (0.1 micron to 1000 microns). This selectivity and control of beam annealing makes completely new types of
processing possible, and the purpose of this paper is to examine
these new possibilities in the light of their usefulness for device
fabrication and particularly for silicon integrated circuit processing. New developments are occurring monthly, however, and this
overview can only hope to be an accurate snapshot of a rapidly moving
field of work.

SOLID, LIQUID AND VAPOUR PHASE PROCESSES

We can change the properties of the silicon, or of films on the
silicon surface, either by solid or liquid phase processes. There
are also vapour phase processes, which are important in I.C. applications, which are used for removing material from the semiconductor
surface (e.g. programmable metallization links) but these will not

be further described here.   The characteristics of the structures
produced by solid and liquid phase processes are very different and
it is useful to list the new capabilities offered by beam-annealing
under these two headings:

New capabilities offered by solid state processes:

(1)   Complete anneal of implantation damage without redistrib-
       ution of dopant.

(2)   Controlled reaction of surface films to form equilibrium
       phases.

New capabilities offered by solid state processes:

(3)   Complete anneal of all damage with novel redistribution of
       dopant.

(4)   Formation of novel phases and solid solutions.

(5)   Epitaxial regrowth of deposited polycrystalline silicon.

(6)   Direct formation of ohmic contacts.

(7)   Localisation of heat treatment to a micron laterally and
       in depth.

(8)   Shaping of surface steps.

(9)   Formation of gettering sites.

Because of the very different recrystallization times for solid
state (greater than 10^{-3} secs) and liquid state (less than 200 nanosecs) processes, the equipment required is completely different.   The
longer times are generated by scanning a focussed spot from a continuous beam, the shorter times from a pulsed beam.   Examples of both
types of equipment are available using laser and electron beams, and
the results from both will be compared in the following section, where
items 1-9 are discussed in more detail.

CAPABILITIES OFFERED BY BEAM ANNEALING

Complete Anneal of Implantation Damage Without Redistribution of
Dopant

When the silicon surface temperature is raised to about $1300^{\circ}C$
for a few milliseconds, complete recrystallization of implant damage
can occur, but diffusion rates of dopants in the solid state at this
temperature are such that no significant redistribution of dopant
occurs.   Such annealing has been achieved using scanned laser beams

[1] and electron beams[2,3,4]. The electrical properties of p-n
junctions made by this technique are reported[4] to be as good as
those thermally annealed; an additional low temperature (450°C)
anneal in hydrogen was needed to achieve this. The same technique
has been used to electrically activate and anneal the boron implantation used for threshold voltage control under the gate of an integrated MOS transistor[5]. A scanned focussed Krypton laser produced
full electrical activation of the dopant (an improvement on the
partial activation achieved thermally) and no redistribution, allowing
the use of lower doses to achieve the same threshold shift. The
scanned millisecond pulse technique has been observed to create solid
solutions of arsenic at concentrations greater than those obtainable
by furnace techniques[43]. Subsequent thermal annealing at temperatures over 400°C caused precipitation of this excess solution.

A disadvantage of this technique is the requirement to control
the beam size and the back temperature of the substrate very accurately. Both these parameters have a strong influence on the transient temperature reached by the silicon surface as the scanned spot
passes over it[7]. Both are also difficult to control. An alternative technique, still utilising solid state recrystallization, is
to raster the beam over a large area, with a high scan rate, so that
the entire sample rises in temperature uniformly, and no thermal
gradients exist either laterally or in depth. It is reported[6]
that for a heating cycle of a few seconds a high percentage electrical activity is produced in both boron and arsenic implants with
little redistribution of the implanted profile. Satisfactory diodes
and bipolar transistors have also been fabricated. Because the
scanned beam essentially generates a large area uniform pulse of a
few seconds, the size of the focussed spot can vary over a wide range:
slice temperature is determined only by beam current, voltage, scan
time and scan area, which are all well controlled parameters.
Because of the relaxed requirement on beam focussing off-axis aberrations are unimportant, and large area scans are feasible (e.g. to
cover a 3 inch or 4 inch slice). This technique is likely to find
a place in I.C. processing[6].

Controlled Reaction of Surface Films to Form Equilibrium Phases

Solid state interdiffusion induced by both scanned laser[9] and
electron[10] beams has been used to form silicide layers from layers
of the appropriate metals (Pd, Pt, Mo, W, Nb) on single crystal
silicon. The composition (and hence conductivity) of the films can
be controlled by adjusting either the beam power or the number of
scans used. Silicide films are of great interest for interconnection
metallization in VLSI devices because of their combination of high
conductivity and compatibility with high processing temperature.
Use of this technique in fabricating devices has not yet been reported.

Complete Anneal of All Implantation Damage With Novel
Redistribution of Dopant

The melting of the surface layer of silicon and subsequent recrystallization of the liquid gives an annealed layer with characteristics quite different from those described for solid state annealing
above.   The most important difference for a device engineer is that
diffusion rates in molten silicon are so high (10^{-4} cm^2 sec^{-1}) that
even in the short pulse times (50 nanosecs) used, significant redistribution can occur in the melt.   Diffusion rates in the adjacent
solid are, of course, much too low (10^{-12} cm^2 sec^{-1}) for any redistribution in such short times.   Thus redistributed profiles not
normally obtainable can be obtained.   In particular, multiple pulses
can result[11] in a thick doped layer (up to 1 micron) with a uniform
doping and a very sharp boundary, determined by the deepest depth of
melting.   Thus, for instance, heavily doped regions with sharp
boundaries can be fabricated using dopants whose solid state diffusion
properties would preclude their use in furnace processes (e.g.
phosphorus).   The segregation coefficient between liquid and solid
can also be utilised to obtain novel dopant redistribution.   Dopants
with very small segregation coefficient, e.g. copper and iron, are
transported to the surfaces[12,13] where it is possible to etch them
off.   This offers the possibility of removing such dopants (known
to be deleterious to both MOS and bipolar device operation) from the
active regions of devices.

The electrical properties of p-n junctions fabricated by liquid
regrowth of doped implanted layers have been studied by a number of
workers[19,23].   The achievement of diode characteristics as good as
those obtained in furnace anneals is, of course, essential if beam
annealing is to be used for fabrication of the active regions of
I.C.S.   Early results were disappointing, but the high leakage currents observed habe been shown to originate from damage occurring
under the passivating oxide overlayer[19].   A systematic study of
the sources of damage and non-uniformity in silicon-silicon dioxide
structures has shown[24] that protection of the oxide from the irradiation is essential and is the key to successful diode fabrication.
A metal-masking process is described which protects oxide during
annealing of implanted regions.   An alternative process using the
self-aligned polysilicon-gate material for protection, has been
used to anneal the source and drain regions in MOSFET structures[20].
Good diode characteristics were obtained and good MOS operation.
In attition, the lateral spread of source and drain under the gate
was much reduced and this is a significant advantage in short
channel devices.

Formation of Novel Phases and Solid Solutions

The rapid recrystallization rates (10 metres/sec) and the high
cooling rates (10^{10} ^oC/secs) involved in regrowth from the liquid

enables non-equilibrium phases and solid solutions to be formed.
One area that is receiving a lot of attention[14],[16] is silicide formation because of the capability of forming high temperature phases
without significantly heating the immediately underlying silicon.
The restriction for integrated circuit metallization on silicides
with formation temperatures below 500°C, which applies to furnace
processes, is overcome by pulsed radiant annealing. Temperatures
in excess of 1000°C can be generated in the metal-silicon layer without raising the substrate surface temperature above 400°C. The
silicides formed have conductivities almost as high as those formed
in equilibrium, and are stable under subsequent processing at temperatures up to 500°C (16). No device applications are reported as yet.

     High solid solubilities of all substitutional dopants are
obtained after liquid phase regrowth[17]. This may enable dopants
not normally useful in silicon planar technology to be incorporated
(e.g. aluminium, indium) or much higher concentrations of familiar
dopants to be used (e.g. boron in p-n-p bipolar transistors). These
supersaturated solutions are reported [18] to be stable up to 500°C.

Crystalline Regrowth of Deposited Amorphous Layers

     Since once the surface layer is melted, the information as to
its previous structure is lost, deposited amorphouse surface layers
can be regrown in the same way as implanted amorphous layers. If
single crystal substrates are used, then a defect free epitaxial
layer results[25],[27]. The attraction of this in relation to VLSI
is the possibility of converting low pressure deposited amorphous
silicon (which is a well-controlled large volume process) into high
quality epitaxial silicon (at present epitaxy is a low-volume poorly
controlled process). One obvious problem is the incorporation of
dopant from the heavily doped substrate (bipolar collector) into the
lightly doped epitaxial layer. This problem is not yet solved. No
device fabrication utilising such regrown layers is yet reported.
Heteroepitaxy is also reported, the interesting case for VLSI being
silicon on sapphire[28],[29]. The strains due to the latticed mismatch in the original deposition, which give rise to leakage currents
in devices, can be removed by regrowing the $Si-Al_2O_3$ interface with
a pulsed laser anneal[29]. 30% increase in channel mobility of MOS
is reported[34]. A development which may have far-reaching implications for VLSI technology is the growth of single crystal regions
on amorphous or polycrystalline substrates. The simple increase in
grain size of polysilicon on oxide has been utilised[30] to reduce
the resistance of doped polysilicon interconnections. Active
regions of devices have also been fabricated in laser regrown polysilicon[31] to have channel mobility within a factor of two of those
fabricated in single crystal substrates. It has been shown that
this process can be further improved to produce single crystal orient-
ated silicon on polycrystalline silicon oxide, either by defining the
deposited polysilicon into small isolated islands [32] or by con-

touring the oxide before deposition[33]. Thus the real possibility
exists of making devices on a cheap substrate material (e.g. oxidised
polysilicon) with the speed and low power advantages at present only
obtainable from silicon on sapphire substrates.

Direct Formation of Ohmic Contacts

The localisation of the reaction zone to the near-surface region
permits the doping and metallization of a contact region in one operation by pulsed beam anneal. Diodes with good ohmic contacts have
been fabricated by direct irradiation of aluminium films on silicon
using a single laser pulse[35,36]. Development of this technique
may permit the omission of the separate doping step at present
required to form degenerate regions for contacting.

Localisation of Heat Treatment

The short heat-flow distance associated with 50 nanosecond beam
pulses (about 0.5 micron) allows localisation of heating both laterally and in depth to the region where the beam is absorbed. The
most precise localisation of the beam is obtainable using photo-
engraved reflection layers[20,24] and selective heat treatment with
a resolution of 0.5 micron is shown to be possible. As has been
seen in the section entitled Formation of Novel Phases and Solid
Solutions, such selectivity is essential for the successful implem-
entation of liquid phase regrowth in integrated device structures.
The implications of this selectivity are not yet explored but they
are likely to be very significant.

Surface Contouring

The capability of pulse heating to only melt the surface of
deposited layers allows these surfaces to be contoured without
affecting the underlying active devices. The surface asperities on
deposited polysilicon, which give rise to breakdown in subsequently
grown oxides, can be completely removed by a pulse melt and regrowth[37] without otherwise affecting the MOS device characteristics.
Step edges in polysilicon can also be rounded, for high yield metallization steps on both sapphire and double-level polysilicon devices[38]. It seems that heavily doped silica can also be countoured,
since the absorption bands caused by the doping can absorb sufficient
power from the laser beam to melt[39].

Formation of Gettering Sites

This technique, unlike all the previous ones, involves causing
as much damage to the silicon as possible. The back of the slice
is covered with arrays of deeply laser melted regions, around the
periphery of which high dislocation densities occur. These damage
regions are very effective gettering sites for heavy metals during

subsequent processing[39,40]. The technique is clean, can be repeated
as often as required, and does not heat the active regions of the
device at all. The trend to lower processing temperatures in VLSI
may make this gettering technique very attractive.

ELECTRON VS LASER BEAM PROCESSING

Nearly all the processes described above can be carried out with
either electron or laser beams of the correct pulse length.   In
practice, there are reasons for preferring particular beam sources
for different applications.   For the liquid state processes, apart
from some specialised epitaxy applications, the most suitable equipment available would appear to be the repetitively pulsed, focussed
and scanned Nd-Yag Q-switched laser with a frequency doubler to
introduce a component of green light (which is strongly absorbed in
silicon) to the major infra-red component of the beam.   In this way,
the high reliability, conversion efficiency and power output of the
NdYag laser can be coupled effectively into the silicon surface.
The processing costs per wafer are calculated to be the same as for
furnace processing, and throughput also equal[41].   The pulsed elec-
tronbeam equivalent does not appear to be as accurately controllable, or as cheap, as the NdYag laser.   There is also the advantage
using the laser that the incident beam can be reflected away where
heating is not required: this is not possible with the electron beam.
Where truly uniform heating is required, rather than the averaged
effect of overlapping pulses given by the repetitively pulsed and
scanned system, then the large area pulse from a ruby laser, homo-
genised to achieve uniformity[24,42], amy need to be used.

For solid state processes, the focussed scanned electron beam
has many advantages.   The conversion efficiency is high, and the
power output is also high.   The visible light continuous lasers are
very inefficient (0.05%) and of limited power (about 20 Watts).
Because of the longer times (greater than 1 millisecond) required for
solid state processes, spatial selectivity of heating is not possible
on a device scale, and so inability to reflect away the beam energy
is not in this case a disadvantage.   Unless the implanted profile
must be retained with no redistribution at all, the multiple-scan
technique[6] would seem to offer the cheapest and highest throughput technique where complete anneal with little redistribution is
required.

REFERENCES

1.   A. Gat, J.F. Gibbons, T.J. Magee, J. Peng, V.R. Deline,
         P. Williams and C.A. Evans Jr., Appl. Phys. Letts. 32
         (1978) 276-278 and Appl. Phys. Letts. 33 (1978) 389-391.
2.   J.L. Regolini, J.F. Gibbons, T.W. Sigmon, R.F.W. Pease,
         T.J. Magee and J. Peng, Appl. Phys. Letts. 34 (1979)
         410-413.

3. K.N. Ratnakumar, R.F.W. Pease, D.J. Bartelink, N.M. Johnson and
    J.D. Meindl, Appl. Phys. Letts. $\underline{35}$ (1979) 463-466.
4. H. Boroffka, E.F. Krimmel, M. Linder and H. Runge, Proceedings
    of the Symposium on Laser and Electron Beam Processing of
    Electronic Materials, Electrochem. Soc. Proc., Vol 80-1
    (1980) 178-186.
5. G. Zimmer, Electronics Letts. (1979) $\underline{15}$ No.6, 184-186.
6. R.A. McMahon and H. Ahmed in Ref.4, p.123-140.
7. J.F. Gibbons in Ref.4, p.1-25.
8. By kind permission of R.A. McMahon and H. Ahmed.
9. T. Shibata, T.W. Sigmon and J.F. Gibbons in Ref.4, p.520-530.
10. T.W. Sigmon, J.L. Regolini, J.F. Gibbons in Ref.4, p.531-536.
11. C.W. White, W.H. Christie, R.E. Eby, J.C. Wang, R.T. Young,
    G.J. Clark and R.F. Wood, to be published.
12. See Ref.36.
13. A.G. Cullis, H.C. Webber, J.M. Poate and N.G. Chew, J.Microsc.
    $\underline{118}$ (1980).
14. M. Wittmer and M. von Allmen, J. Appl. Phys. $\underline{50}$ 4786 (1979).
15. J.M. Poate, H.J. Leamy, T.T. Sheng and G.K. Celler, Appl. Phys.
    Letts. $\underline{33}$ 918 (1978).
16. Proceedings of the Laser Solids Interactions and Laser
    Processing Symposium Ed. S.D. Ferris, H.J. Leamy, J.M. Poate,
    A.I.P. Conf. Proceedings No.50 (1979).
17. C.W. White, S.R. Wilson, B.R. Appleton, F.W. Young Jr., J.Appl.
    Phys. $\underline{51}$ (1980), 738-749.
18. J. Narayan, R.T. Young, C.W. White, J. Appl. Phys. $\underline{49}$ (1978),
    3912-3917.
19. M. Linder, Phys.Stat.Sol. (a) $\underline{57}$ (1980) 263.
20. M. Koyanagi, H. Tamura, M. Miyao, N. Hashimoto and T. Tokuyama,
    Appl. Phys. Letts. $\underline{35}$ (1979) 621-623.
21. J. Narayan, R.T. Young, R.F. Wood, W.H. Christie, Appl. Phys.
    Letts. $\underline{33}$ (1978) 338-340.
22. K.L. Wang, Y.S. Liu and C. Burman, Appl. Phys. Letts. $\underline{35}$ (1979)
    263-265.
23. R. Stuck, E. Fogarassy, A. Grob, J.J. Grob, J.C. Muller and
    P. Siffert in Ref.4, pp.193-203.
24. C. Hill in Ref.4, p.26-43.
25. J.C. Bean, H.J. Leamy, J.M. Poate, G.A. Rozgonyi, J.P. van der
    Ziel, J.S. Williams and G.K. Celler, J. Appl. Phys. $\underline{50}$
    (1979) 881.
26. P. Revesz, G. Farkas, G. Mezey and J. Gyulai, Appl. Phys. Letts.
    $\underline{33}$ (1978) 431-433.
27. D. Hoonhout, C.B. Kerkdigk, F.W. Saris, Phys. Lett. A66 (1978)
    145.
28. G.A. Sai-Halasz, F.F. Fang, T.O. Sedgwick Armin Segmüller,
    Appl. Phys Letts., $\underline{36}$ (1980) 419-422.
29. M.E. Roulet, P. Schwob, K. Affolter, W. Lüthy, M. von Allmen,
    M. Fallavier, J.M. Mackowski, M.A. Nicolet and J.P. Thomas,
    J. Appl. Phys. $\underline{50}$ (1979) 5536-5538.
30. P. Shah, R. Shah, L. Crosthwait in Ref.4, 227-234.

31. K.F. Lee, J.F. Gibbons, K. Saraswat, T.I. Kamins, Appl. Phys.
        Letts. 35 (1979) 173-175.
32. J.F. Gibbons, K.F. Lee, T.J. Magee, J. Peng and R. Ormond,
        Appl. Phys. Letts. 34 (1979) 831.
33. M.W. Geis, D.C. Flanders, H.I. Smith, Appl. Phys. Letts. 35
        (1979) 71.
34. G. Yaron and La Verne D. Hess, Appl. Phys. Letts. 36 (1980)
        220-222.
35. F.E. Harper and M.I. Cohen, Sol. State Elec. 13 (1980) 1103.
36. C.W. White, J. Narajan and R.T. Young, Science 204 (1979)
        461-468.
37. G. Yaron, L.D. Hess and S.A. Kokorowski, IEDM (1979) Abs.9.4.
.38. C.P. Wu and G.L. Schnable, RCA Review 40 (1979) 339-344.
39. H. Runge, private communication.
40. Y. Hayafuji, T. Yanada and Y. Aoki, Elec. Chem. Soc. Meeting
        (Fall 1979), Abs.485 Abstracts Vol.79-2.
41. R.A. Kaplan, M.G. Cohen, K.C. Liu, Ref.4, 58-82.
42. A.G. Cullis, H.C. Webber and P. Bailey, J. Phys. E. 12 (1979)
        688-689.
43. A. Lietoila, J.F. Gibbons, T.J. Magee, J. Peng and J.D. Hong,
        Appl. Phys. Letts. 35 (1979) 532-534.

HANDLING OF HIGH POWER BEAMS FOR

LASER METAL WORKING SYSTEMS

V. Bartiromo, A. Cutolo, and S. Solimeno

Istituto Elettrotecnico, Università di Napoli

G.N.E.Q.P., C.N.R.

1. INTRODUCTION

A small and well shaped focal spot remains the ultimate goal
for a good design of a beam handling system. The thermal deformations
of the mirrors used for focusing on or simply directing a laser beam
cause distortions of the wavefronts which result in an increased
focal spot. This circumstance reduces the peak irradiance with
prejudice for the efficiency and the overall performance of the
laser apparatus. In some systems the laser beam is used in stations
which are separated several meters from the source. In these cases
some effects due to the air turbulence and the thermal blooming can
be of some relevance.

A way to overcome some of the problems posed by the free space
propagation of the laser beam has been indicated in the use of
flexible metallic waveguides. These devices have a reasonably good
transmission efficiency (>90% per meter) and can have a helical
shape in order to provide the flexibility needed for directing
and scanning the laser spot, thus relieving from the use of complex
tables for positioning the workpiece.

This paper is aimed to review the main problems connected
with the transfer of the laser power from the source to the working
station. Some analytic preliminaries are premised in Sec.2 in order
to provide the basic tools needed for evaluating the effects of the
beam aberrations on the focal spot irradiance. Section 3 is dedicated

to the evaluation of the aberrations brought about by thermal
deformation of the mirrors. Thermal blooming and turbulence effects
are briefly reviewed in Sec.4. Metallic waveguides are discussed
in Sec.5 with particular emphasis on the design of the input sections.

2. DIFFRACTION ANALYSIS OF BEAM PROPAGATION
2.1 Aberration Function W and Circle Polynomials

The deviation of the wavefront from a reference sphere can be
represented by the so-called aberration function W which measures
the distance between these two surfaces (Fig.1). If we introduce
polar coordinates (ρ,θ) on the reference sphere and normalize   to
the radius of the wavefront aperture, supposed circular, we can
express W in the form[1]

$$W(\rho,\theta) = \sum_{n=0}^{\infty} \sum_{m=0}^{n}{}' A_n^m R_n^m(\rho)\cos(m\theta) \tag{2.1}$$

where the prime is intended to indicate that when writing this series
out in full every A_n^o is to be raplaced by $2^{-\frac{1}{2}}A_n^o$. The functions R_n^m
are the Zernike polynomials which form an orthogonal basis on the
interval $(1,0)$ with respect to the weight function ρ^2, i.e.

$$\int_0^1 \rho R_n^m(\rho)R_{n'}^{m'}(\rho)d\rho = \frac{1}{2(n+1)} \delta_{nn'}\delta_{mm'} \tag{2.2}$$

If we choose the normalization constant so that for all n and m
$R_n^m(1) = 1$, then

$$R_n^m(\rho) = \sum_{s=0}^{\frac{1}{2}(n-m)} (-1)^s \frac{(n-m)!}{s!(\frac{n+m}{2} - s)!(\frac{n-m}{2} - s)!} \rho^{n-2s} \tag{2.3}$$

Nijober has proved that

$$\int_0^1 R_n^m(\rho)J_m(v\rho)\rho d\rho = (-1)^{\frac{1}{2}(n-m)} \frac{J_{n+1}(v)}{v} \tag{2.4}$$

This relation proves particularly useful in the evaluation of the
effect of W on the focal plane field distribution U.

Let NA be the wavefront numerical aperture and A the field
amplitude, in proximity of the focus we have

$$U(P) = U(u,v,\psi) = -\frac{i}{\lambda} \, NA \, \exp(iu/NA^2)$$

$$\int_0^1 \int_0^{2\pi} A\exp\left[i(kw - v\rho\cos(\theta-\psi) - \tfrac{1}{2}u\rho^2)\right]\rho \, d\rho \, d\theta \qquad (2.5)$$

where u and v are the two <u>optical</u> <u>coordinates</u> of P,

$$u = kzNA^2 \qquad v = k\rho NA \qquad\qquad (2.6)$$

and ψ is the anomaly.

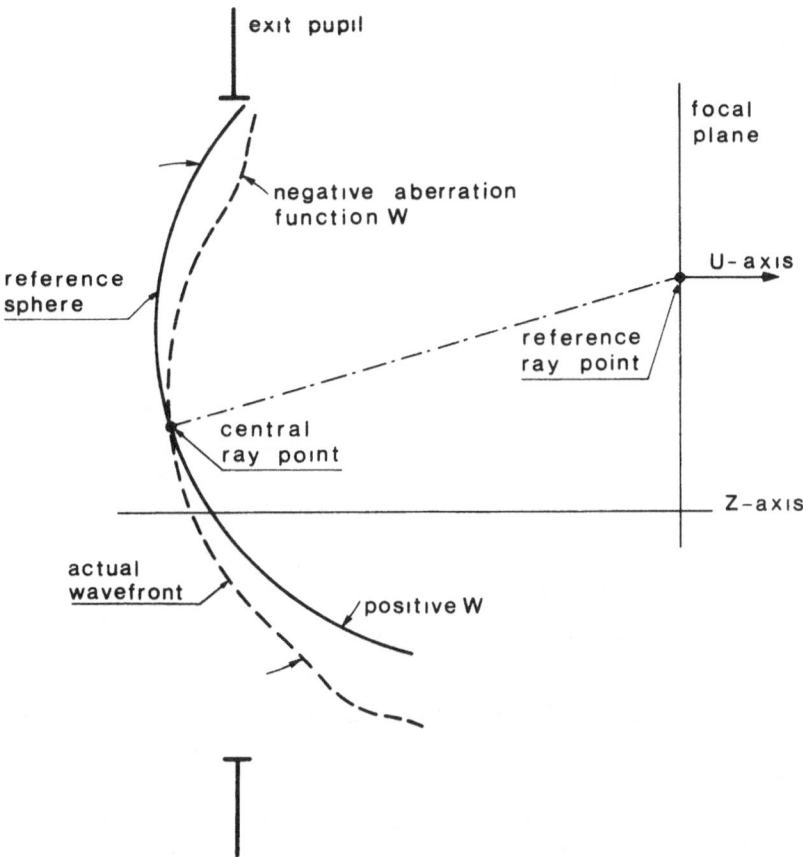

Fig. 1.   Cross section of actual and reference spherical wavefronts.
The aberration function W measures the distance between
these two surfaces. The exit pupil is shown since it defines
the domain of definition of the function W.

The intensity is given, for constant A, by

$$I(P) = I_o \, i(P) \qquad\qquad\qquad (2.7)$$

where

$$I_o = \tfrac{1}{2}k^2A^2NA^2 \qquad\qquad\qquad (2.8)$$

is the peak intensity in absence of aberrations, and i(P) is the normalized intensity

$$i(P) \equiv \left| \frac{1}{\pi} \int_o^1 \int_o^{2\pi} \exp\left[i(kW - v\rho\cos(\theta-\psi) - \tfrac{1}{2}u\rho^2)\right]\rho d\rho d\theta \right|^2 \qquad (2.9)$$

It follows that the normalized intensity at the Gaussian focus is

$$i(F) = \left| \frac{1}{\pi} \int \int \exp(ikW)\rho d\rho d\theta \right|^2$$

$$\simeq \left| 1 + ik\langle W\rangle - \tfrac{1}{2}k^2\langle W^2\rangle \right|^2$$

$$\simeq 1 - k^2\langle\Delta W^2\rangle \qquad\qquad\qquad (2.10)$$

where the bracket $\langle...\rangle$ indicates the average value on the wavefront and $\Delta W = W - \langle W\rangle$. Accordingly, for small aberrations i(F) is proportional to the mean-square deformation of the wavefront. Using the expansion (2.1) and taking into account the orthogonality of the functions $R_n^m\cos(m\theta)$ on the circular domain (see eq. 2.2) we can write i(F) in the form

$$i(F) = 1 - k^2 \sum_{n=1}^{\infty} \sum_{m=o}^{n} \frac{(A_n^m)^2}{n+1} \qquad\qquad (2.11)$$

In general,we can expand W in a power series in ρ and $\cos\theta$ containing only even powers

$$W(\rho,\theta) = W^{(0)} + W^{(2)} + W^{(4)} + W^{(6)} + \ldots \qquad (2.12)$$

where $W^{(2k)}$ is a polynomial of degree 2k in the variables ρ and θ. It is easy to show that the term $W^{(0)} + W^{(2)}$ does not change the intensity distribution apart from a displacement of the whole distribution. The term $W^{(4)}$ represents the Seidel's primary aberrations

$$W^{(4)} = -\frac{1}{4} B\rho^4 - C\rho^2\cos^2\theta - \frac{1}{2}D\rho^2 + E\rho\cos\theta + F\rho^3\cos\theta \quad (2.13)$$

the term proportional to B represents the <u>spherical aberration</u>. The term containing C is due to the <u>astigmatism</u>. The <u>field curvature</u> is represented by the factor D. The terms proportional to E and F represent respectively the <u>distortion</u> and the <u>coma</u> (Fig.2). In view of the aberration invariance with respect to $W^{(2)} + W^{(0)}$ (2.13) can be rewritten as

$$W^{(4)} = \frac{1}{24} BR_4^0 - \frac{1}{2} CR_2^2\cos 2\theta - \frac{1}{4} DR_2^0 + ER_1^1\cos\theta + \frac{1}{3}FR_3^1\cos\theta \quad (2.14)$$

The diffraction image on the focal plane in the presence of a single primary aberration is represented by the integral (2.5) with W replaced by one of the terms of (2.14) and u = 0, i.e.

$$U(u,v,\psi) = \frac{A}{\pi}\int_0^1\int_0^{2\pi} \exp\left[i(a_n^m R_n^m\cos(m\theta) - v\rho\cos(\theta-\psi))\right]\rho d\rho d\theta$$

$$= A\left[U_0 + ia_n^m U_1 + (ia_n^m)^2 U_2 + (ia_n^m)^3 U_3 + \ldots\right] \quad (2.15)$$

where U_0 represents the unperturbated diffraction image of the circular pupil (the so-called <u>Airy</u> spot),

$$U_0 = 2\int_0^1 J_0(v\rho)\rho d\rho = \frac{2}{v} J_1(v) \quad (2.16a)$$

while the other terms are due to the aberration contributions,

$$U_1 = 2(-i)^m\cos(m\theta)(-1)^{\frac{1}{2}(n-m)} \frac{J_{n+1}(v)}{v} \quad (2.16b)$$

$$U_2 = \frac{1}{2}\int_0^1 R_n^{m2} J_0(v\rho)\rho d\rho$$

$$+ \frac{1}{2} i^{2m}\cos(2m\theta)\int_0^1 R_n^{m2} J_{2m}(v\rho)\rho d\rho \quad (2.16c)$$

For U_3 and U_4 see Ref.1. Nijboer has shwon that for a_n^m of the order of unity, the four terms in the expression (2.15) suffice to give the intensity to within a few per cent.

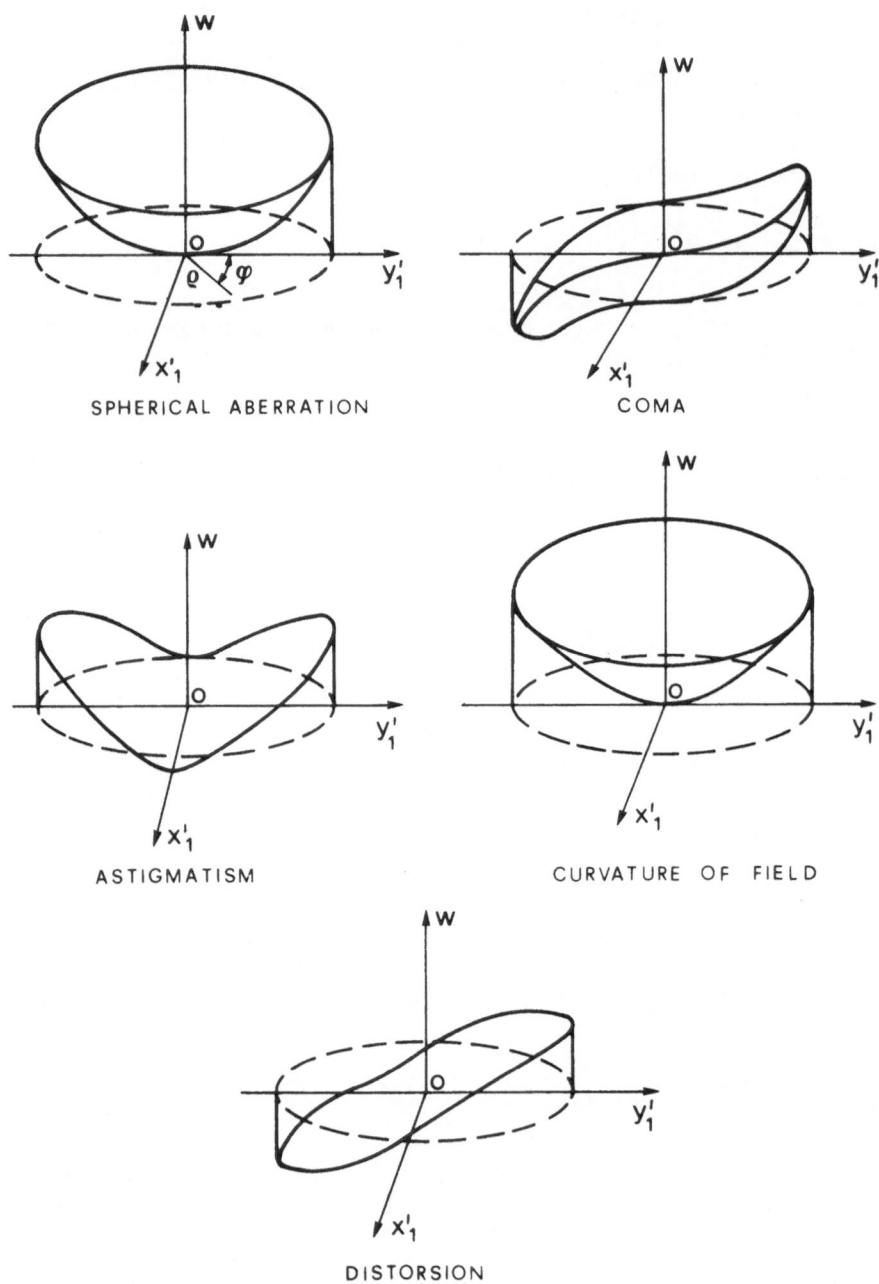

Fig. 2 W-functions for the primary Seidel's aberrations

For <u>primary</u> <u>coma</u>

$$U_1(0,v,\psi) = \frac{2i\cos\psi}{v} J_4(v) \qquad (2.17a)$$

$$U_2(0,v,\psi) = \frac{1}{2v} \left[\frac{1}{4} J_1 - \frac{1}{20} J_3 + \frac{1}{4} J_5 \right.$$

$$\left. - \frac{9}{20} J_7 - \cos(2\psi) \frac{2}{5} J_3 + \frac{3}{5} J_7 \right] \qquad (2.17b)$$

the argument of the Bessel functions having been omitted for nota-
tional simplicity. When the aberration grows up the image becomes of
the forms specified by geometrical optics broken up by a series of
dark bands. For <u>primary</u> <u>astigmatism</u>:

$$U_1(0,v,\psi) = - \frac{4\cos(2\psi)}{v} J_3 \qquad (2.18a)$$

$$U_2(0,v,\psi) = \frac{1}{2v} \left(\frac{1}{3} J_1 - \frac{1}{2} J_3 + \frac{1}{6} J_5 + \cos(4\psi) J_5 \right) \qquad (2.18b)$$

In the ray optical limit the above listed aberrations give rise to
a <u>blur</u> <u>spot</u> whose diameter B is given for a spherical mirror by the
following relations

$$B = 7.8 \ F \ NA^3 \ 10^{-3} \qquad \text{spherical} \qquad (2.19a)$$

$$B = 62.5 \ \theta \ F \ NA^2 \ 10^{-3} \qquad \text{sagittal coma} \qquad (2.19b)$$

$$B = 500 \ \theta^2 \ F \ NA \ 10^{-3} \qquad \text{astigmatism} \qquad (2.19c)$$

θ being the angle in radians that the object and image are off axis.

2.2 Numerical Evaluation of the Aberrated Beam in the Focal Zone

Essential to the evaluation of the field distribution in proximity of the focal plane is the numerical calculation of the diffraction intergal (2.5). Once known the aberration function W (2.5)
reduces to the two-dimensional Fourier transform of the function

$$A \ \exp i(kW - \tfrac{1}{2}u\rho^2) \qquad (2.20)$$

Such a calculation can be performed very quickly by resorting to the FFT alogrithm. The function W can be calculated analytically in the simple cases of a system containing very few mirrors. In most systems the laser beam undergoes many reflections before being focused. In all these cases it is necessary to evaluate W numerically by calculating the optical length of several rays traced starting from an assigned initial wavefront and intercepting the ideal sphere centered on the exit pupil.

The ray tracing for the optical systems adopted in metal working apparatus is made particularly complicated by the presence of articulate arms, elbows, periscopes, collimators, telescopes, etc.. Appendix A describes a program, called POSE-I, specially designed for the ray tracing when the mirrors are disuniformly distributed in space.

In conclusion, the field calculation has to be carried out in two steps: first, the aberration function W has to be calculated numerically by ray tracing procedures; second, the diffraction integral can be calculated by using the FFT algorithm for the function (2.20) defined on the exit pupil of the optical system.

2.3 Media and Power Driven Aberrations

The above discussed wavefront aberrations that one commonly encounters in high power laser systems is what is commonly known as media aberrations[3]. Since all high average power lasers have a flowing gas as their gain media, shock waves, weaks or heat release act as sources of refractive index variations. These give rise to the primary aberrations such as tilt, defocusing and astigmatism.

The next class of aberrations, known as power driven, occur when a mirror surface deforms due to a small absorption of optical energy. This results in an aberration function which is a function of the mirror thermal load. Two classes of such aberrations can be generally found. The former one is characterized by low spatial frequencies and is due to slow intensity variation over the beam diameter. The second class results from the so-called Fresnel ripples characterized by relatively high spatial frequencies and well described by the normalized intensity (2.10).

A third type of aberration is the so-called <u>thermal blooming</u>.
If the medium is initially in thermal equilibrium and the laser
is turned on at t=0, the absorption will give rise to a local heating
which in turn will produce a small pressure increase. The medium
then expands at the speed of sound so as to restore pressure balance
leaving behind a small decrease in the density. The local refractive
index will decrease in proportion to the local density change. If
the cumulative effect of these negative refractive index changes
is large enough, the shape of the laser beam profile will change
as it propagates away from the source.

3. WAVEFRONT ABERRATIONS CAUSED BY THERMAL DEFORMATIONS OF MIRRORS
3.1 Introduction

We will discuss now some analytical results obtained solving
exactly the thermoelastic equations for a water-cooled finite-size
mirror illuminated by a c.w. doughnut shaped laser beam. No restrictive
hypothesis will be made on the intensity distribution and full account
will be taken of the cooling process together with the elastic
constraints imposed on the mirror. The mirror will be modelled as
a cylinder of finite size (radius R and thickness d) cooled by forced
convection as a result of the water flowing at temperature T_c inside
a rigid support to which the inner face of the mirror is stiffly
constrained. The thermoelastic equation will be solved and the
results of some numerical calculations will be presented, together
with the quantitative evaluation of the deforations on the peak
irradiance.

3.2 Temperature Field

We shall condider an analytical approach to the problem of
evaluating the temperature field in a mirror irradiated by an intense
laser beam. It is wortwhile to mention that the analytical solutions
are not always possible to obtain and, indeed, in many istances are
very cumbersome and difficult to use. In these cases numerical tech-
techniques are used to advantage. The analysis carried out by Cutolo
et al.[4][5] differs from the one discussed by Apollonov et al.[6][7] by
the consideration of a finite size mirror and by accounting for the
forced convection cooling.

Consider a metallic mirror (Fig.3) exposed on the outer face
to a laser beam producing an illumination $I(r,\phi)$. In many cases the
beam is collimated and axial-symmetric. We shall ignore temperature

transients by considering steady-state distributions. Accordingly, thetemperature field inside the mirror $T(r,z,\phi)$ will be found among the solutions of the heat equation

$$\nabla^2 T(r,z,\phi) = 0 \qquad\qquad (3.1)$$

subject to the boundary conditions

$$k\frac{\partial}{\partial z} T = AI(r,\phi) = AI_{max}I(r,\phi) \qquad \text{as } z = 0 \qquad (3.2a)$$

$$k\frac{\partial}{\partial z} T = h(T-T_c) \qquad\qquad \text{as } z = -d \qquad (3.2b)$$

$$\frac{\partial}{\partial r} T = 0 \qquad\qquad \text{as } r = R \qquad (3.2b)$$

with A the absorption coefficent of the mirror (1-5%), K the thermal conductivity (380 Wm$^{\circ}$C for copper), h the thermal convection coeffic- ient ($10^2 - 10^4$ W/m^2°C), T_2 the temperature of the cooling water. To solve the system (3.1)-(3.2) the separation of variable method can be used together with a cylindrical coordinate system with origin at the mirror outer face center and the positive axis pointing outwards. To this end we can make use of a _Dini series_ expansion thus obtaining after some involved algebra

$$T - T_c = \sum_{m,n=0}^{\infty} \cos(2n\phi)J_{2n}(a_{2n,m}r)$$

$$\cdot \left[D_{2n,m}\exp(a_{2n,m}z) + D_{2n,-m}\exp(-a_{2n,m}z) \right] \quad (3.3)$$

with $a_{2n,m}$ a zero of the derivative of the Bessel function of order 2n. Now, by expanding also $\hat{I}(r,\phi)$ in a _Fourier-Dini_ double series

$$\hat{I}(r,\phi) = \sum_{n=0}^{\infty} \cos(2n\phi)\hat{I}_{2n}(r)$$

$$= \sum_{m,n=0}^{\infty} \cos(2n\phi)\hat{I}_{2n,m}J_{2n,m}(a_{2n,m}r) \qquad (3.4)$$

with

$$\hat{I}_{2n,m} = \frac{2a_{2n,m}^2}{(a_{2n,m}^2 R^2 4n^2)} \frac{\int_0^R r\hat{I}_{2n}(r)J_{2n}(a_{2n,m}r)dr}{J_{2n}^2(a_{2n,m}R)} \qquad (3.5)$$

and plugging (3.3) and (3.4) into (3.2a) we finally obtain

$$T(r,z,\phi) - T_c = \frac{A}{k} I_{max} \sum_{m,n=0}^{\infty} \hat{I}_{2n,m} \cos(2n\phi) R_{2n,m}(r) Z_{2n,m}(z)$$

$$= \sum_{m,n=0}^{\infty} T_{2n,m}(r,z) \cos(2n\phi) \qquad (3.6)$$

being

$$R_{2n,m}(r) = J_{2n,m}(a_{2n,m}r) \qquad (3.7a)$$

$$Z_{2n,m}(z) = \frac{a_{2n,m}\cosh(a_{2n,m}(d+z)) + g\sinh(a_{2n,m}(d+z))}{a_{2n,m}\sinh(a_{2n,m}d) + g\cosh(a_{2n,m}d)} \qquad (3.7b)$$

and $g = h/k$ (≈ 10 m^{-1}). Expansion (3.6) and (3.4) have been used for calculating the temperature field in a copper mirror illuminated with an annular 15 kw beam. We have assumed a beam of intensity $I = I_{max}\exp(-(r-r_o)^2 k_o)$, with $r_o = 5$cm and $k_o = 10^4$cm^{-2}. The results are plotted in figs.4a and b.

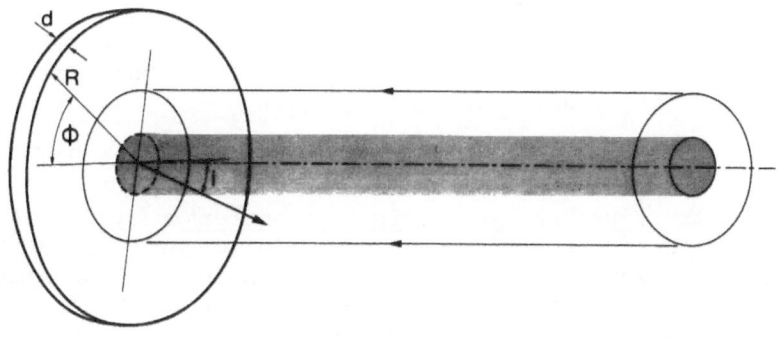

Fig.3   Annular laser beam impinging on a metallic mirror stiffly held and water cooled on the basis opposite to exposed surface.

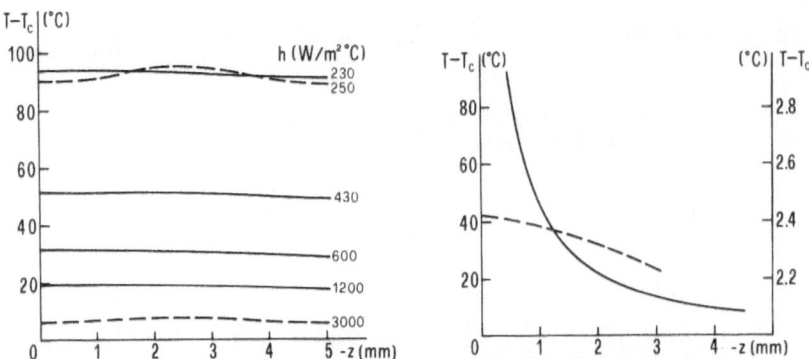

Fig.4   (a) Plot of the temperature rise T-T$_c$ of the exposed face
        of a mirror with R=10cm, d=5mm, A=.05, for different values
        of h and k=384 W/m°C(copper) (continuous line) and 3.84 W/m°C
        (dashed lines). Note that for copper T is almost uniform
        radially. A 15 KW laser beam with I(r)∝exp(-β(r-r$_o$)^2),
        β=10^{4}m^{-2} and r$_o$=5cm, has been considered. (b) Plot of the
        temperature rise of the center of the exposed face for the
        same conditions of fig.4a and K=384W/m°C (dashed line). Plot
        of the rise T(0,z)-T$_c$ versus z for h=9000 W/m °C (dashed line).

3.3 Mirror Deformations and Related Beam Aberrations
Displacement Vector Field

     The temperature field T(r,z,ϕ) induces a deformation of the
mirror, described by a displacement vector field $\underline{u}$(r,z,ϕ) =
$\underline{u}$(r,z,ϕ,T) - $\underline{u}$(r,z,ϕ,T$_c$). The following analysis will be limited
to normal incident beams so as to get rid of ϕ..In the limit of
small deformations, $\underline{u}$ satisfies the equation

$$(\lambda+2\mu)\nabla\nabla \cdot\underline{u} - \mu\nabla\times\nabla\times\underline{u} + \alpha(2\mu+3\lambda)\nabla T = 0 \qquad (3.8)$$

with λ and μ the <u>first</u> and <u>second Lamé parameters</u> respectively, α
the thermal expansion coeficent. The strain tensor $\underline{\underline{\varepsilon}}$ is related
to $\underline{u}$ through the equation $\underline{\underline{\varepsilon}}$ = sim $\underline{u}$, sim standing for the symmetric
part of the tensor. Next, imposing the vanishing of the forces
applied externally on the front and lateral surfaces of the mirror,
and supposing the rear face stiffly constrained to a rigid support
results in a set of boundary conditions which can be satisfied by
the vector field

$$\underline{u}(r,z) = \sum_m \left[\hat{r} J_1(a_{om}r)\alpha_m(z) + \hat{z} J_o(a_{om}r)\beta_m(z) \right] \quad (3.9)$$

where, on account of (3.8),

$$\frac{d^2}{dz^2}\alpha_m - a_{om}^2(\lambda+2\mu)\alpha_m - a_{om}(\lambda+\mu)\frac{d}{dz}\beta_m$$

$$= \alpha(2\mu+3\lambda)a_{om}I_{max}\uparrow_{om}\frac{A}{k}Z_{om} \qquad (3.10a)$$

$$(\lambda+2\mu)\frac{d^2}{dz^2}\beta_m - a_{om}^2\mu\beta_m + a_{om}(\lambda+\mu)\frac{d}{dz}\alpha_m$$

$$= -\alpha(2\mu+3\lambda)\frac{d}{dz}Z_{om}I_{max}\frac{A}{k}\uparrow_{om} \qquad (3.10b)$$

the coefficents α_m and β_m satisfying the boundary conditions

$$- a_{om}\beta_m(0) + \alpha_m'(0) = 0 \qquad (3.11a)$$

$$(\lambda+2\mu)\alpha_m'(0) + a_{om}\mu\beta_m(0) + \alpha(2\mu+3\lambda)\frac{A}{k}Z_{om}I_{max}\uparrow_{om} = 0 \quad (3.11b)$$

$$\alpha_m(-d) = \beta_m(-d) = 0 \qquad (3.11c)$$

The coefficents α_m and β_m have been calculated numerically for a copper mirror and used to calculate u_z on the outer face, as shown in fig.5.

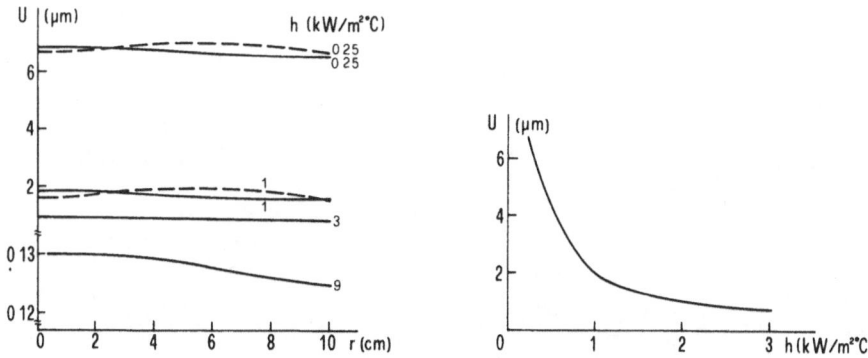

Fig.5 (a) Normal displacement of the outer face of the mirror
        described in fig.4 as a function of h for k=384 W/m°C(continuos
        line) and k=3.84 (dashed lines). (b) Normal displacement of
        the center of the exposed face versus h.

Reflected Beam Aberration

A collimated beam reflected by the deformed mirror is characteriz by an aberration function $W = 2u_z(r,0)$. By using (3.9) and (2.1) we have

$$u_z(r,0) = \sum_m J_o(a_{om} r)\beta_m(0) = \sum_m A_{2m} R_{2m}^o(r/R) \qquad (3.12)$$

with

$$A_{2n} = (4n+2)\int_o^1 u_z(\rho R)R_{2n}^o \rho d\rho$$

$$= (-1)^n(4n+2)\sum_{m=1}^\infty \beta_m(0) \frac{J_{2n+1}(a_{om} R)}{a_{om} R} \qquad (3.13)$$

As a consequence of (2.11)

$$i(F) = 1 - (\frac{8\pi}{\lambda})^2\sum_{n=1}^\infty (2n+1) \sum_{m=1}^\infty \frac{\beta_m(0)J_{2n+1}(a_{om} R)}{a_{om} R} \qquad (3.14)$$

In particular, in case $d/R \ll 1$ a simple expression of $\beta_m(0)$ can be found

$$\beta_m(0) \simeq \frac{A}{k} I_{max} \alpha d \frac{2\mu+3\lambda}{\lambda+2\mu} \hat{I}_{om} \frac{a_{om}+3g/8}{a_{om}(a_{om}^2 d+g)} \qquad (3.15)$$

For a gaussian beam with $I(r) = I_{max} \exp(-\beta r^2)$ and $\beta R \gg 1$

$$\hat{I}_{om} \simeq \frac{\exp(-a_{om}^2/4\beta)}{R J_o(a_{om} R)} \qquad (3.16)$$

and

$$i(F) \simeq 1 - 4 \frac{A}{k} \frac{d}{R} \frac{P_z}{R^2} \frac{2\mu+3\lambda}{\lambda+2\mu}$$

$$\cdot \sum_{n=1}^\infty (2n+1) \sum_{m=1}^\infty \frac{(a_{om}+3g/8)\exp(-a_{om}^2/4\beta)}{a_{om}^2(a_{om}^2 d+g)J_o(a_{om} R)} J_{2n+1}(a_{om} R)$$

4. THERMAL BLOOMING, TURBULENCE AND KINETIC COOLING[10]

To understand how the density modification induced by the absorption of a laser beam might change the laser propagation, let us consider a Gaussian beam. The heating profile shows a maximum on the axis of the beam while the corresponding refractive index exhibits a minimum. Thus the wavefront moves with slightly greater speed on the beam axis and the rays of the beam can thus be thought of as diverging outward from their original direction[11]. The entire beam will bloom as the heating of the medium will continue.

For what concerns the kinetics of the absorption process, it has been found for the 10.6μ case that the atmosphere undergoes first a cooling process, which is then followed by heating. This was predicted by Wood et al.[12] and verified experimentally by Gebhardt and Smith[13] When a CO_2 laser beam travels through the atmosphere, a fraction is absorbed by the excited $CO_2(100)$ molecules thus inducing a transition to the level $CO_2(001)$ which is in turn in resonance with the level $N_2(v=1)$. Consequently, each excited $CO_2(001)$ molecule tends to trasfer its energy to the nitrogen. As a result of this process the $CO_2(100)$ molecules loose their energy returning to the fundamental level. To restore the population of $CO_2(100)$ molecules some unexcited molecules are collisionally raised to the level 100 by detracting from the gas translational energy. The net result of this process is a cooling of the gas.[14] This phenomenon is the opposite of the thermal blooming and tends to favor the propagation of high power beams.

A way to contrast the thermal blooming is to use pulsed lasers[15] In this way even if we assume that the absorption process occurs instantly and is immediately converted into translational energy, there exists a finite time needed to convert the air density along the beam path. These density changes propagate with the sound velocity. This implies that the beam distortion can occur only if the atmosphere is irradiated for a time no less than the time needed by an acustic perturbation to cross the beam.

Due to the spreading of the beam the peak irradiance undergoes some reduction which can be accounted for by introducing a distortion parameter N given by[16]

$$N_0 = \frac{n_T \alpha_{abs} P_l z^2}{\pi np C_p v \omega^3} \qquad (4.1)$$

where n, n_T, α_{abs}, ρ and c_p are respectively the refractive index,
the coefficent of index change with respect to the temperature, the
absorption coefficent, density and specific heat at constant pressure
P and z are the laser power and range respectively, ω stands for
the spot size. The parameter N has been generalized to include the
effects of focusing and nonuniform velocity profile along the beam
axis and is given by[17]

$$N = N_o \frac{2}{z^2} \int_0^z \frac{\omega_o}{\omega(z')} \, dz' \int_0^{z'} \frac{\omega_o^2 U \exp(-\alpha z'')}{\omega^2 v} \, dz'' \qquad (4.2)$$

When the velocity profile varies uniformly with z, $U(z') = U_o + \omega z$,
N can be written in the form[18]

$$N = N_o f(\alpha z) q(\omega_o/\omega) s(\omega z/v_o) \qquad (4.3)$$

where the three correction factors f, q and s are given by

$$f(x) = \frac{2}{x} (x - 1 + e^{-x}) \qquad (4.4a)$$

$$q(x) = \frac{2x}{x-1} (1 - \frac{\ln x}{x-1}) \qquad (4.4b)$$

$$s(x) = \frac{2}{x} ((x+1)\ln(x+1) - x) \qquad (4.4c)$$

The peak irradiance i(F) with blooming normalized by the undistorted
value is approximatively given by the simple empirical relationship

$$i(F) = \frac{1}{1 + 0.0625 \, N} \qquad (4.5)$$

An important feature of the thermal blooming is its dependence on
the beam power. This can be considered as a potential limitation
for very high power beam propagation in stagnant atmosphere. The
heated atmosphere acts like a thin lens with aberrations. Accordingly,
the propagation effects can be compensated by imposing to the beam
an appropriate phase compensation. This might be done by using
active optics (COAT) or compensating through phase adaption by non
linear techniques (f.i. PANT).[18]

Another factor affecting the regular motion of wavefront is the atmospheric turbulence. This process induces a spreading ω_t of the beam which can be expressed by the approximate formula[19]

$$\omega_t = \frac{2}{k\rho_o} \tag{4.6}$$

being

$$\rho_o = (0.545k^2C_N^2z)^{-3/5} \tag{4.7}$$

the so-called <u>lateral</u> <u>coherence</u> <u>length</u> of a spherical wave. C_N is the <u>refractive</u> <u>index</u> <u>structure</u> <u>constant</u>. This form of the coherence length is valid for $z > z_c$ being

$$z_c = (0.4k^2C_N^2(L_o/2\pi)^{5/3})^{-1} \tag{4.8}$$

and L_o is the <u>outer</u> <u>scale</u> <u>length</u> <u>of</u> <u>the</u> <u>turbulence</u>.

5.   METALLIC WAVEGUIDES
5.1 Modes and Relative Losses of Straight and Helical Waveguides

When the atmosphere becomes an unsuitable medium, metallic wave-guides can be used for transmitting, shaping and directing high power beams. The waveguide most studied by the Center for Laser Studies, U.S.C. (see E. Garmire et al. Refs. 20-24) is formed of two metallic strips separated by shimstocks whose edges form the sidewalls (see Fig.6). It allows the propagation of TE and TM modes.

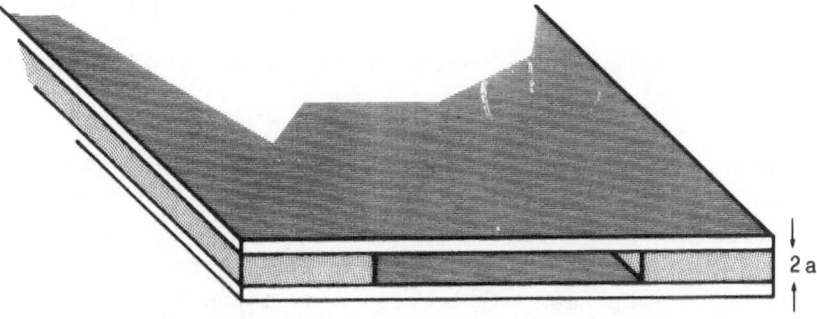

Fig.6   Metallic waveguide formed of two metallic strips separated by shimstocks whose edges form the sidewalls.

Customarily the waveguide height 2a is much smaller than the width so that the propagation is insensitive to the presence of the sidewalls and occurs as through two parallel infinite plates. The modes, shown in fig.7, are given by

$$\underline{E} \propto \hat{y} \, \sin\left[\frac{m\pi}{2a}(x-a)\right]\exp(-i\beta_m z) \qquad TE_{mo} \text{ mode} \qquad (5.1a)$$

$$\underline{E} \propto \hat{x} \, \cos\left[\frac{m\pi}{2a}(x-a)\right]\exp(-i\beta_m z) \qquad TM_{mo} \text{ mode} \qquad (5.1b)$$

with $\beta_m = \left[k^2 - (m\pi/2a)^2\right]^{\frac{1}{2}}$.

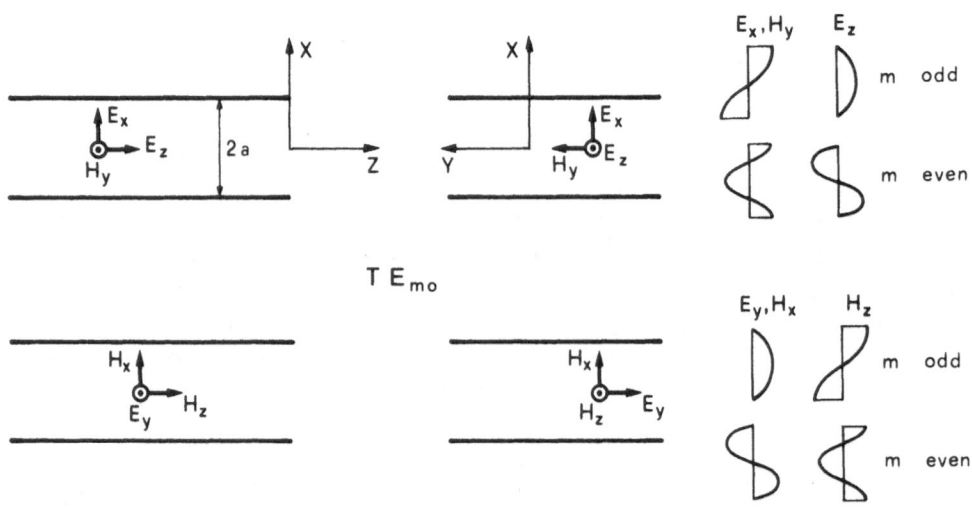

Fig.7 Field distributions for the TE and TM modes of two parallel metallic plates.

These waveguides exhibit strong attenuations of the higher modes. The relative loss factors can be calculated by a simple ray-optical analysis[25],[26]. In fact, a mode of index m is formed by two plane waves travelling at angle $\theta_m = m\pi/2a$ with respect to the waveguide axis. Then each ray travels a distance $d_m = a/tg\theta_m$ between two reflections on the upper and lower plates. Let $A(\theta)$ the loss per reflection at the incidence angle $i = \pi/2 - \theta_m$, the mode undergoes an attenuation per unit length equal to

$$\alpha_m = \frac{m\lambda A(\theta)}{2a^2} \tag{5.2}$$

for grazing incidence A is given by

$$A^{TE}(\theta) = 4Re(n^{-1})\sin\theta \simeq 4\cdot0Re(n^{-1}) \tag{5.3a}$$

$$A^{TM}(\theta) = \frac{4Re(n)\sin\theta}{1 + 2Re(n)\sin\theta + |n|^2\sin\theta} \tag{5.3b}$$

n being the complex refractive index of the metal. Consequently

$$\alpha_m^{TE} = \frac{m^2\lambda^2}{a^3} Re(n^{-1}) \tag{5.4a}$$

$$\alpha_m^{TM} = \frac{m^2\lambda^2}{a^2} Re(n) \tag{5.4b}$$

A transmission of more than 95% per meter is a straight waveguide having a cross section of .5x7 mm has been measured[19]. When rectangular waveguide are bent, propagation takes place mainly by means of surface waves along the outer walls[27].

The fact that a waveguide affords only one degree of freedom limits its practical usefulness. Some authors[28-30] have proposed the use of helical-circular waveguides which are quite flexible while the curvature of the cross section would balance diffraction (fig.8).

The modes of a waveguiding strip which is curved in the transverse direction have been extensively studied by Casperson et al.[30] and by Marhic et al.[28,29] starting from the expression of the component E_z parallel to the metallic strip and perpendicular to the helix.

$$E_z(r,z,\phi) = A(r,z,\phi) \exp[-ik(r_o\phi\cos\theta + z\sin\theta)] \tag{5.5}$$

(r,z,ϕ) being cylindrical coordinates, θ and r_o the pitch angle and the radius respectively of the helix. Casperson and Garfield have put A in the form

$$A(r',z',\phi') = B(r',z',\phi') \exp\left[-i\frac{q}{2}(z'-d_a)^2\right] \tag{5.6}$$

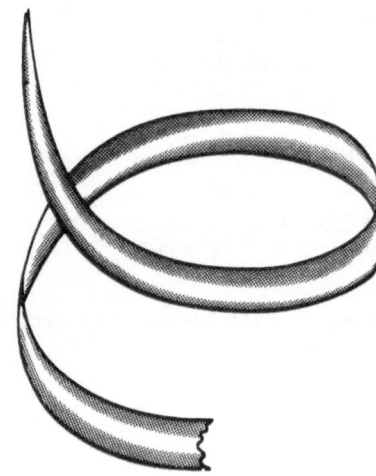

Fig.8 Flexible waveguide with curved cross section

where z' and φ' are coordinates representing displacements parallel
and perpendicular to the helix trajectory (see fig.9)

$$\phi' = \phi\cos\theta + \frac{2}{r}\sin\theta$$

$$z' = -r\phi\sin\theta + 2\cos\theta$$

(5.7)

The factor q is related to the wavefront curvature and the spot
size by the standard relation $q^{-1} = R^{-1} - i\pi/\lambda\omega^2$. Introducing the
distance $x = (\cos\theta/r_oR_o)^{\frac{1}{2}} r_o\phi'$, with R_o the curvature radius across
the waveguide strip, q is expressed by

$$\frac{q(r_o,\phi')}{k} = -\frac{(\cos\theta/r_oR_o)^{\frac{1}{2}}\sin x - (q(0)/k)\cos x}{\cos x + (q(0)/k)(r_oR_o/\cos\theta)^{1/2}\sin x}$$

(5.8)

Accordingly, the spot size oscillates periodically according to the
real part of (5.8) and exhibits a periodic pinching of the beam
waist. In addition, due to the presence of the factor $d_a(\phi')$ at
exponent of (5.6), the beam center is displaced with respect to
the center of the guide of the quantity $d_a(r_o,\phi')$ which varies
with φ' according to the sinusoidal law

$$d_a(r_o,\phi') = d_o(0)\cos x + d_a'(0)(r_oR_o/\cos\theta)^{\frac{1}{2}}\sin x$$

(5.9)

The factor B can be expressed in the form

$$B(r',\phi',z') \quad H_m(\frac{z'-d_a}{\omega}) \; Ai\left[(\frac{2k^2\cos^2\theta}{r_o})(r'-r_n')\right| \tag{5.10}$$

where H_m is the Hermite polynomial of order m and Ai is the Airy function. $r'(2k^2\cos^2\theta/r_o)^{1\ 3}$ is the nth zero of the Airy function. The losses of this waveguide are almost coincident with those of a single curved metallic surface. Then it turns out that

$$\alpha = \frac{1}{r_o} \; Re(n^{-1}) \tag{5.11}$$

Accordingly, the attenuation is independent of the mode order but depends on the angle of the bend arc. A loss of 5% per radian of bend arc has been measured by Garmire et al.[19]

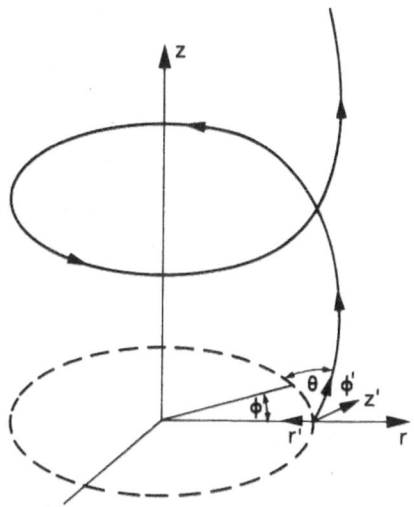

Fig.9 Geometry relative to the helical waveguide

5.2 Launching of a Gaussian Beam into a Waveguide

Consider a current sheet $\underline{J}^t$ inside the waveguide at a distance z_t from the end, having a distribution such as to excite in the straight waveguide infinitely long a single TE_{mo} or TM_{mo}. If we apply the reciprocity theorem to the couples of currents $\underline{J}^t$ and $\underline{J}^s = \hat{x}\delta(r-r_g)$, we obtain

$$\int_{-a}^{a} \underline{J}^{t}(x,z_{t}) \cdot \underline{E}^{s}(x,z_{t}) dx = \underline{J}^{s} \cdot \underline{E}^{t}(S) \qquad (5.12)$$

$\underline{J}^{s}$ is a <u>source</u> <u>current</u> dipole having complex coordinates $S(x_{s},z_{s})$ in order to produce a Gaussian beam (see fig.10).

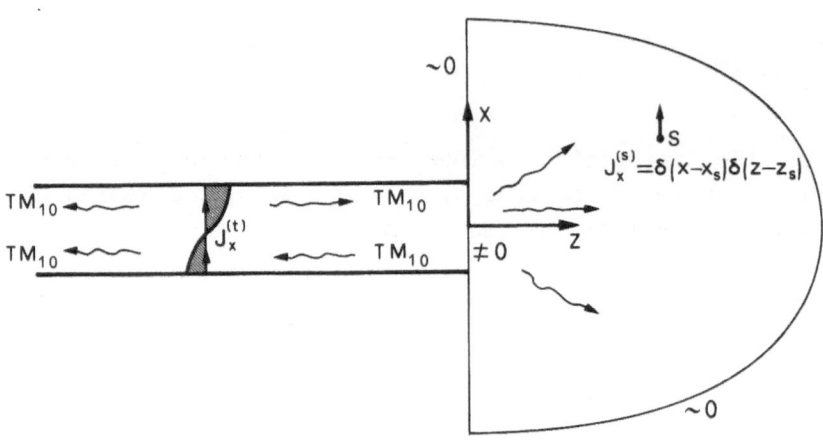

Fig.10   Schematic of the test and source currents used for calculating
         the amplitude of the modes excited in the waveguide.

Choosing suitable test currents $\underline{J}^{t}$ distributions we will get

$$\underline{J}^{t} = \hat{x}\ 2\cos\left[\frac{m\pi}{2a}(x-a)\right]\delta(z_{t}-z)\exp(i\beta_{m}z_{t})$$

$$\rightarrow TM_{mo} \rightarrow H^{t} = \hat{y}\ \cos\left[\frac{m\pi}{2a}(x-a)\right]\begin{cases} \exp(i\beta_{m}z) & z>z_{t} \\ \\ \exp(-i\beta_{m}z) & z<z_{t} \end{cases} \qquad (5.13a)$$

$$\underline{J}^{t} = \hat{y}\ 2\sin\left[\frac{m\pi}{2a}(x-a)\right]\delta(z_{t}-z)\exp(i\beta_{m}z_{t})$$

$$\rightarrow TE_{mo} \rightarrow \underline{E}^{t} = \hat{y}\ \frac{\beta_{m}}{\omega\varepsilon}\sin\left[\frac{m\pi}{2a}(x-a)\right]\begin{cases} \exp(i\beta_{m}z) & z>z_{t} \\ \\ \exp(-i\beta_{m}z) & z<z_{t} \end{cases} \qquad (5.13b)$$

Expanding the electric field produced inside the waveguide by the source $\underline{J}^s$

$$E_x^s = E^s(w) \sum_n C_n^{TM} \cos\left[\frac{n\pi}{2a}(x-a)\right]\exp(-i\beta_n z)$$

$$\text{(5.14)}$$

$$E_y^s = E^s(w) \sum_n C_n^{TE} \sin\left[\frac{n\pi}{2a}(x-a)\right]\exp(-i\beta_m z)$$

$E^s(w)$ being the field value at the waist point w of the Gaussian beam generated by $\underline{J}^s$, (5.12) fields the coupling coefficents

$$c_m = \frac{1}{2aE^2(w)} \underline{J}^s \cdot \underline{E}^t(S) \qquad (5.15)$$

This relation allows us to calculate the amplitude of a mode excited by $\underline{J}^s$ by using the field $\underline{E}^t$ diffracted outside the waveguide by the mode itself, having amplitude complying with (5.13). Asymptotic expressions of these diffracted fields can be found in the literature. They are

$$H_y^t = 4iG^{(2)}(k\rho)F_h(\theta_{2m},\theta+\pi) \frac{\cos\frac{1}{2}\theta_m}{\cos\frac{1}{2}\theta}$$

$$\cdot \frac{G_+(k\cos\theta_{2m})}{G_+(k\cos\theta)} \sin(ka\sin\theta) \qquad TM_{2m,o} \quad (5.16a)$$

$$H_y^t = 4iG^{(2)}(k\rho)F_h(\theta_{2m+1},\theta+\pi) \cos(ka\sin\theta)$$

$$\cdot \frac{L_+(k\cos\theta_{2m+1})}{L_+(k\cos\theta)} \qquad TM_{2m+1,o} \quad (5.16b)$$

$$E_y^t = -4iG^{(2)}(k\rho) \frac{\beta_{2m}}{\omega\epsilon} F_e(\theta_{2m},\theta+\pi) \sin(ka\sin\theta)$$

$$\cdot \frac{G_+(ka\cos\theta_{2m})}{G_+(ka\cos\theta)} \frac{\cos\frac{1}{2}\theta_{2m}}{\cos\frac{1}{2}\theta} \qquad TE_{2m,o} \quad (5.16c)$$

$$E_y^t = -4G^{(2)}(k\rho) \frac{\beta_{2m+1}}{\omega\varepsilon} F_e(\theta_{2m+1}, \theta+\pi)\cos(kasin\theta)$$

$$\cdot \frac{L_+(kcos\theta_{2m+1})}{L_+(kcos\theta)} \qquad\qquad TE_{2m+1,o} \qquad (5.16d)$$

where $\cos\theta_m = \beta_m/k$. The functions F_e and F_h are the directivity factors defined by

$$F_h(\theta_m, \theta+\pi) = \frac{\cos(\tfrac{1}{2}\theta_m)\sin(\tfrac{1}{2}\theta)}{\cos\theta - \cos\theta_m} \qquad\qquad (5.17a)$$

$$F_e(\theta_m, \theta+\pi) = \frac{\sin(\tfrac{1}{2}\theta_m)\cos(\tfrac{1}{2}\theta)}{\cos\theta_m - \cos\theta} \qquad\qquad (5.17b)$$

$G_+(\alpha)$ and $L_+(\alpha)$ are defined by the functional equations

$$G_+(\alpha)G_+(-\alpha) = \exp(-\gamma a)\frac{\sinh(\gamma a)}{\gamma a} \qquad\qquad (5.18a)$$

$$L_+(\alpha)L_+(-\alpha) = \exp(-\gamma a)\cosh(\gamma a) \qquad\qquad (5.18b)$$

together with the condition of being regular and non-zero in the upper α-plane, $\gamma^2 = (\alpha^2-k^2)$. The factor $G_+(\alpha)$ can be expressed as

$$G_+(kcos\theta) = (-ika)^{-\frac{1}{2}} \frac{1}{\cos\tfrac{1}{2}\theta} \quad \exp \frac{\exp(2ika-3i\pi/4)}{(ka)^{05}\cos\theta}$$

$$\cdot \phi(e^{2ika}, \frac{3}{2}, 1) \qquad\qquad (5.19)$$

where ϕ is defined by

$$\phi(z,u,1) = \sum_0^\infty \frac{z^n}{(1+n)^u} \qquad |z|<1,\ Arg(z)\neq 0 \qquad (5.20)$$

Similar expression holds for L_+. It can be shown[32] that $|\phi|$ is of the order of unity. Then, neglecting the exp factor on the right of (5.19) and using (5.16) fields

$$H_y^t \simeq 4iG^{(2)}(k\rho)F_h(\theta_{2m},\theta+\pi)\sin(ka\sin\theta)\exp\left|\frac{\exp(2ika-3i\pi/4)}{(\pi ka)^{05}}\right.$$

$$\left.\cdot\left(\frac{1}{\cos\theta_{2m}} - \frac{1}{\cos\theta}\right)\right] \tag{5.21}$$

When the Gaussian source is no too far from the waveguide termination, (5.21) must be replaced by

$$H_y^t \simeq 2iG^{(2)}(k\rho)F_h(\theta_{2m},\theta_1+\pi) + (-1)^m 2iG^{(2)}(k\rho)F_h(\theta_{2m},\theta_2+\pi) \tag{5.22}$$

where $\rho_1(\rho_2)$ and $\theta_1(\theta_2)$ refer to the edge 1 (2) of the waveguide (Fig.11). Plugging (5.22) into (5.15) yelds the coupling coefficents for the TM and TE modes

$$c_m^{TM} \simeq \frac{1}{ka}\left[\frac{E^S(\rho_1)}{E^S(w)} F_h(\theta_m,\theta_1+\pi) + (-1)^m \frac{E^S(\rho_2)}{E^S(w)} F_h(\theta_m,\theta_2+\pi)\right]$$

$$c_m^{TE} \simeq -\frac{1}{ka}\left[\frac{E^S(\rho_1)}{E^S(w)} F(\theta_m,\theta_1+\pi)\frac{\cos\theta_m}{\cos\theta_1}\right. \tag{5.23}$$

$$\left. - (-1)^m \frac{E^S(\theta_2)}{E^S(w)} F_e(\theta_m,\theta_2+\pi)\frac{\cos\theta_m}{\cos\theta_2}\right]$$

When the angles θ_m and $\theta_{1,2}$ are small the diffraction factors must be replaced by [33]

$$F_{h,e}(\theta_m,\theta+\pi) = -\frac{1}{2}\frac{F\left[\frac{1}{2}k\rho\theta^2(\theta_m/\theta-1)^2\right]}{\theta_m-\theta}$$

$$\pm\frac{1}{2}\frac{F\left[\frac{1}{2}k\rho\theta^2(\theta_m/\theta+1)^2\right]}{\theta_m+\theta} \tag{5.24}$$

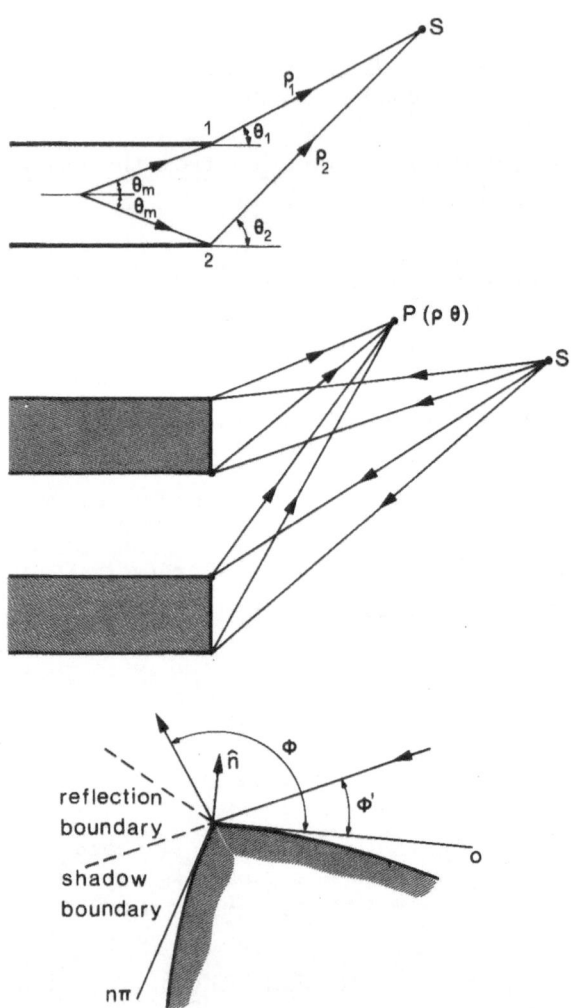

Fig.11   Geometry relevant to the rays diffracted by two half-planes
          and two square edges.

where the transition function is defined by

$$F(x) = 2ix^{\frac{1}{2}}e^{ix} \int_{x^{\frac{1}{2}}}^{\infty} \exp(-iz^2)dz \qquad (5.25)$$

which for $x > .7$ can be expressed in the polynomial form

$$F(x) \simeq 1 + \frac{A}{x} + \frac{B}{x^2} \qquad (5.26)$$

with

$$A = -0.14 + 0.48i \quad \text{and} \quad B = -0.05 - 0.25i$$

For a Gaussian beam having the waist on the input section of the waveguide, (5.23) and (5.24) give

$$c_m^{TE} = \frac{1}{4} e^{-\eta^{-2}} \left[\frac{F\left[-i\eta^{-2}(\frac{1}{2}m\pi\eta^2-1)^2\right]}{1-\frac{1}{2}im\pi\eta^2} + \frac{F\left[-i\eta^{-2}(\frac{1}{2}im\pi\eta^2+1)^2\right]}{1+\frac{1}{2}im\pi\eta^2} \right]$$

where $\eta = w_o/a$.

5.3 Design of the Input Section

For exciting the fundamental mode of the guide, which exhibits the lowest transmission losses, the waist of the beam must have a radius comparable with the waveguide height, as can be shown by calculating (5.27). This condition is difficult to meet in most practical cases, specially when using very intense beams. In fact, the beam aberrations are so strong as to produce a large focal spot, however larger than 0.5 mm, the typical height of these metallic waveguides[19]. To cope with this problem a flared input section can be used (Fig. 12). The modes of this section are described by

$$E_y = k^2 H^{(1)}_{\frac{m\pi}{\phi_o}}(k\rho)\sin(\frac{m\pi\phi}{\phi_o})$$

$$\simeq k^2 (\frac{2}{\pi k\rho})^{\frac{1}{2}}\exp(ik\rho)\exp\left[-i\frac{\pi}{4}(\frac{2m\pi}{\phi_o} - 1)\right]\sin(\frac{m\pi\phi}{\phi_o}) \quad (5.28)$$

ϕ_o being the flare angle. In horns of small flare angles the wave-fronts are virtually plane and so the fields near the mouth and the junction with the straight waveguide. Accordingly, the coupling co-efficients with a Gaussian beam can be calculated by a simple gen-eralization of the above illustrated procedure. The mismatch between the wavefronts of the laser beam and the cylindrical wavefronts of the flared section produces a reduction of the coupling as can be seen from Fig. 13.

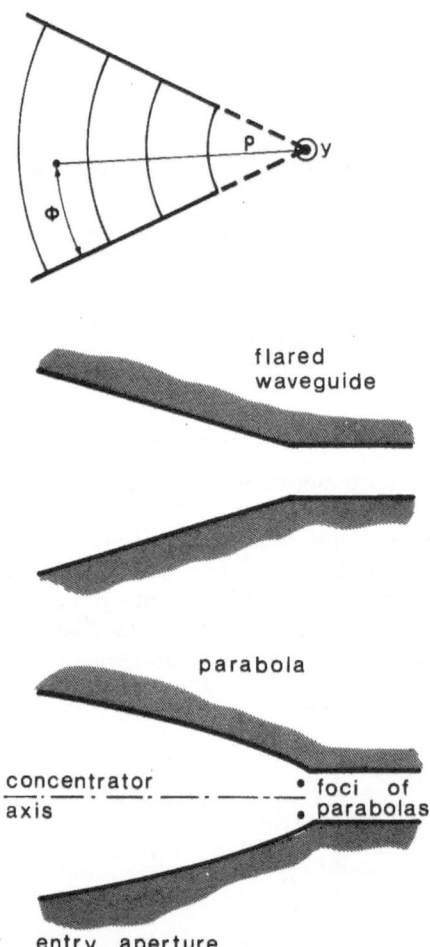

Fig.12   Geometry of flared waveguides and compound parabolic
         concentrators proposed as input sections for FIT waveguides.

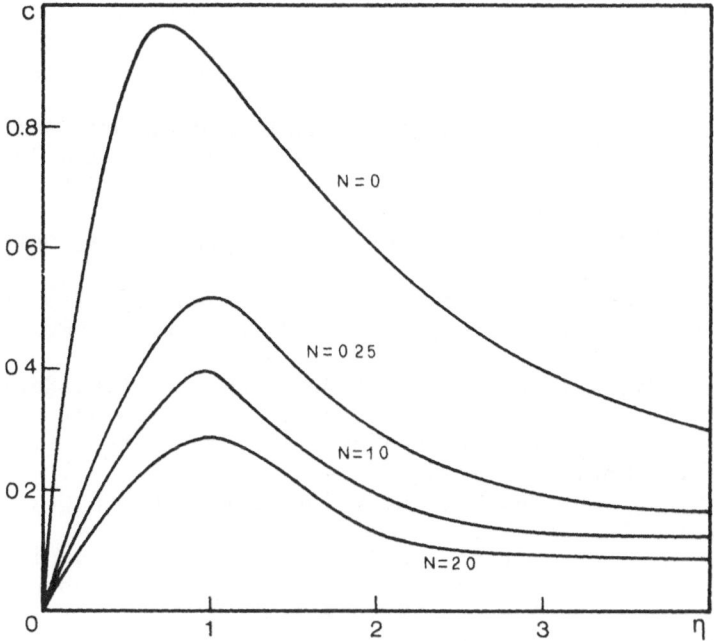

Fig.13   Coupling coefficents for different modes as functions of
$\eta = w_o/a$.

Acknowledgements
     This work was supported by the C.N.R. under contract no.80.
00028.02 and the Fiat Research Center, Orbassano, Torino. The
fruitful collaboration of E.Pastore, E.Maiorino, E.Di Stasio and
G.Angelo is gratefully acknowledged.

REFERENCES

1. M. Born and E. Wolf, Principles of Optics, Pergamon Press,
   Oxford (1975)
2. S. Cornbleet, Microwave Optics, Academic Press, New York (1976)
3. P.V.Avizonis, in Adaptive Optics and Short Wavelength Sources,
   eds. S.F. Jacobs, M. Sargent III and M.O. Scully, Addison-
   Wesley, Reading (1978) pp. 1-54
4. A. Cutolo, P. Gay and S. Solimeno, Lasersymposium 2, Delft (1979)
   paper H11
5. A. Cutolo, P. Gay and S. Solimeno, Optica Acta $\underline{27}$, 1105 (1980)
6. V.V. Apollonov, A.I. Barchukov and A.M.Prokhorov, IEEE J. Quantum
   Electron. $\underline{QE-10}$, 505 (1974)
7. V.V. Apollonov, A.I. Barchukov, A.M. Prokhorov, A.M. Karlov and
   E.M. Shefter, Sov. J. Quant. Electron. $\underline{5}$, 216 (1975)
8. G.N. Watson, A Treatise on the Theory of Bessel Functions, Cambridg
   Univ.Press, Cambridge (1966)
9. B.A. Boley and J.H. Weiner, Theory of Thermal Stresses, J. Wiley
   and Sons, New York ((1963)
10. J.L. Walsh and P.B. Ulrich, in Laser Beam Propagation in the
    Atmosphere, ed. J.W. Strohben, Topics in Applied Physics vol.25
    Springer-Verlag, Berlin (1978) pp.223-320
11. R.C.C. Leite, R.S. Moore and J.R.Whinnery, Appl.Phys.Lett. $\underline{5}$,
    141 (1964)
12. A.D. Wood, M. Camac and E.T. Gerry, Appl.Optics $\underline{10}$, 1877 (1971)
13. F.G.Gebhardt and D.C. Smith, Appl.Phys.Lett. $\underline{20}$, 129 (1972)
14. D.C. Smith, Proc.IEEE $\underline{65}$, 1679 (1977)
15. ——— , IEEE J. Quant.Electron. $\underline{QE-5}$, 600 (1969)
16. F.G. Gebhardt and D.C. Smith, IEEE J. Quantum Electron. $\underline{QE-7}$,
    63 (1971)
17. F.G. Gebhardt, Appl.Optics $\underline{15}$, 1479 (1976)
18. A. Yariv, in Adaptive Optics and Short Wavelength Sources, eds.
    S.F. Jacobs, M.Sargent III, M.O. Scully, Addison Wesley, Reading
    (1978) pp.175-216
19. E. Garmire, T. McMahon and M. Bass, IEEE J. Quantum Electron.
    $\underline{QE-16}$, 23 (1980)
20. ——— , Appl.Opt. $\underline{15}$, 145 (1976)
21. ——— , Appl.Phys.Lett. $\underline{29}$, 254 (1976)
22. ——— , Appl.Phys.Lett. $\underline{31}$, 92 (1977)
23. ——— , Appl.Phys.Lett. $\underline{34}$, 35 (1979)
24. E. Garmire, Appl.Opt. $\underline{15}$, 3037 (1976)
25. N. Nishihara, T. Inare and J.Koyama, Appl.Phys.Lett. $\underline{25}$, 291
    (1974)
26. T. McMahon, E.Garmire and M. Bass, Opt.Lett. $\underline{3}$, 235 (1978)
27. H. Krammer, Appl.Opt. $\underline{17}$, 316 (1978)
28. M.E. Marhic, L.I.Kwan and M. Epstein, Appl.Phys.Lett. $\underline{33}$, 609
    (1978)
29. ——— , Appl.Phys.Lett. $\underline{33}$, 874 (1978)
30. L.W. Casperson and T.S. Garfield, IEEE J.Quant.Electr. $\underline{QE-15}$,
    491 (1979)

31.F. Crescenzi, A. Cutolo, G. Panariello, R. Pierri and S. Solimeno
    2nd Nat.Conf. on Appl.Electromagn., June (1980) Bari
32.F. Crescenzi, A. Cutolo, P. Gay, I. Pinto and S. Solimeno, Europ.
    Conf. on Opt.Syst. and Applications, Ultrecht, Sept. (1980)
33.R.G. Kouyoumjan and P.H.Pathak, Proc. IEEE 62, 1448 (1978)

Appendix A. Listing of the program POSE I

```
C         IT CALCULATES THE FOCAL RADIANCE OF AN OPTICAL SYSTEM STARTING
C         FROM THE INCIDENT FIELD FOR ASSIGNED GEOMETRY AND INCIDENT FIELD
          REAL C(15),U(15),T(15),R(15),DX(15),DY(15),TX(15),TY(15),TZ(15)
          REAL CO,LAMBDA,WW(64,64)
          INTEGER N,P(64,64)
          LOGICAL L
          COMPLEX UU(64,64),UI
          COMMON C,U,T,R,DX,DY,TX,TY,TZ,CO,N
          COMMON L1,CX1,CY1,CZ1,L2,XO,YO,L3,TO,L4,F7,L7,A,B,L10,LAMBDA,INIZ
          COMMON DELTA,FFT1,IBACK,F2,WW,P,F,XR,YR,ZR,COR,CXR,CYR,CZR,F1,UU
          EXTERNAL UI
    1     FORMAT (I2)
    2     FORMAT(F20,6)
    3     FORMAT(2F20,6)
    4     FORMAT(3F20,6)
    5     FORMAT(2F20,6)
    6     FORMAT(I3)
   20     FORMAT(1HO,30X,'DESIGN PARAMETERS OF THE OPTICAL SYSTEM')
   21     FORMAT(1HO,'INDEX',10X,'CURVATURE',4X,'REFRACTIVE INDEX',4X,
          'AXIAL-DISTANCE',5X,'APERTURE RADIUS',2X,'X-DECENTRALIZATION',
          2X'Y-DECENTRALIZATION)
   22     FORMAT(1X,I2,6F20,6)
   23     FORMAT(1HO,'INDEX',9X,'X-TILT',14X,'Y-TILT',14X,'Z-TILT')
   24     FORMAT(1X,I2,3F20,6)
   25     FORMAT(1HO,'AXIAL DISTANCE OF THE PROPAGATING RAY',13X,'RADIA-
          TION-WAVELENGTH')
   26     FORMAT(1X,'TO=',F20,6,30X,'LAMBDA=',F20,6)
   27     FORMAT(1HO,'DIRECTION COSINES OF RAY CNGRUENCE')
   28     FORMAT(1X,'CX1=',F20,6,'CY1=',F20,6,'CZ1=',F20,6)
   29     FORMAT(1HO,'OBJECT POINT COORDINATES',)
   30     FORMAT(1X,'XO=',F20,6;10X,'YO=',F20,6)
   31     FORMAT(1HO,'NUMBER OF RAYS IN ONE DIMENSION')
   32     FORMAT(1X,'NU=',I3)
   33     FORMAT(1HO,'REFERENCE RAY FAILS TO HIT EXIT PLANE')
   34     FORMAT(1HO,'FP INDETERMINATE')
C         READS DESIGN PARAMETRS OPTICAL SYSTEM
          READ(5,I)N
          READ(5,2)(C(I),I=1,N),(U(I),I=1,N),(T(I),I=1,N),(DX(I),I=1,N)
          -(DY(I),I=1,N),(TX(I),I=1,N),(TY(I),I=1,N),(TZ(I),I=1,N)
          WRITE(6,20)
          WRITE(6,21)
          WRITE(6,22)  (I,C(I),U(I),T(I),DX(I),DY(I),I=1,N)
```

```
        WRITE(6,23)
        WRITE(6,24) (I,TX(I),TY(I),TZ(I),I=1,N)
        WRITE(6,25)
        WRITE(6,26) TO,LABDA
        WRITE(6,27)
        WRITE(6,28) CXI,CYI,CZI
        GO TO 200
  100   READ(5,5) XO,YO
        WRITE(6,29)
        WRITE(6,30) XO,YO
  200   READ(5,6) NU
        WRITE(6,31)
        WRITE(6,32)NU
C       REFERENCE RAY TRACING
        CALL RIFRAY (L,N,R)
        IF(.NOT.L) GO TO 300
C       EVALUATES THE FOCAL POINT
        CALL RICPF(XR,ZR,YR,CXR,CZR,XP,YP,ZP,L)
        IF(.NOT.L) GO TO 400
C       COMPUTES PUPIL AND ABERRATION FUNCTION
        CALL TOT(RSR,RU,XP,YP,ZP,R,NU)
C       EVALUATES THE FIELD ON THE EXIT PLANE
        CALL CPU(UI,NU)
        CALL FPI(RSR,RU,NU)
        STOP
  300   WRITE(6,33)
        STOP
  400   WRITE(6,34)
        END

        SUBROUTINE RIFRAY(L,N,R)
        REAL VR(5),R(15)
        INTEGER IR(15)
        LOGICAL L
        COMMON L1,CX1,CY1,CZ1,L2,XO,YO,L3,TO,F,XR,YR,ZR,COR,CXR,CYR,CZR
        COMMON O1(15),O2(15),O3(15),O4(15),O5(15),O6(15),O7(15),CO
    1   FORMAT(1HO,'REFERENCE RAY ON THE EXIT PLANE')
    2   FORMAT(1X,'XR=',F20.6,5X,'YR=',F20.6,5X,'ZR=',F20.6)
    3   FORMAT(1HO,'DIRECTION COSINES OF THE REFERENCE RAY AT THE
       -EXIT PLANE')
    4   FORMAT(1X,'CXR=',F20.6,5X,'CYR=',F20.6,'CZR=',F20.6)
    5   FORMAT(1HO,'REFERENCE RAY OPTICAL PATH')
    6   FORMAT(1X,'COR=',F20.6)
        I=1
```

```
      DO 100, J=1,N
      IF(R(J).GT.O) GO TO 100
      VR(I)=VR(J)
      IR(I)=0
      R(J)=1.E18
      I=I+1
100   CONTINUE
      X=0.
      Y=0.
      Z=0.
      IF(TO.GT.O) GO TO 200
      CX=CXI
      CY=CYI
      CZ=CZI
      CO=CXxX+CYxY+CZxZ
      GO TO 300
200   ALFA=(X-XO)/TO
      BETA=(Y-YO)/TO
      CZ=1./SQRT(1+ALFA²+BETA²)
      CX=ALFAxCZ
      CY=BETAxCZ
      CO=SQRT((X-XO)²+(Y-YO)²+TO²)
300   CAL RAY(X,Y,Z,CX,CY,CZ,L)
      IF(.NOT.L) RETURN
      XR=X
      YR=Y
      ZR=Z
      CXR=CX
      CYR=CY
      CZR=CZ
      COR=CO
      WRITE(6,1)
      WRITE(6,2) XR,YR,ZR
      WRITE(6,3)
      WRITE(6,4) CXR,CYR,CZR
      WRITE(6,5)
      WRITE(6,6) COR
      IJ=I-1
      DO 400 J=1,IJ
      IRR=IR(J)
400   R(IRR)=VR(J)
      RETURN
      END
```

```
      SUBROUTINE RAY(X,Y,Z,CX,CY,CZ,L)
      INTEGER N
      LOGICAL L
      REAL C(15),U(15),T(15),R(15),DX(15),DY(15)
      COMMON C,U,T,R,DX,DY,TX,TY,TZ,CO,N
      DO 1500 I=1,N
      X=X-DX(I)
      Y=Y-DY(I)
      Z=O.
      IF(TX(I).EQ.O) GO TO 100
      YTT=Y
      Y=YxCOS(TX(I))
      Z=YTTxSIN(TX(I))
      CYTT=CY
      CY=CYxCOS(TX(I))-CZxSIN(TX(I))
      CZ=CYTTxSIN(TX(I))+CZxCOS(TX(I))
      DD=Z/CZ
      X=CX-XxDD
      Y=Y-CYxDD
      Z=O.
      CO=CO-ABS(U(I-1))xDD
200   IF(TZ(I).EQ.O) GO TO 300
      XTT=X
      X=XxCOS(TZ(I))-YxSIN(TZ(I))
      Y=XTTxSIN(TZ(I))+YxCOS(TZ(I))
      CXTT=CX
      CX=CXxCOS(TZ(I))-CYxSIN(TZ(I))
      CY=CXTTxSIN(TZ(I))+CYxCOS(TZ(I))
300   T1=C(I)
      T2=CZ-C(I)x(XxCX+YxCY)
      T3=C(I)x(XxX+YxY)
      T4=1-T1xT3/(T2xT2)
      IF(T4.LT.O) GO TO 1600
      IF(C(I).EQ.O) GO TO 600
      Q=T3/T2x(1+SQRT(T4))
      IF(I.EQ.1) GO TO 400
      CO=CO+ABS(U(I-1))xQ
      GO TO 500
400   CO=CO+Q
500   X=X+CXxQ
      Y=Y+CYxQ
      Z=CZxQ
      CXN=-C(I)xX
      CYN=-C(I)xY
```

```
        CZN=1
  700   SPH=(XxX+YxY+ZxZ)xC(I)-2xZ
        RR=SQRT(XxX+YxY)
        IF(R(I)) 900,800,800
  800   IF(R(I)-RR) 1600,1000,1000
  900   IF(-R(I)-RR) 1000,1600,1600
 1000   IF(I.EQ.1) GO TO 1100
        V=U(I-1)/U(I)
        GO TO 1200
 1100   V=1/U(I)
 1200   IF(V.EQ.1) GO TO 1400
        G=SQRT(CXNxCXN+CYNxCYN+CZNxCZN)
        IF(V.EQ.-1) GO TO 1300
        T5=1-(GxGx(VxV-1))/(VxVxT2xT2xT4)
        IF(T5.LT.0) GO TO 1600
        W=VxT2xSQRT(T4xT5)
        A=(W-VxT2xSQRT(T4))/GxG
        CX=VxCX+AxCXN
        CY=VxCY+AxCYN
        CZ=VxCZ+AxCZN
        GO TO 1400
 1300   CX=CX-(2xT2xSQRT(T4)/(GxG))xCXN
        CY=CY-(2xT2xSQRT(T4)/(GxG))xCYN
        CZ=CZ-(2xT2xSQRT(T4)/(GxG))xCZN
 1400   SS=CXxCX+CYxCY+CZxCZ
        IF(I.EQ.N) GO TO 1500
        D=(T(I)-Z)/CZ
        CO=CO+ABS(U(I))xD
        X=X+CXxD
        Y=Y+CYxD
        Z=0
 1500   CONTINUE
        L=.TRUE.
        RETURN
 1600   L=.FALSE.
        RETURN
        END

        SUBROUTINE RICPF(XR,YR,ZR,CXR,CYR,CZR,XP,YP,ZP,L)
        LOGICAL L
        COMMON C,U,T,R,DX,DY,TX,TY,TZ,CO,N
        COMMON L1,CX1,CY1,CZ1,L2,XO,YO,L3,TO,L4,FS
    1   FORMAT(1X,'COORDINATES OF THE FOCAL POINT')
    2   FORMAT(1X,'XP=',F20.6,5X,'YP',F20.6,5X,'ZP=',F20.6)
    3   FORMAT(1X,'THE RAY IS LOST')
```

```
  4   FORMAT(1X,'THE RAYS DO NOT INTERSECT, FP INDETERMINATE')
      I=1
      DO 100, J=1,N
      IF(R(J).GT.0) GO TO 100
      VR(I)=R(J)
      IR(I)=J
      R(J)=1.E18
      I=I+1
100   CONTINUE
      X=0
      Y=0.01
      Z=0
      IF(TO.GT.0) GO TO 200
      CX=CX1
      CY=CY1
      CZ=CZ1
      CO=XxCX+YxCY+ZxCZ
      GO TO 300
200   ALFA=(X-XO)/TO
      BETA=(Y-YO)/TO
      CZ=1/SQRT(1+ALFA²+BETA²)
      CX=ALFAxCZ
      CY=BETAxCZ
      CO=TO/CZ
300   CALL RAY(X,Y,Z,CX,CY,CZ,L)
500   IF(.NOT.L) GO TO 600
      FS=Y/0.01
      CALL INTER(XR,YR,ZR,CXR,CYR,CZR,X,Y,Z,CX,CY,CZ,XP,YP,ZP,L)
      IF(.NOT.L) GO TO 700
      WRITE(6,1)
      WRITE(6,2) XP,YP,ZP
      GO TO 800
600   WRITE(6,3)
      GO TO 800
700   WRITE(6,4)
      IJ=I-1
800   DO 900 J=1,IJ
      IRR=IR(J)
      R(IRR)=VR(J)
900   CONTINUE
      RETURN
      END
```

```
      SUBROUTINE INTER(X1,Y1,Z1,CX1,CY1,CZ1,X2,Y2,Z2,CX2,CY2,CZ2,XO,YO,Z)
      LOGICAL L
1     FORMAT(1X,'THE TWO RAYS ARE OBLIQUE',10X,'COMP=',F20.6)
2     FORMAT(1X,'THE TWO RAYS ARE PARALLEL')
3     FORMAT(1X,'CX1=',F20.6,5X,'CY1=',F20.6,5X,'CZ1=',F20.6)
4     FORMAT(1X,'CX2=',F20.6,5X,'CY2=',F20.6,5X,'CZ2=',F20.6)
      COMP=(X1-X2)x(CY1xCZ2-CY2xCZ1)-(Y1-Y2)x(CX1xCZ2-CX2xCZ1)-
     -(Z1-Z2)x(CX1xCY2-CX2xCY1)
      IF(COMP.LE.0) GO TO 200
      IF(CX1.EQ.CX2).AND.(CY1.EQ.CY2).AND.(CZ1.EQ.CZ2) GO TO 200
      PUN=CY2xCZ1-CZ2xCY1
      IF(PUN.EQ.0) GO TO 200
      L=.TRUE.
      NUM=CY2x(Z2-Z1)-CZ2x(Y2-Y1)
      PAR=NUM/PUN
      XC=X1+CX1xPAR
      YC=Y1+CY1xPAR
      ZC=Z1+CZ1xPAR
      RETURN
100   L=.FALSE.
      WRITE(6,1) COMP
      RETURN
200   L=.FALSE.
      WRITE(6,2)
      WRITE(6,3) CX1,CY1,CZ1
      WRITE(6,4) CX2,CY2,CZ2
      RETURN
      END

      SUBROUTINE CPU(UI,NU)
      COMPLEX UU(64,64),FP,C
      DIMENSION S1(15),S2(15),S3(15)
      COMMON S1,S2,S3,R,L10,LAMBDA,L7,A,B,INIZ,DELTA,F2,WW,P,F1,UU
      AB=3.14/BxB)
      G=3.14xB
      DD=2xG/NU
      DO 200  I=1,N
      Y=G-(I-1.5)xDD
      Y1=R(1)-(I-0.5)xDELTA
      DO 100  J=1,NU
      X=G-(J-0.5)xDD
      RIAL=FLOAT(P(I,J))xCOS(6.28xWW(I,J)/LAMBDA)
      AIMM=FLOAT(P(I,J))xSIN(6.28xWW(I,J)/LAMBDA)
      FP=CMPLX(RIAL,AIMM)
      BB=ABx(XxX+YxY)
```

```
      C=CMPLX(0,BB)
      UU(I,J)=FPxUI(X1,Y1)xCEXP(C)
100   CONTINUE
200   CONTINUE
      RETURN
      END

      SUBROUTINE TOT(RSR,RU,XP,YP,ZP,R,NU)
      LOGICAL L
 1    FORMAT(1H0,'REFERENCE SPHERE RADIUS',5X,'RSR=',F20.6)
 2    FORMAT(1H0,'RAY GRID SPACING',5X,'DELTA=',F20.6)
 3    FORMAT(1H0,'SCALE FACTOR',5X,'FS=',F20.6)
 4    FORMAT(1H0,'I','J',14X,'X1','Y1','XU','YU','CX','CY','CZ','WWL')
 5    FORMAT(1X,I2,1X,I2,8F20.6)
 6    FORMAT(11X,'NUMBER OF TRACED RAYS','NUMR=',I8)
 7    FORMAT(1H0,'ABERRATION FUNCTION VARIANCE',5X,'VARWW=',F20.6)
 8    FORMAT(1X,'AVARAGE VALUE OF THE ABERRATION FUNCTION',WWM='F20.6)
 9    FORMAT(1H0,'RMS=',F20.6)
10    FORMAT(1H0,'OPTICAL SYSTEM IS ABERRATED')
11    FORMAT(1H0,'OPTICAL SYSTEM IS DIFFRACTION LIMITED')
12    FORMAT(1H0,'EXIT PUPIL VANISHES')
13    FORMAT(1H0,'RU=',F20.6)
14    FORMAT(1H0,'FNUM=',F20.6)
      RSR=SQRT((XR-XP)**2+(YR-YP)**2+(ZR-ZP)**2)
      NUMR=0
      DELTA=2xR(1)/NU
      DO 200  I=1,NU
      DO 100  J=1,NU
      WW(I,J)=0
100   CONTINUE
200   CONTINUE
      WRITE(6,1) RSR
      WRITE(6,2) DELTA
      WRITE(6,3) FS
      WRITE(6,4)
      DO 700 I=1,NU
      Y=R(1)-(I-0.5)xDELTA
      DO 600 J=1,NU
      XR=R(1)-(I-0.5)xDELTA
      Z=0
      RAG=SQRT(X**2+Y**2)
      IF(RAG.GT.33) GO TO 300
      CX=CX1
      CY=CY1
      CZ=CZ1
```

```
          CO=CXxX+CYxY+CZxZ
          GO TO 400
   300    COMMON ALFA,BETA,CX,CY,CZ,CO
   400    XT=X
          YT=Y
          CALL RAY(XT,YT,Z,CX,CY,CZ,L)
          IF(.NOT.L) GO TO 500
          NUMR=NUMR+1
          P(I,J)=1
          WW(I,J)=CO-COR-RSR+SQRT((XT-XP)²+(YT-YP)²+ZP²)
          GO TO 600
   500    P(I,J)=0
   600    CONTINUE
   700    CONTINUE
          WWM=0
          DO 800  I=1,NU
          DO 800  J=1,NU
          WWM=WWM+WW(I,J)
   800    CONTINUE
          WWM=WWM/NUMR
          VARWW=0
          DO 1000 I=1,NU
          DO 1000 J=1,NU
          VARWW=VARWW+(WW(I,J)-WWM)²/NUMR
  1000    CONTINUE
          RMSWW=SQRT(VARWW)
          WRITE(6,6) NUMR
          WRITE(6,7) WWM
          WRITE(6,8) VARWW
          WRITE(6,9)RMSWW
          GO TO 1300
  1300    DO 1500 I=1,NU
          DO 1500 J=1,NU
          IF(P(I,J).EQ.1) GO TO 1600
  1500    CONTINUE
          WRITE (6,12)
          STOP
  1600    RU=ABS((R(1)-(I-1)xDELTA)xFS))
          FNUM=RSR/(2xRU)
          WRITE(6,13) RU
          WRITE(6,14) FNUM
          A=-RU
          B=RU
          RETURN
          END
```

METAL PROCESSING AT CULHAM

I.J. Spalding

UKAEA Culham Laboratory
Abingdon, Oxon, OX14 3DB

LASERS

The Laser Applications Group at Culham utilizes a wide range of commercially-available and in-house CO_2 laser systems, covering the power range 5-15,000 Watts (continuous). The sub-kilowatt systems utilize stable optical resonators, providing Gaussian-mode outputs, and some of these have high mean-power pulsing facilities at repetition rates up to 1 kH_z. The highest power in routine use is the 5-6kW unstable-cavity (M=2 $\sim$3) output from the transverse-flow CL5 laser, discussed in the first lecture. Detailed numerical simulations of the behaviour of this laser have been undertaken: Figure 1 illustrates the good agreement obtained between the computed small-signal gain α_o (full curve) and that measured experimentally as a function of gas flow velocity and the distance z downstream of the electrode (Armandillo and Kaye, 1980). To obtain this agreement CO_2 dissociation of $\sim$50% has been assumed. Lower dissociation (and higher gain) are observed immediately after switching on; here the Culham code agrees with the numerical predictions and experimental observations reported by Yoder et al (1978). Representative near and far-field burns from this laser are shown in Figure 2. (More quantitative measurements are made using high-speed IR cameras on irradiated targets. With f/4 spherical mirror optics not less than 80% of the power is focussed through an aperture of diameter 0.3mm onto a calorimetric detector such as a 'Joule Stick'). Three of these systems are being used within the UKAEA; they have operated routinely for a cumulative total of $\sim$ 7000 hours.

LASER SYSTEMS

A wide range of complete (prototype) laser-systems have also been built to demonstrate important aspects of the technique, and to meet specific industrial requirements. Typical examples are:

i.   a pipe-cutter (in which the cutting head rotates about the stationary workpiece, (Figure 3);
ii. an articulated arm, coupling (manually or by robot) a stationary laser to a three-dimensional workpiece (Figure 4);
iii. the CL5 laser-processing centre (Figure 5).

The CL5 system was demonstrated as a fully-operational two work-station welding system at the International Welding Exhibition (WELDEX) in Birmingham, September 1979. It is now used daily at Culham as a three workstation facility, using a mini computer to control beam-switching, high-speed materials-processing, and to provide a full range of management information and systems-status monitoring. Considerable attention has been paid to personnel-safety, to conform with statutory UK Health and Safety requirements.

LASER APPLICATIONS

A wide range of industrially-funded work on non-metallic work-pieces is undertaken. An example of our sub-kilowatt work on non-metals is the automatic production of rotogravure (printing) cylinders, in collaboration with Crosfields Ltd. At these powers thin metals can also be welded, such as silicon steel transformer laminations and the stainless-steel tube to tube-plate orbital welds shown in Figure 6. At higher powers much thicker samples can, of course, be welded out of vacuum with very low distortion. A cross-section of a double-sided 5kW weld in 16mm titanium is illustrated in Figure 7, and a pressure-vessel test on single-pass welds in 8-10mm stainless-steel in Figure 8. Laser-cut, or cropped, samples occasionally require filler-wire, such as the mild-steel samples illustrated in Figure 9.

Perhaps one of the clearest contrasts between electron-beam and laser techniques is provided by work we have pioneered in collaboration with the Dounreay Nuclear Power Development Establishment (DNE) over the past ten years or so (Higginson and Campbell, 1980). The laser-beam can be 'piped' over long distances in air, into relatively inaccessible and hostile environments to remotely machine components (with minimal mechanical stress). Figure 10 shows schematically one (of two) 450W CO_2 lasers being used for this purpose outside a post-irradiation cell at DNE. The beam is transmitted through the 2m lead-concrete cell walls via special seals, and is then focused onto the 3m long hexagonal outer wrapper of a Prototype Fast Reactor (PFR) fuel sub-assembly, thus providing access to the spent fuel elements, which are continuously cooled with molten sodium. A wide variety of specialized heat-treatment work is also undertaken (Hill et al 1974, Trafford et al 1979). In conclusion, the development of reliable multikilowatt CO_2 lasers consuming relatively little He has led to

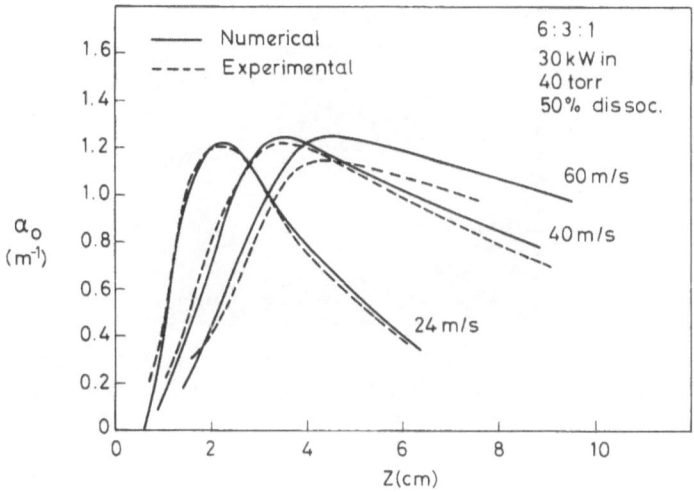

Fig 1.   Small-signal gain (α_0) as a function of flow velocity and
distance downstream (z); gas mix 6:3:1 ($He/N_2/CO_2$), at
50% dissociation.   Full curve, numerical; broken curve,
experimental (after Armandillo and Kaye, 1980).

their increasing use in certain high-technology or mass-production
applications in industry. The concept of the flexible (laser)
machine tool seems likely to find increasing application in most
industrialized countries over the next decade.

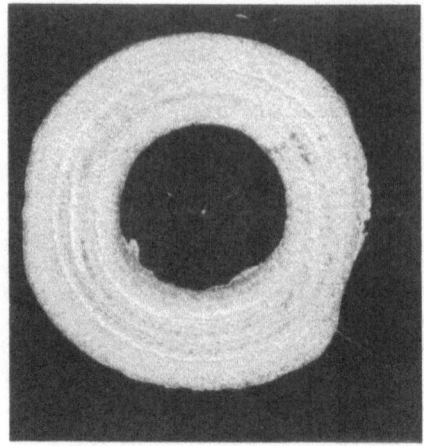

Fig 2(a)   Near-field burn pattern in lucite; produced by short ($\sim\frac{1}{2}$s)
           exposure to Culham 5kW CL5 CO_2 laser, using a positive-
           confocal (M = 2) 'unstable' resonator.

Fig 2(b)   Far-field burn pattern produced by laser-beam of Fig 1(a).
           The axis of symmetry is vertical.

Fig 3. Laser pipe-cutting machine, handling tube diameter of 12–150mm with wall thickness up to 4–5mm. (Lengths cut from few mm to ∿ 500mm to accuracy of ± 0.25mm with cut width of ∿ 0.1mm, burr-free).

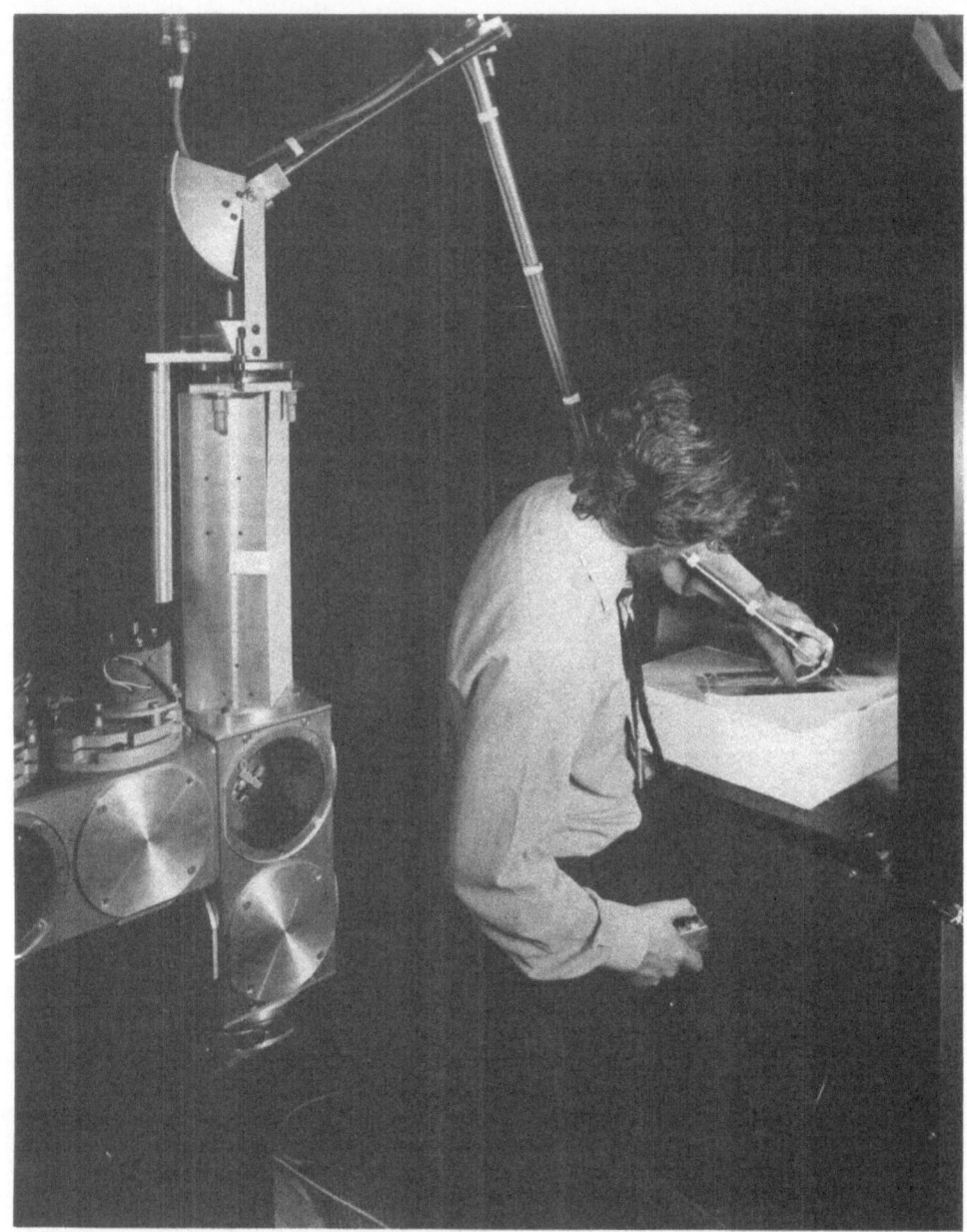

Fig 4.   400 W flexible laser beam guide (12mm bore).

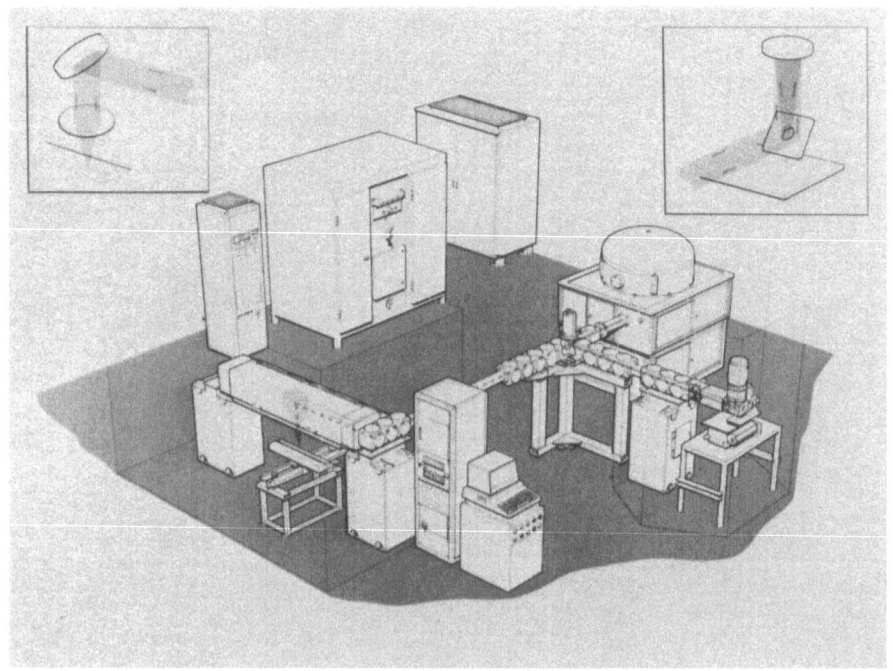

Fig 5.   Culham CL5 laser-processing centre. (Two work-stations
         illustrated, with 'Digital LSI11' mini computer, VDU,
         1 mega-byte floppy disc storage, and Camac interface).

Fig 6.   Tube to tube-plate weld at 400 W.

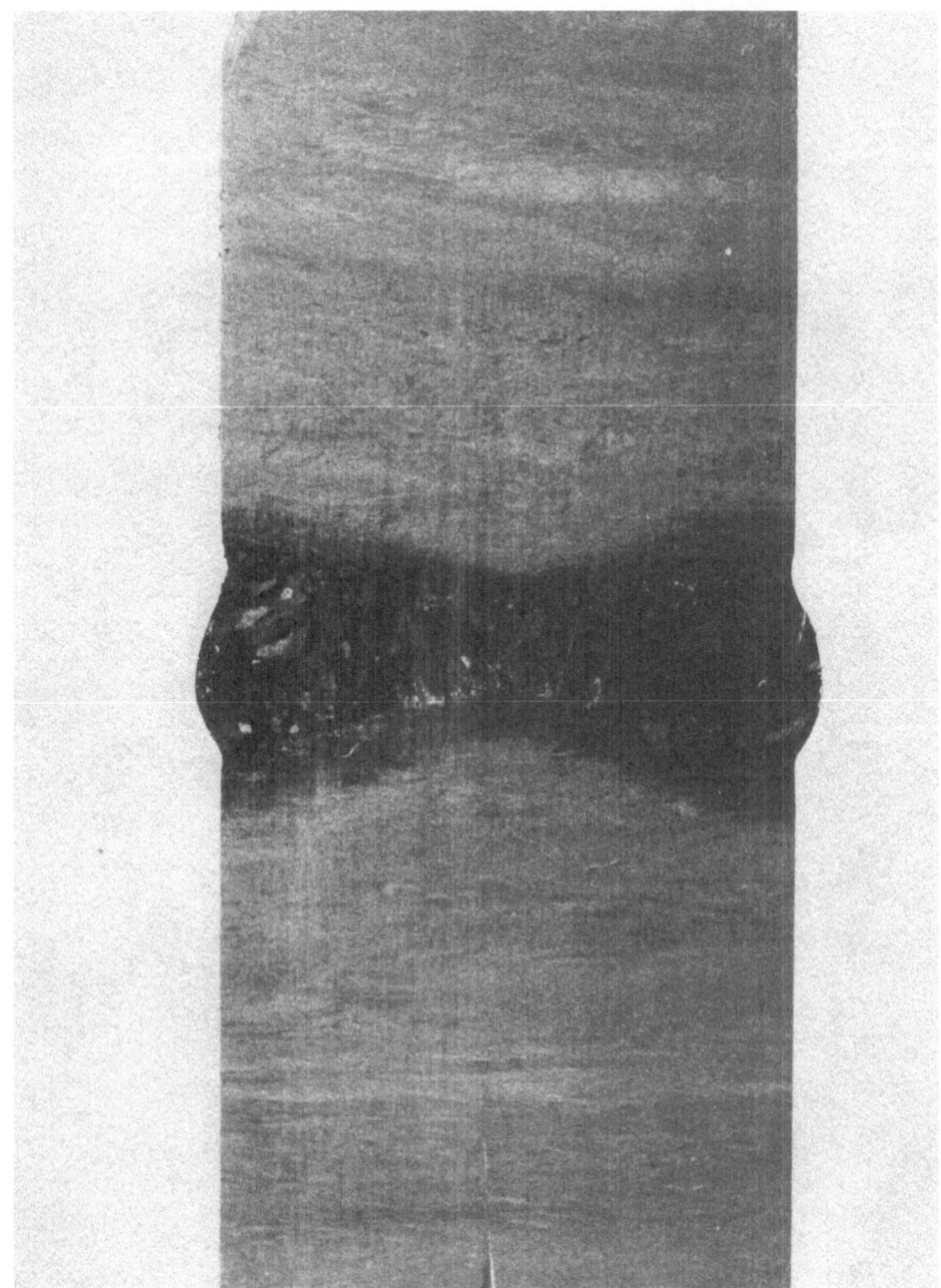

Fig 7.   Cross section of double-sided weld in 16mm T_i, made with
         the Culham CL5 laser.

Fig 8.   Pressure-vessel test on CL5 laser-weld in 9mm stainless-
         steel: the crack initiates at a stress-raiser and runs
         cleanly across the weld.

(a)

(b)

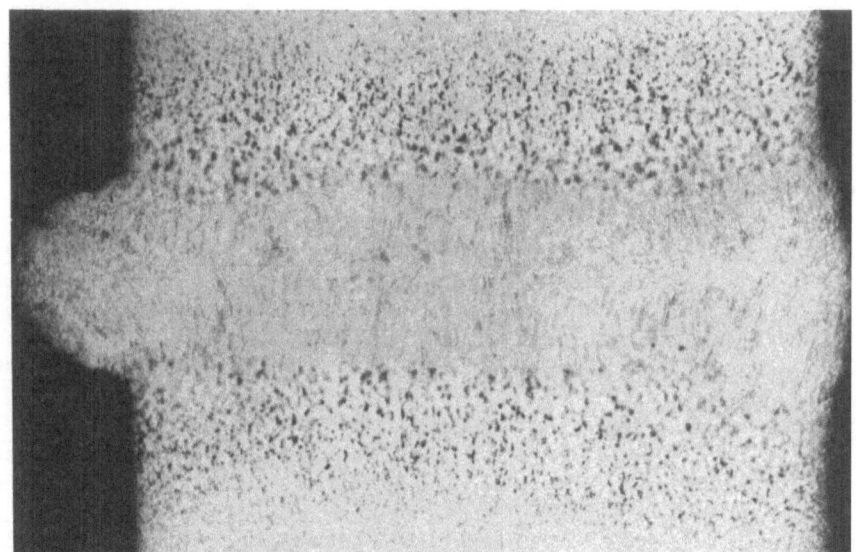

Fig 9.   CL5 weld on 6mm mild steel (a) uncropped, no filler,
         (b) cropped, with filler.

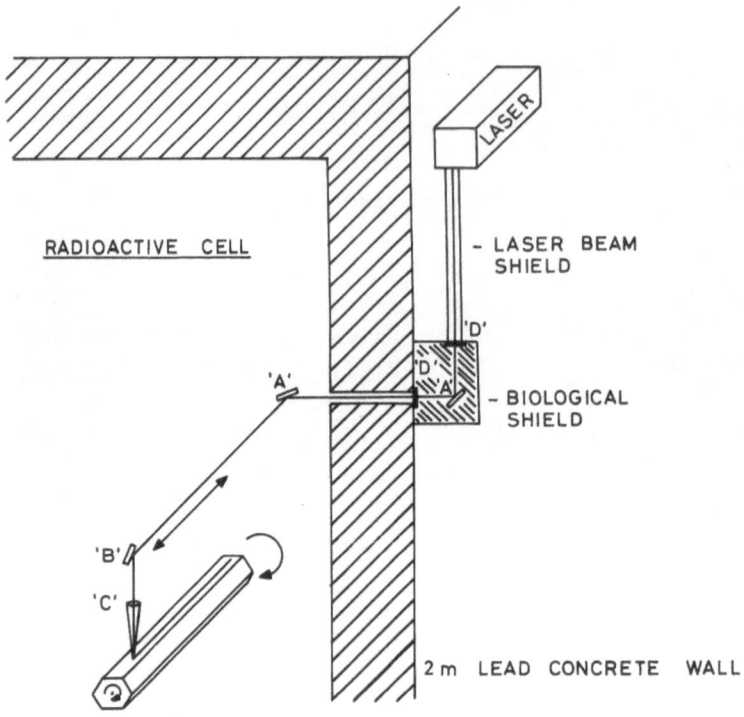

Fig 10.   Schematic diagram of laser cutting system at DNE.

REFERENCES

Armandillo, E. and Kaye, A.S., 1980, J. Phys. D:Appl. Phys. 13, 321.
Higginson, P.R. and Campbell, D.A., 1980, Paper 21 in Proceedings of
        B.N.E.S. Conference on Post Irradiation Examination,
        Grange-over-Sands, UK, May 1980.
Hill, J.W., Lee, M.J. and Spalding, I.J., 1974, Optics and Laser
        Technology, 6, 276.
Trafford, D.N.H., Bell, T., Megaw, J.H.P.C. and Bransden, A.S.,
        1979, Proceedings of 'Heat Treatment 1979' Conference,
        Birmingham 22-24 May 1979, Metals Society/Am. Soc. for
        Metals.
Yoder, M.J., Legner, W.H., Jacob, J.H. and Abhouse, D.R., 1978,
        J. Appl. Phys. 49, 3171.

TIME DEPENDENT EFFECTS IN LASER HEATING OF ORGANIC MATERIALS

F. T. Arecchi, C. Castellini, J. Tredicce

Istituto Nazionale di Ottica

Largo Enrico Fermi, 6 - Arcetri -Firenze,Italia

INTRODUCTION

During the last years, a special interest in heating, drilling and cutting of solids by medium and high power lasers was shown. From the theoretical point of view, the exact solution of the basic heat equation

$$\nabla^2 T - 1/K_e \frac{\partial T}{\partial t} = - \frac{A(x,y,z,t)}{K}$$

(1)

where   T = temperature
K_e = thermal diffusivity
K = thermal conductivity
t = time
A(x,y,z,t) : rate of heat per unit time per unit volume supplied to the solid

presents very difficult problems and can only be obtained in an analytic form when a variety of assumptions concerning the spatial and temporal dependance of the laser radiation and the geometry of the workpiece are made [1-2].

However, in many cases, those calculations can predict some experimental results, but only in a small range of laser intensities and for some metallic materials [3-4].

Our experimental work, described below, consists in a very accurate measurement of the necessary energy and time to drill different depths of non-metallic materials

like plexiglas and cotton and our results are not explained by the presently available simplified theoretical models.

EXPERIMENTAL ARRANGEMENT

In Fig. 1 we show the experimental set-up. Two beams,
one of which provided by a CO_2 laser (100 W nominal power,
TEM_{oo}) and the other by a He-Ne laser (5 mW) go through
a shutter. When the shutter is open the He-Ne beam is in
part sent over the photodiode 1, the signal of which sets
the time counter, and in part goes to the back-surface
of the workpiece. The photodiode 2 detects the reflection on the back-surface. The time counter consists of
an oscillator and a pulse counter.

The CO_2 laser beam, going through a 20 cm focal
length lens, drills the workpiece. At the moment when
the material is drilled, the signal of the photodiode 2
falls down and stops the time counter.

This system measures the time necessary to drill a
width "e" with an accuracy determined by the oscillator;
in our case, of about 1 μsec.

The CO_2 laser power is measured by a bolometer,
before the beam arrives to the shutter, with a relative
precision better than 1% .

The results are shown in the graph of Fig. 2. It is
possible to clearly see the existence of two energy minima which are not predicted by theoretical results.
Furthermore, these minima are aligned on two constant
power lines.

With this set-up, measurements were performed for
different widths of plexiglas and cotton, with a N_2 jet
and without, and results qualitatively similar were
found.

TENTATIVE EXPLANATION

We attempt here some qualitative interpretations
of our plots. We expect that for a given energy deposition at an "optimal" time, there is a local temperature
increase at the focal spot up to the melting point (or
to the transition, responsible for bonds breaking, which

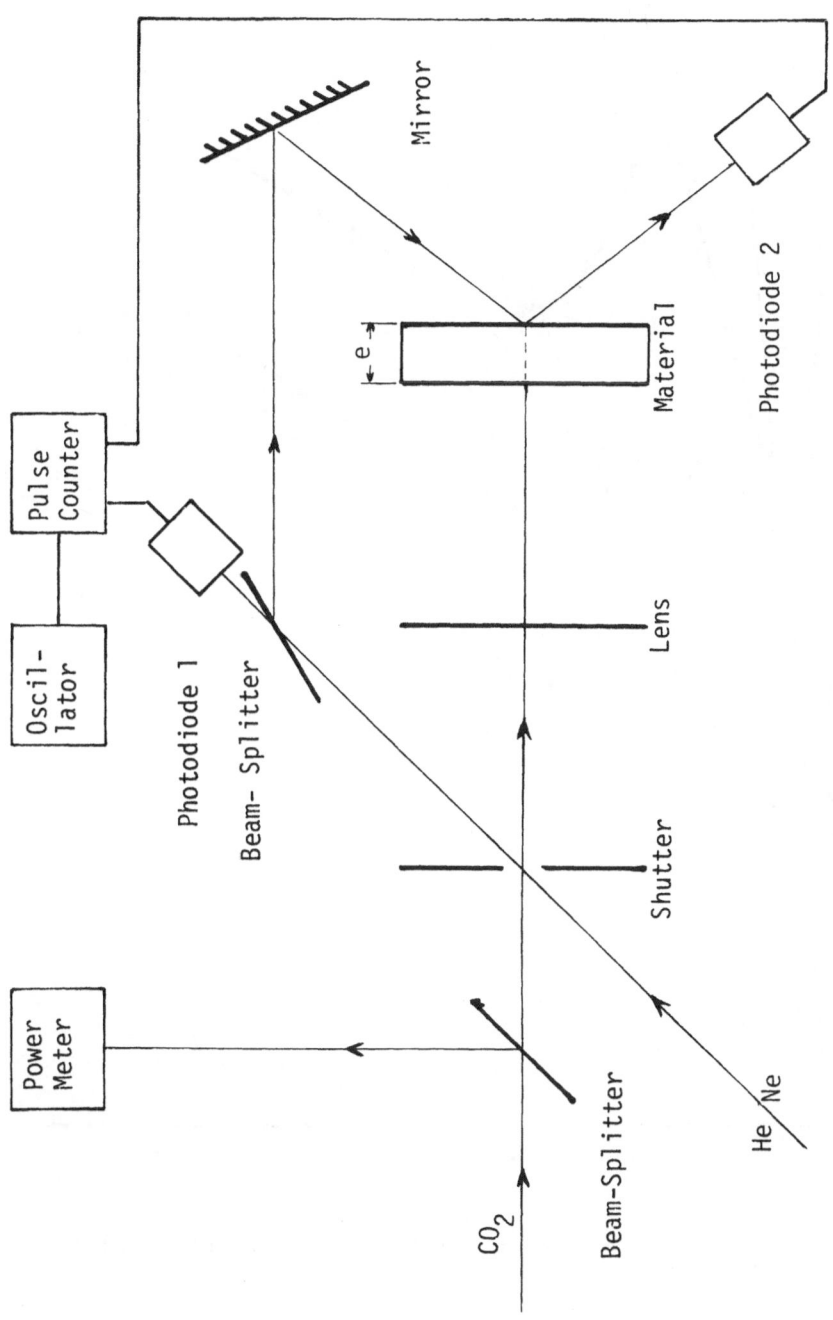

Fig.1 : Experimental arrangement

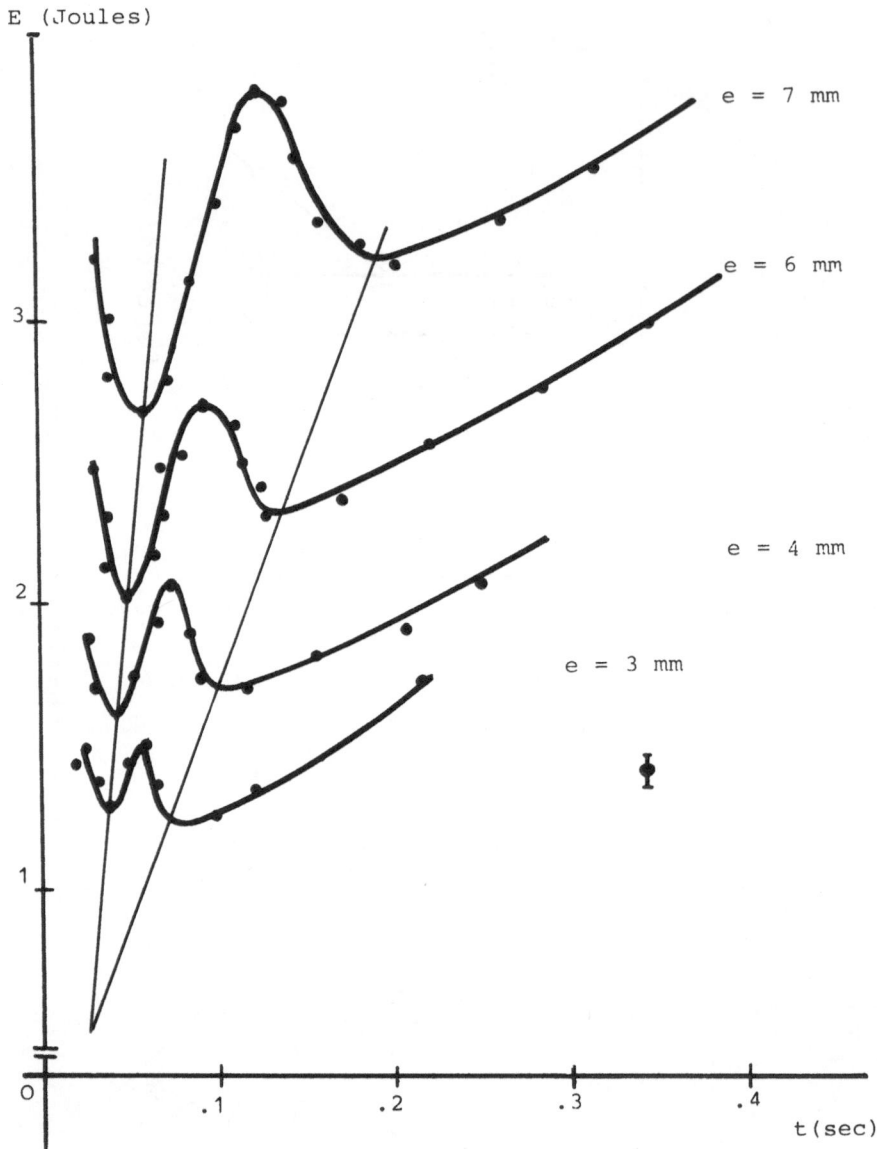

Fig.2 : Energy vs. time curves for different widths of plexiglas.

may not be necessarily a solid-liquid transition when the sample is made of long polymeric molecules) with the maximum temperature reached right at the end of the time pulse (*).

For longer times, we expect that the competitive heat diffusion process away from the focal region reduces the local temperature raise, hence requiring a higher energy deposition.

This might explain the long time branch of our plots.

For shorter times, probably, the high temperature frontier has not arrived at the end of the material, and when the laser pulse has passed by, the material is left with a drop of liquid phase, which has to find its way through the material by advancing with the liquid-solid frontier up to the end of the material. During this process however, part of the hot drop vaporizes bringing with itself some latent heat of vaporization, and hence increasing the energy requirement for drilling the hole.

This qualitative model would explain a single minimum, and would also justify the alignment of minima corresponding to different thicknesses over a constant power line. However, it does not absolutely explain why there are two minima.

A tentative explanation, which is being explored, is that the two minima correspond to the passage of two moving frontiers vapour-liquid and liquid-solid, and the liquid phase has optical and thermal parameters(absorption coefficient and thermal conductivity) sufficiently different from the solid one to justify a new optimum energy and hence a new minimum in the plot.

CONCLUSIONS

Observing the graphs, a different situation for very thin layers of materials can be predicted. In that sense, the intersection of the power lines, where the minima are aligned determines a point below which it is proba-

---

(*) For sake of simplicity in this model we think of a standard continuous material undergoing the ordinary phase transition solid to liquid to vapour.

ble to find only one minimum, due to the lateral diffusion of heat into the material.

       As the relative error on the energy is better than
5% and the present state of the theory cannot interpret
our measurements, a new model of the laser drilling is
necessary to understand them. However, the experimental
results are very important to make an appropriate use of
the C.W. lasers for heat treatment of organic materials.

       It is important to notice that, independently from
the interpretation, the presence of such effects has a
high relevance for laser use in surgery, since similar
time dependent energy effects must appear in IR laser
exposure of biological material.

REFERENCES

1. H. S. Carslaw and J. C. Jaeger, "Conduction of Heat
      in Solids", 2nd ed. Oxford Univ.Press (Clarendon),
      London and New York (1959).
2. U. C. Paek and F. P. Gagliano, IEEE J. Quantum Elec-
      tronics, QE-8, 112 (1972).
3. S. I. Anisimov, A. M. Bonch-Bruevich, M. A. El'yash-
      evich, Ya. A. Imas, N. A. Pavlenko, and G. S. Ro-
      manov, Sov. Phys.-Tech. Phys., 11, 945 (1967).
4. V. B. Braginskii, I. I. Minakova, and V. N. Rudenko,
      Sov. Phys.-Tech. Phys., 12, 753 (1967).

ION BEAM ANALYSIS OF NEAR SURFACE REGIONS

Salvatore Ugo Campisano

Istituto di Struttura della Materia dell'Università

Corso Italia 57 I95129 Catania,Italy

INTRODUCTION

The ion-atom collision is used as a tool to investigate so-lid surfaces[1-5] . Composition and its depth dependence can be determined up to a thickness of about 1 μm with a resolution of the order of 0.01 μm using Rutherford backscattering (RBS) technique. The reduction of the yield due to the alignment of a low index axis of a single crystal target with the analyzing beam direction [1,6] can be used to determine the presence of defects, their natu re and amount. Other effects of the ion-atom collision[1,5] as the emission of characteristics X-ray or of the product of a nuclear reaction, are used for analytical purposes too.

A typical experimental set-up[2] consists of a Van de Graaf electrostatic generator with 3 MV accelerating voltage and beam intensity of the order of $1-10^3$ nA with beam size of the order of 1 mm in diameter. Accelerated ions are generally H^+ or He^+. For channeling measurements the single crystal target is mounted on a goniometer system allowing its orientation with an accuracy of about 0.05°. Reverse biased p-n junctions are used to measure the energy of both particles and X-ray impinging on the detector. Ty pical energy resolutions are 15 keV for 5.5 MeV He^+ particles and 200 eV for 6 keV X-ray.

In the following we will illustrate briefly RBS and channe-ling effect techniques and will show their applications.

BACKSCATTERING

     The RBS technique is based on the detection of particles scat-
tered at large angles, after having experienced   elastic collision
with a target atom. Due to the conservation of energy and momentum, the energy of the scattered ion is related to its mass, energy before impact, scattering direction and mass of the scattering
atom[1] . The ratio of recoil energy E_1 to impinging energy E_o is
reported in Fig.1 as a function of the atomic mass of the target
atoms for proton helium and oxygen ions and for 165° scattering

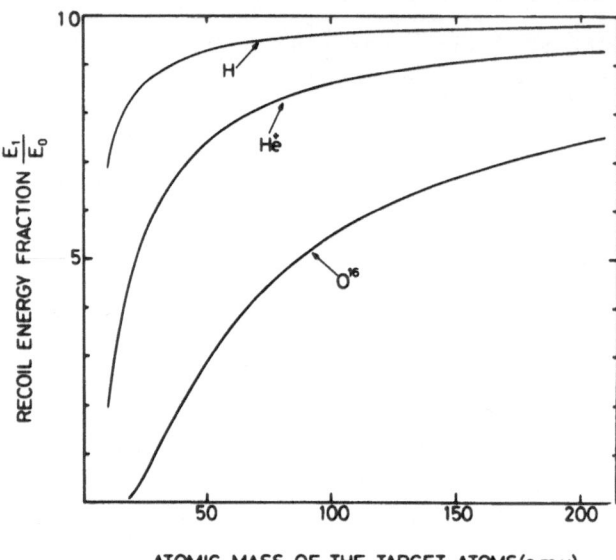

ATOMIC MASS OF THE TARGET ATOMS(amu)

Fig. 1. Recoil energy
        fraction as
        a function
        of the ato-
        mic mass of
        the target
        atom and for
        several ana-
        lyzing beam
        particles.
        The scattering
        angle is 165°.

angle. For He[+] ions the energy ratio function depends strongly on
the atomic mass of the target atom up to a value of about 75 (corresponding to arsenic). For heavier target atoms the energy ratio
saturates toward unity. For example in Fig. 2 is reported the energy spectrum of 2.5 MeV He[+] ions scattered from a target consisting
of an equal number of copper, silver and gold atoms. The three
peaks correspond to He[+] ions scattered for the different target
atoms. The peak labelled Cu is splitted into two sub-peaks corresponding to the isotopes of mass 63 and 65 a.m.u. In a thin target
then the measurement of the scattering energy allows the determination of the mass of the target atoms.

     The scattering yield increases with the square of the atomic

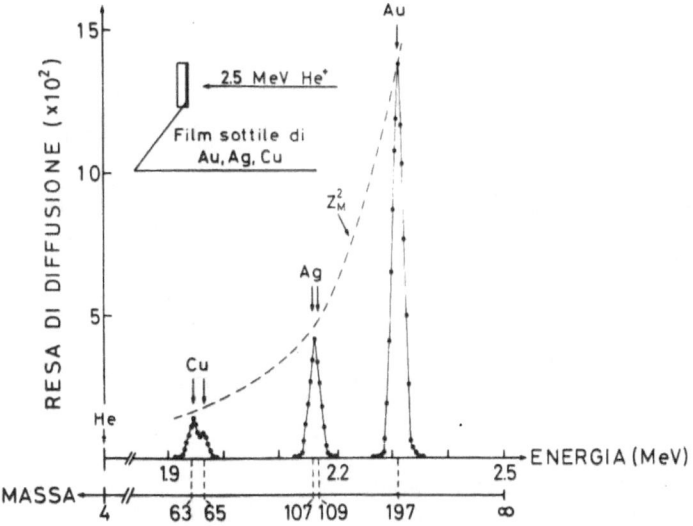

Fig. 2. Backscattering energy spectra of 2.5 MeV He⁺ beam from a
thin target containing an equal number of Cu, Ag and Au
atoms. The reading of the figure is in Italian.

number of the scattering atom as shown in the same Fig. 2 where the
peak due to scattering on Au atoms is larger than that of Ag, although the number of atoms in the target is the same.

The fraction of beam particles which experience a large angle
scattering is very small and most of the ions penetrate inside the
target. They suffer collisions with electrons and atoms and then
lose their energy before they can experience a small impact parameter collision. Such particles will be revealed then at energies
smaller than those scattered from surface atoms as shown in Fig. 3.
Being the stopping cross section a known function of the velocity
and atomic number of the ion and of atomic number of the target
atoms, this energy difference can be converted into a depth difference. RBS technique is then sensitive to the depth at which the
scattering event occurs. In Fig. 3 is shown schematically the energy spectrum of particles from a thin layer and the analytical relationships between energy width and thickness. The experimental
dependence of the energy spectrum on the film thickness is shown
in Fig. 4. The Pt film thickness ranges from 125 to 4000 Å and
the corresponding energy width ranges from 19 to 650 keV respectively.

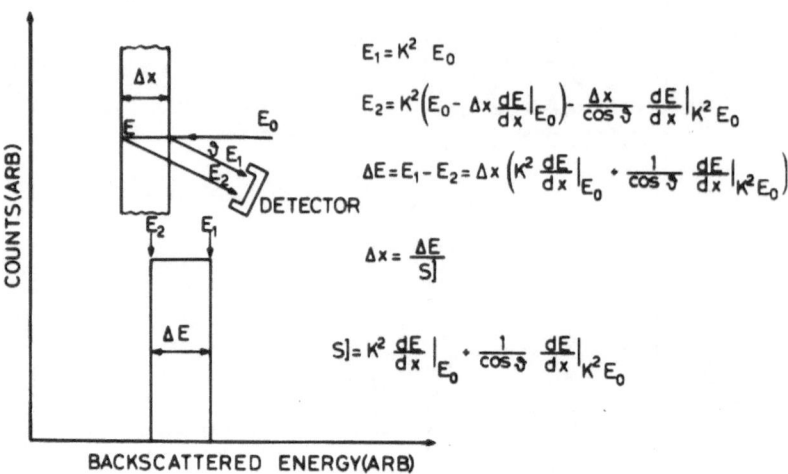

Fig. 3. Schematic backscattering energy spectrum from athin elemental target.   The analytical relationship between energy
width and film is reported.

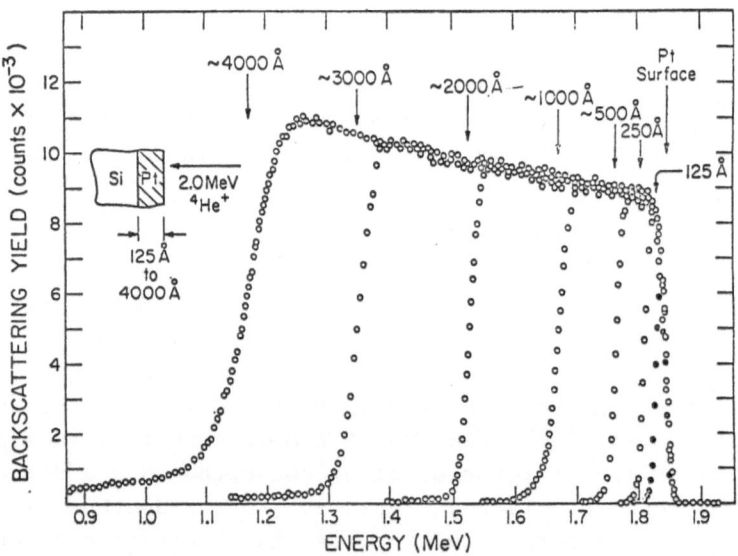

Fig. 4. Backscattering energy spectra of 2.0 MeV He[+] from Pt films
of various thicknesses ranging from 125 to 4000 Å.

In the case of atomic mixture of elements A and B, having different atomic mass the energy spectrum shows two different peaks,
as shown schematically in Fig.5. Particles can be scattered either by A or by B atom. The energy width is related to the total
thickness through the weighted average energy loss whilst the

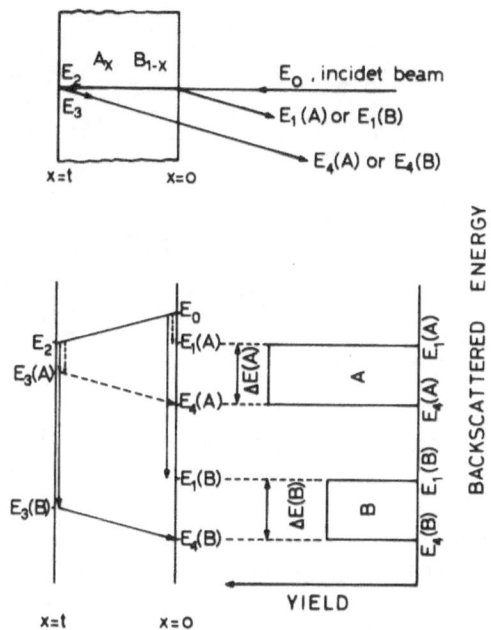

Fig. 5. Schematic backscattering
        spectrum from a thin
        composite target.

yield ratio is proportional to the atomic ratio of the corresponding elements. The technique is thus suitable for quantitative
measurements of atoms ratio.

A typical application of the RBS technique is the investigation of mass transport in thin film structures[3],[4]. In Fig. 6 is
reported an application to the Au-Al system[7]. The dashed line
in the left hand side spectrum corresponds to the as deposited bilayer structure and from the corresponding energy width the measured thickness were 900 Å of Al deposited onto 1100 Å of Au. After
annealing for 15 min at 140°C the RBS spectrum modifies as shown

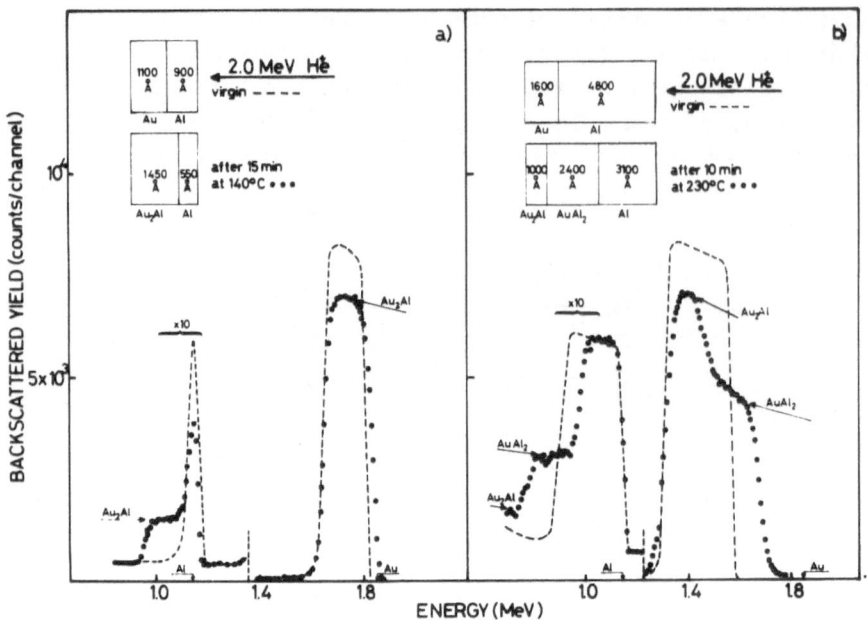

Fig. 6. Backscattering energy spectra of 2.0 MeV He⁺ from Au-Al
thin film bilayer after various thermal treatment.

by the dotted spectrum. The Au signal becomes wider and shorter
and the Al signal exhibit a plateaux in its low energy side. From
the thickness and height ratio we can determine the presence of
1450 Å of Au_2Al under 550 Å of unreacted Al. For higher temperature annealing $AuAl_2$ is formed from Au_2Al and unreacted Al, as
shown in the right hand side spectrum. The formed phases must
be, in any case, identified by a structure sensitive technique,
as X-ray diffraction, because RBS is sensitive only to atoms depth
location but not to their structure within a lattice.

CHANNELING

An energetic ion, entering a crystal in a direction parallel
to a low index axis, is gently channeled in its motion by the potential of the atomic row[8,9]. The yield for processes requiring
a small impact parameter collision are then strongly reduced. The
backscattering yield in aligned condition is 3-4% of the yield
corresponding to a misoriented crystal for Si along one open direction (<100>, <111> or <110> axes) and for 2 MeV He⁺ analyzing
beam, as shown schematically in Fig. 7. Atoms which are displaced
from their lattice sites into random position can instead be hit

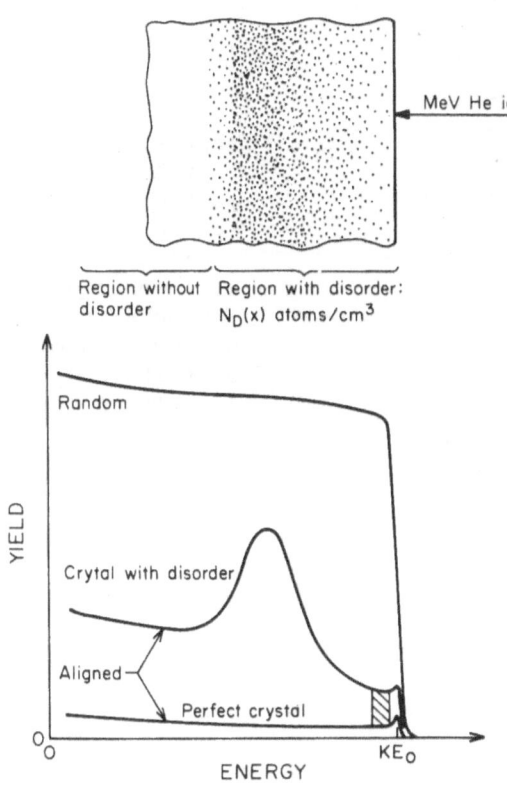

Fig. 7. Schematic energy spectra of particles impinging along a random and a low index axis directions of a perfect or disordered crystal.

by the beam particles. These scattering events increase the aligned yield in a quantity which is, in a zero order approximation, proportional to the density of displaced target atoms[6]. These defects can be moreover distributed in a non uniform profile along the sample. The yield increase is then a function of the revealed particle energy. The shape of the aligned spectrum can then give a measurement of the amount and depth distribution of displaced atoms.

Informations on displacements of atoms from their lattice sites can be argued from the angular yield profile. In Fig. 8 is reported the normalized yield measured around a low index axis. As it appears the angular profile shows a dip with a minimum at the parallel incidence of the analyzing beam. The dip is charac-

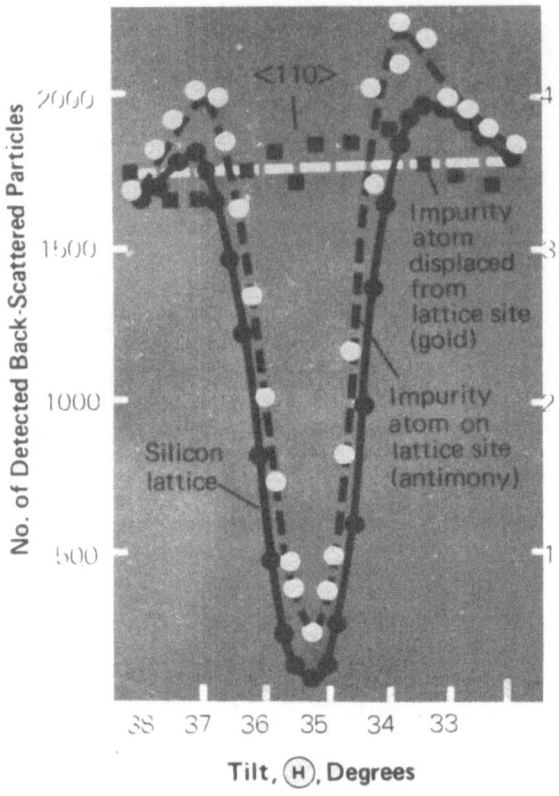

Fig. 8. Angular yield profile of particles scattered from a silicon crystal containing substitutional or randomly distributed impurities.

terized by a FWHM which depends[8], for a perfect crystal, on beam energy and atomic number and on the crystal direction and atomic number. For a low index axis of a Si crystal and 2.0 MeV He$^+$ the FWHM is of the order of 1°. The narrowing of the dip in a damaged crystal can be related to the average displacement of atoms from their lattice sites[6] and this method gives an accuracy less than 0.1 Å.

Channeling effects have been used to determine the presence of extended defects[6,10] in a single crystal. The capability the method has been demonstrated and applied to detect dislocations[10], twins[11] and stacking faults[12]. However the sensitivity is quite poor and complementary techniques are required.

In addition the angular yield profiles can be used to locate foreign particles within a crystal lattice[9,13]. In the same Fig. 8 are shown the angular profiles obtained from substitutional Sb atoms and from randomly distributed Au atoms in the Si crystal.

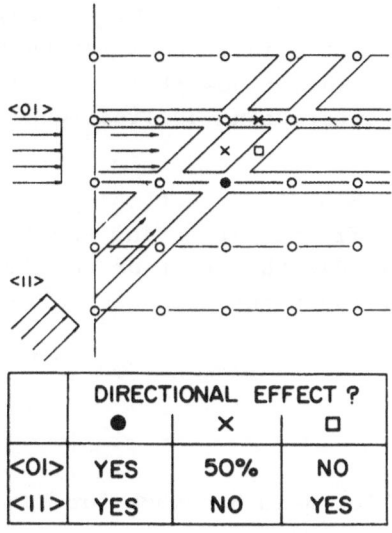

Fig. 9. Schematic diagram
showing the capability
of channeling effect
technique to locate
impurities within a
lattice·

	DIRECTIONAL EFFECT ?		
	●	x	□
<0I>	YES	50%	NO
<II>	YES	NO	YES

The substitutional atoms give rise to a dip similar to that of Si
whilst Au atoms do not give rise to any angular dependence of the
scattering yield. If the impurities are located in a perfect substitutional site, as shown schematically in Fig. 9 by (o), the backscattering yield for aligned incidence is reduced for both host
crystal and impurity. Other lattice sites, as (x) or (□), show
instead yield reduction only along particular analyzing direction.
Measurements performed along several crystal axes give then a precise lattice location of the foreign atoms. This method has been
used almost extensively to locate ion implanted dopant in semiconductors, after annealing with furnace[14], laser[15], electron[16]
and ion beams[17].

CONCLUSIONS

We have shown some of the principles and capabilities of ion
beam analysis. These techniques have been currently used in the
last ten years in pure and applied research areas and new understanding in materials science and surface physics have been achie-

ved. As for other analytical techniques, as electron microscopy,
Auger electron spectroscopy etc, most of the early work refers to
the development of the technique itself whilst now it is used as
a routine tool. For this reason many interesting applications are
not mentioned nor quoted.

I am indebted with my collegues E. Rimini, G. Foti and P. Baeri
for their encouragements and for their friendship during the last ten
years of group work. I wish to acknowledge the staff of the Van
deer Graaff accelerator in Catania for their enthusiastic collaboration and technical assistance.

REFERENCES

1. J.W.Mayer and E.Rimini (editors): "Ion Beam Handbook for Material Analysis" (Academic Press, N.Y. 1977)

2. J.F.Ziegler (editor): "New Uses of Ion Accelerators"
   (Plenum Press, N.Y.1975)

3. O.Meyer,G.Linker and F.Kappler (editors): "Ion Beam Surface
   Layer Analysis" (Plenum Press,N.Y.1976)

4. S.T.Picraux,E.P.EerNisse and F.L.Vook (editors): "Application
   of Ion Beams to Metals" (Plenum Press,N.Y.1974)

5. S.U.Campisano, Le Scienze 143,76(1980)

6. E.Rimini: "Analysis of defects by channeling", in "Material
   Characterization Using Ion Beams" ed. by J.P.Thomas and A.
   Cachard,(Plenum Press,N.Y.1978)

7. S.U.Campisano,G.Foti,E.Rimini,S.S.Lau and J.W.Mayer;
   Phil.Mag.31,903(1975)

8. J.F.Lindhard; Dan.Vidensk Selsk.Mat.Fys.Medd. 28,n.8(1954)

9. D.S.Gemmell; Review of Modern Physics 46,129(1974)

10. S.T.Picraux,E.Rimini,G.Foti and S.U.Campisano;
    Phys.Rev.B18,2078(1978)

11. G.Foti,L.Csepregi,E.Kennedy,J.W.Mayer,P.Pronko and M.D.Retchin;
    Phil.Mag. A37,591(1978)

12. S.U.Campisano,G.Foti,E.Rimini and S.T.Picraux;
    Nucl.Inst.Meth.149,371(1978)

13. J.A.Davies, in "The Structure of Nuclei" ed. by I.A.E.A. Vienna (1972)p.457

14. S.U.Campisano,E.Rimini,P.Baeri and G.Foti; Appl.Phys.Lett.51,2680(1980)

15. P.Baeri,S.U.Campisano,G.Foti and E.Rimini; Appl.Phys.Lett.33,137(1978)

16. P.Baeri,G.Foti,J.M.Poate and A.G.Cullis; Proc.of "Laser and Electron Beam Solid Interactions and Materials Processing" Material Research Society Boston, Nov.1980. (to be published)

17. R.T. Hodgson J.E.E.Baglin,R.Pal,J.M.Neri and D.Hammer; Appl.Phys.Lett.37,187(1980).

SHAPING CERAMICS WITH A CONTINUOUS WAVE CARBON DIOXIDE LASER

Russell Wallace, Michael Bass, and Stephen Copley*

Center for Laser Studies
University of Southern California
University Park
Los Angeles, California 90007

INTRODUCTION

The application of the continuous wave carbon dioxide laser to processes such as straight-line cutting and hole drilling is well established.[1] Recently the CW-CO_2 laser has been used to shape ceramics in lathing and milling operations. In this report we present results on such shaping of Si_2N_4, SiAlON and SiC workpieces. A straightforward analysis is presented, which predicts the feed and power corresponding to a specific surface roughness grade and effective material removal rate.

EXPERIMENTAL PROCEDURES

The apparatus for laser machining of ceramic materials is identical with that used in LAM except that no mechanical cutting tool is used. The laser beam is directed along a path that is parallel to the turning axis of the lathe. It is reflected from a mirror mounted on the carriage along a direction perpendicular to the turning axis to a pair of mirrors mounted on the cross-slide. After reflection by the cross-slide mirrors it is focused by a ZnSe lens on the workpiece along a radial direction for turning or parallel to the turning axis for facing. Because the direction of motion of the carriage is parallel to the turning axis and the direction of motion of the cross-slide is perpendicular to the turning axis, the angular relationships among the reflected beams remains constant during normal machining conditions.

---

*Department of Materials Science, University of Southern California, University Park, Los Angeles, CA 90007

This work employed either a 450 W, CW, CO_2 laser (λ = 10.6 μm) operating in the TEM$_{00}$ mode or a 1400 W laser operating in a mixture of spatial modes. The focused beam diameter was 160 μm ($1/e^2$ diameter) for the former while the latter could only be focused to a spot of ~0.3mm in diameter.

A description of the materials which have been shaped in our experiments is given in Table I.

Table 1

Materials Investigated

Material	Description
SiC	NC 435 (Norton Co.) A sintered SiC containing 29 vol pct free Si
Si_3N_4	Nc 132 (Norton Co.) A hot-pressed Si_3N_4.
SiAlON	A hot-pressed SiAlON by General Electric Co.
Alumina	AV30 produced by McDaniel Inc.

Milling

Laser machining in the milling configuration is obtained by employing a scanner to sweep the focused beam back and forth in a direction perpendicular to the direction of workpiece motion. If the advance of the workpeice during one cycle of the scanner is less than the width of the groove, then a nearly continuous swath of material will be removed. Most scanners deflect the light with a sinusoidal motion. As a result the light dwells longer on the material at the end of each stroke than it does on that in the middle. This leads to a nonuniform rate of material removal and a cut bottom with a cross section shown in Fig. 1a. By placing metallic masks over the sample at the ends of the scan the possibility of cutting too deeply is eliminated and a flat bottomed cut as shown in Fig. 1b can be produced.

RESULTS

Lathing

Figure 2 shows the cross section of the grooves of a laser machined spiral in SiC. The grooves were cut in the facing configuration using the TEM$_{00}$ mode laser; the high numbered grooves were closest to the turning axis and thus correspond to the lowest cutting velocities. It is evident that the groove depth increases with decreasing cutting velocity while the groove width decreases slightly, being approximatley equal to the beam diameter.

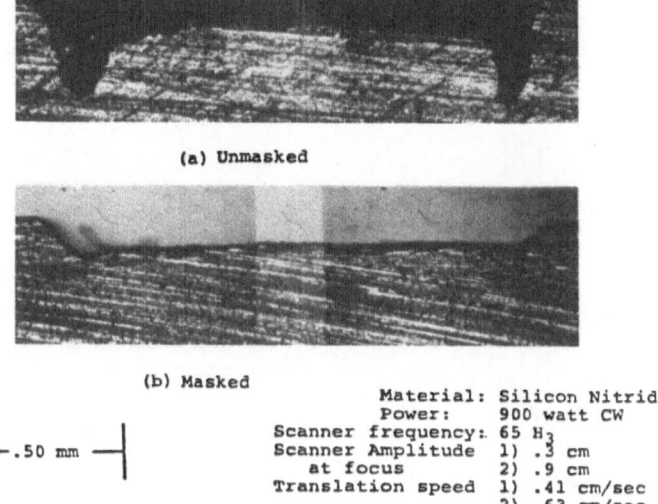

(a) Unmasked

(b) Masked

├─ .50 mm ─┤

Material:	Silicon Nitride
Power:	900 watt CW
Scanner frequency:	65 H$_z$
Scanner Amplitude	1) .3 cm
at focus	2) .9 cm
Translation speed	1) .41 cm/sec
	2) .63 cm/sec
	Two passes

Fig. 1. Cross Section of Milled Grooves

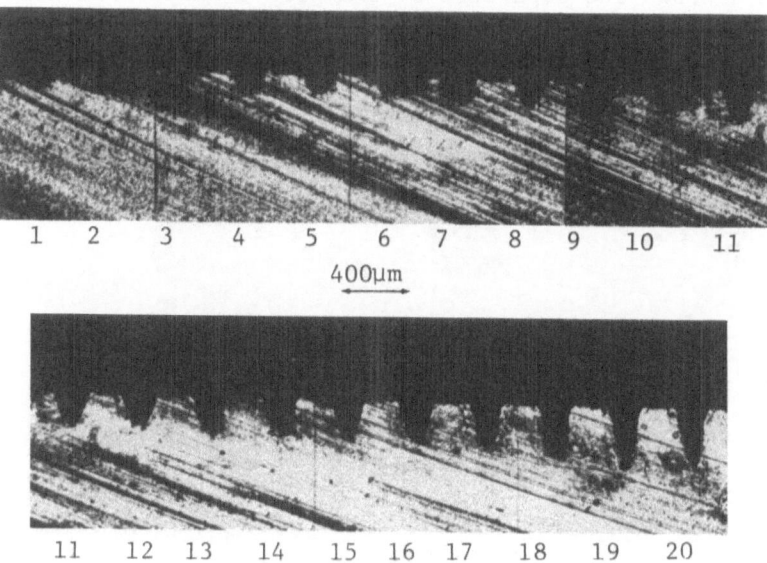

1   2   3   4   5   6   7   8   9   10   11

400μm

11   12   13   14   15   16   17   18   19   20

Fig. 2. Photomicrographs of laser machined spiral
         grooves on SiC

In Fig. 3, the groove depth is plotted as a function of dwell time given by the equation, $\theta = b/v$, where b is the beam diameter and v is the cutting velocity.

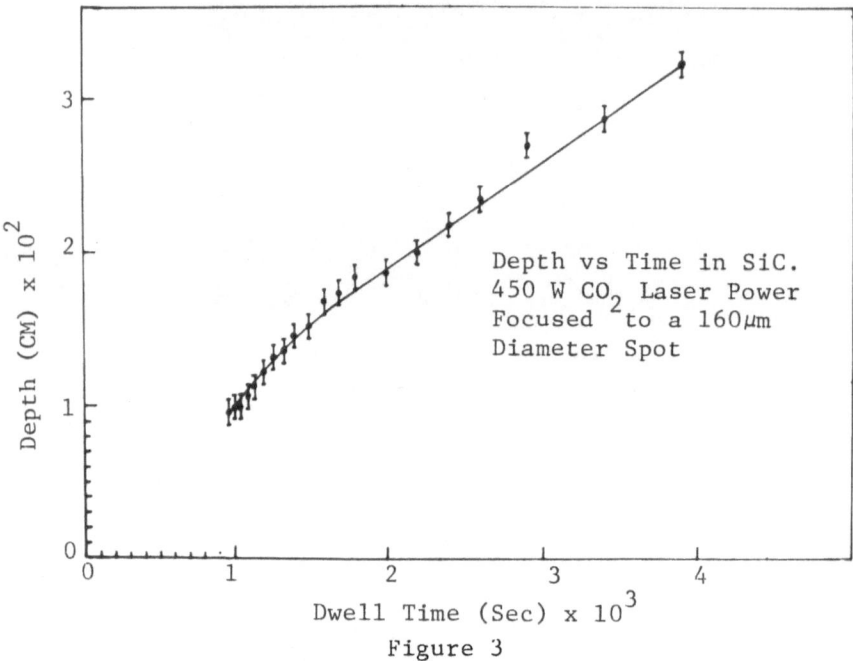

Depth vs Time in SiC.
450 W CO_2 Laser Power
Focused to a 160µm
Diameter Spot

Figure 3

Figure 4 shows the departure of the groove shape from nearly Gaussian which occurs when a multimode laser was used and an O_2 gas jet was added to supplement the burning process.

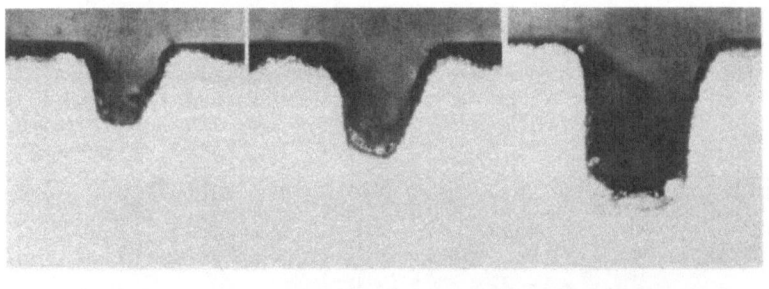

2.9   4.8   14.6
Dwell Time (Sec) x 10^{-3}
├─.50mm─┤   Material: Silicon Nitride
                  Power: 900 watts CW
                  Gas Jet: O_2 coaxial

Fig. 4   Cross section of laser machined grooves
         versus dwell time.

The assymmetry seen in this figure can become extreme as seen in Fig. 5 where for a long dwell time the groove is actually curved. This is thought to be caused by the asymmetric spatial distribution and some light guiding into the most deeply cut part of the groove.

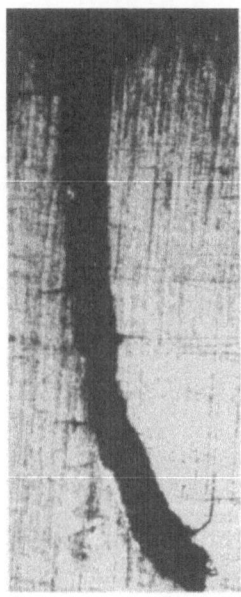

Material: Silicon Nitride
Power: 1300 watts CW
Gas Jet: N_2 large orifice
Scan Speed: 2.7 cm/sec

├─.50mm─┤

Fig. 5.   Cross section of laser cut

When one wishes to obtain faster cutting, the laser induced burning process can be enhanced by a colinear jet of reactive gas. Figure 6 shows the measured cut depth versus dwell time obtained using N_2 and O_2 gas jet assists to cut Si_3N_4. The depth of cut is deeper than without either gas jet assist but the form of the relationship between depth and dwell time is the same as shown in Fig. 3.

A major concern in evaluating the potential of shaping with a laser is the quality of the resulting surface. Figure 7 shows the topography of laser machined grooves in SiC. Although no cracks were observed in the grooves of this material or the others examined, deposits suggestive of a condensate or oxide were observed. The mechanical properties of specimens with laser machined surfaces have not yet been investigated.

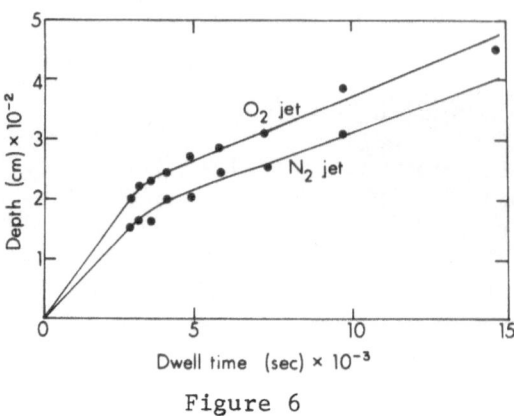

Figure 6

400μm

Fig. 7   Laser machined "0" ring groove
         in SiC illumination incident
         obliquely from the right.

Figure 8 shows the cross sections of grooves 2 and 18 at a higher magnification than in Fig. 2. It appears that their shape is approximately Gaussian. The mechanism of material removal is believed to be evaporation.

A

Fig.   8A.   Cross section of ring #2 in laser
                   machined spiral on SiC.
        B.   Cross section of ring #18 in laser
                   machined spiral on SiC.

100μm

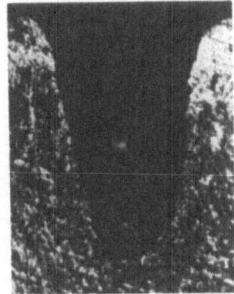

B

Milling

Figure 9 shows three laser machined milled paths in hot pressed Si_3N_4. These data show that increasing the overlap of a pass with the previous pass (decreasing the feed of the work-piece) does not necessarily result in a smoother surface. Figure 10 is a series of profilimeter traces of the cross section of the laser machined surfaces for increasing overlap. The pro-filimeter has scanned perpendicular to the milled grooves (i.e., parallel to the workpiece motion). A ground surface pro-file is included in Fig. 10 to facilitate quick evaluation of the laser machine surface.

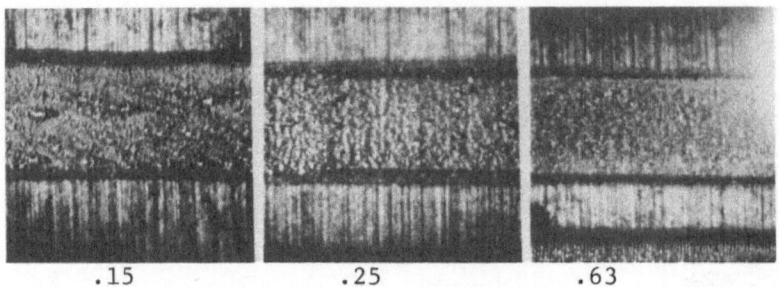

.15   .25   .63

Translation speed (cm/sec)

.42   .29   .19
Depth of groove (mm)
Material: Silicon Nitride
Power: 900 watts CW
Scanner frequency: 65 H$_3$
Scanner Amplitude: .7cm unmasked

Fig. 9   Milled surfaces

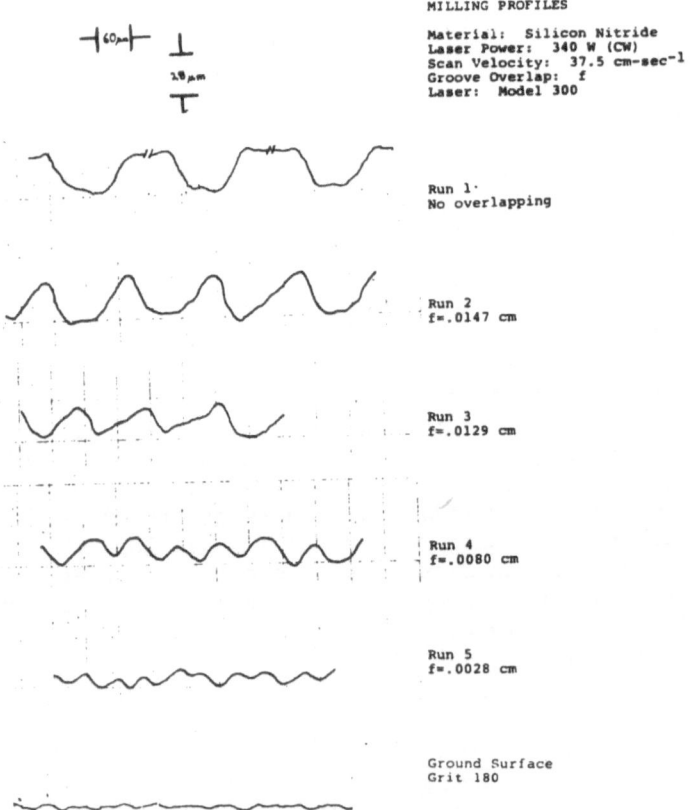

Figure 10

In Fig. 11 similar data is shown but for a higher laser power. It is clear in this case that the surface roughness increases for too small a feed.

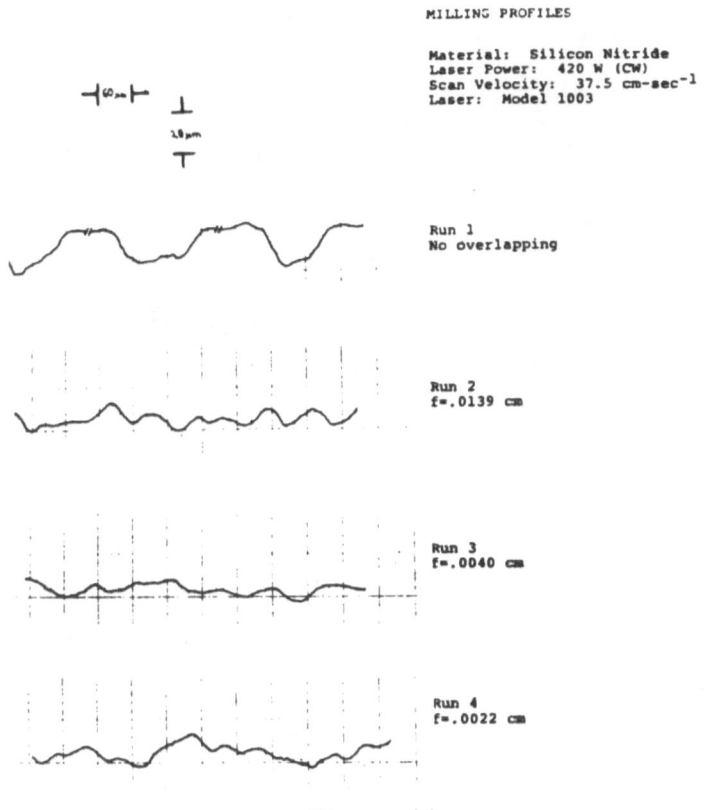

MILLING PROFILES

Material: Silicon Nitride
Laser Power: 420 W (CW)
Scan Velocity: 37.5 cm-sec^{-1}
Laser: Model 1003

Run 1
No overlapping

Run 2
f=.0139 cm

Run 3
f=.0040 cm

Run 4
f=.0022 cm

Figure 11

Photos of the roughened surfaces are shown in Fig. 12. The regular pattern may be the result of incident laser light interacting with material that remains hot following prior passes. This model requires more careful study and is being evaluated both experimentally and theoretically at the time of this writing.

ANALYSIS OF LASER MACHINING

The shape of the laser machine grooves can be described employing the normal curve

$$D(y) = D_o \exp \left(- \frac{y^2}{a^2}\right) \qquad (1)$$

Tracings of grooves 2 and 18 along with normal curves corresponding to the parameters given in Table II are shown in Fig. 13.

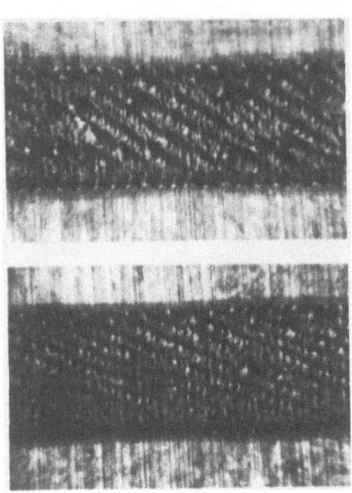

Material: Silicon Nitride
Laser Power: 630 W (CW)
Scan Velocity: 91 cm-sec^{-1}

Run 1
Groove overlap .00154 cm

Run 2
Groove overlap .00215 cm

Fig. 12 Milled surfaces

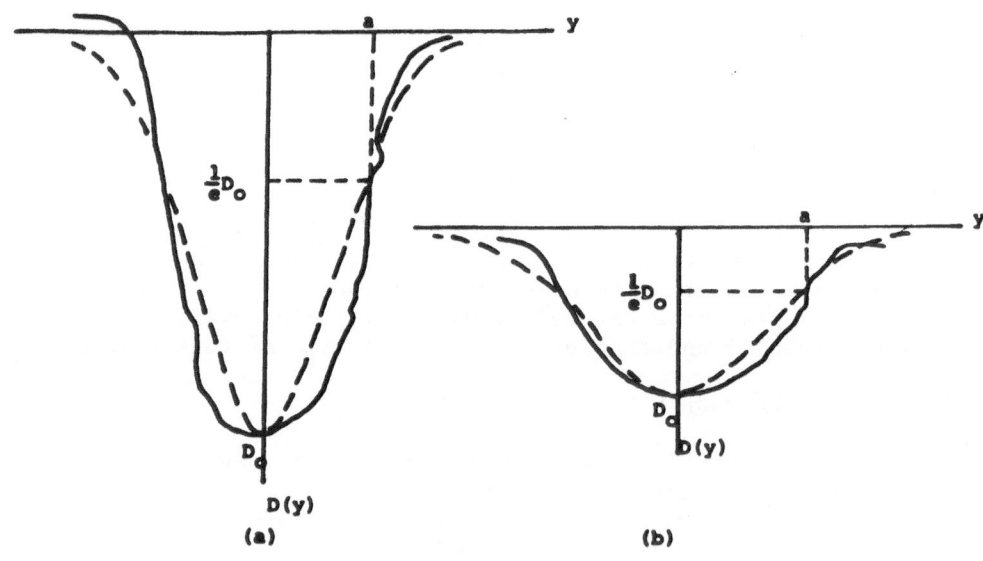

Figure 13

The total area under the normal curve is

$$A = D_o \, a \, \sqrt{\pi} \, .$$ (2)

The material removal rate is

$$Z = Av$$ (3)

and the specific cutting energy neglecting power expended by the lathe is

$$\rho = \frac{P}{Z}$$ (4)

where P is the incident beam power. Values for A, Z and ρ for grooves 2 and 18 are also given in Table II. It is interesting that the cross section of groove 2 is less than that of groove 18 but the material removal rate of groove 2 is the greatest. Since the incident beam power was the same for both grooves, the specific cutting energy the groove 2 must be the least. Thus, the faster the beam scans the surface, the more efficient becomes the laser machining process. Although the origin of this effect has not been determined, it may be due to lower conductive losses or possibly better coupling of the laser beam to the workpiece at high speed. The latter may be caused by excessive ejecta at low speeds blocking the incoming light. This ejecta may even become ionized and then form an opaque plasma. Various models exist in which such a plasma reradiates energy to the surface of a metal and thus increase the coupling. As yet there is insufficient evidence to determine the role of such a process in laser machining of ceramics.

Table II

Parameters Describing Laser Machined Grooves

Groove	D_o (μm)	a (μm)	$A(10^{-8} m^2)$	$Z(10^{-9} m^3 s^{-1})$	(GJm^{-3})
2	125	100	2.22	3.42	131
18	300	75	3.99	2.19	204

Actual turning or facing involves the partial overlapping of laser machined grooves as shown in Fig. 14. In the following analysis, we shall assume that the shape of the groove cut by the laser in turning is the same as in Fig. 2 even though in turning it is cut into the edge of a step. The surface roughness (R), which is equal to the sum of the absolute values of all the areas above and below the mean line divided by the sampling length,[2] is given by the equation

$$R = \frac{\overline{4abc}}{f} \qquad . \qquad (5)$$

The effective material removal rate, which is the material removal rate corrected for the overlapping of grooves is given by

$$\dot{z}' = vD_o a\sqrt{\pi} - v(\overline{geh}) \qquad . \qquad (6)$$

The areas $\overline{abc}$, $\overline{cde}$ and $\overline{geh}$ needed to evaluate the surface roughness and the effective material removal rate can be evaluated with the help of tables giving the ordinates of the normal curve and the areas under the normal curve.[3] Accordingly, the depth of the laser machined groove is given by

$$D(y) = D_o \sqrt{2\pi} \quad (\dot{z}) \qquad (7a)$$

$$\dot{z} = \frac{y\sqrt{2}}{a} \qquad (7b)$$

and the area of the laser machine groove is given by

$$\int_0^y D(y) \, dy = D_0 \, a\sqrt{\pi} \, F(z) \qquad (8a)$$

$$\dot{z} = \frac{y\sqrt{2}}{a} \qquad (8b)$$

where

$$f(z) = \frac{1}{\sqrt{2\pi}} \quad \exp\left(- \frac{z^2}{2}\right) \qquad (9)$$

$$F(z) = \int_0^z f(z) \, dz \qquad .$$

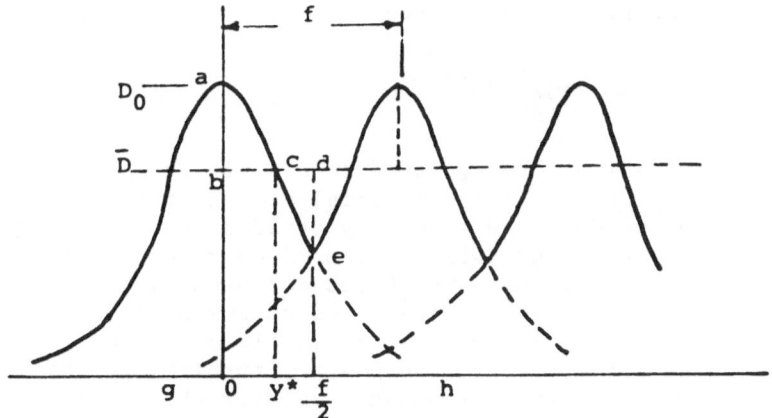

Figure 14

The surface roughness can be determined from Eq. (5) and is given by

$$R = \frac{4}{f} \left[D_0 \; a\sqrt{\pi} \; F(\frac{y^*\sqrt{2}}{a}) - \bar{D} \; y^* \right] \tag{10}$$

where y* is the abcissa value when D(y) equals the mean diameter, see Fig. 14. The mean diameter can be found by equating abc (the quantity in brackets in Eq. (10)) to cde, and solving for $\bar{D}$, where

$$\overline{cde} = \bar{D} \; (\frac{f}{2} - y^*) - \left[D_0 \; a\sqrt{\pi} \; F(f \sqrt{2}/2a) \right.$$
$$\left. - D_0 \; a\sqrt{\pi} \; F(y^* \; \sqrt{2}/a) \right] \; . \tag{11}$$

Following this procedure, we obtain

$$\bar{D} = \frac{2D_0 \; a\sqrt{\pi}}{f} \; F \; (\frac{f\sqrt{2}}{2a}) \; . \tag{12}$$

The parameter y* can be calculated with the equation

$$y^* = \frac{a}{\sqrt{2}} \; F^{-1} \left(\frac{\bar{D}}{D_0 \sqrt{2\pi}} \right) \; . \tag{13}$$

The effective material removal rate is obtained from Eq. (6) and
is given by

$$z' = 2vD_0 \ a\sqrt{\pi} \ F(\frac{f\sqrt{2}}{2a}) \ .$$

(14)

Equations (10) and (14) are employed in Fig. 15 to make a plot of
surface roughness and effective material removal rate for a
groove cut with incident power density of 22.3 GWm^{-3}, beam diameter ($1/e^2$) of 160 μm and dwell time of 1.05 x 10^{-3}s^{-1} (see
groove 2, Table II, Fig. 8). The feeds corresponding to ISO roughness grades N8, N9 and N10 are indicated on the abcissa of Fig. 8.
A feed of 0.05 mm rev^{-1} is normally available on engine lathes.
As a rough guide for comparison purposes[3], turning operations in
metals normally give surface roughness greater than 1 μm.

Consider turning a large radius piece to the shape shown in
Fig. 16 using the focused laser beam as the cutting tool. The
piece of material removed by the laser beam at any time is indicated by the black triangle and has the shape shown in Figs. 8
and 13. Its width is 2a and its depth is D_0. If this is fed
parallel to the axis of rotation by an amount f per turn there
will be

$$N_t = \frac{L}{f}$$

(15)

turns per pass. Since each pass cuts to a depth D_0

$$N_P = \frac{R}{D_0}$$

(16)

passes are required in order to achieve the desired shape. On the
average, each turn takes $2\pi\bar{R}/v$ seconds where v is the speed with
which the surface rotates under the beam. Thus the time required
for cutting is

$$t_c = \frac{\Delta R}{D_0} \ \frac{L}{f} \ \frac{2\pi\bar{R}}{v} \ .$$

(17)

If after each pass one takes a time T_R to reposition laser beam,
then the total time taken in this part of the process is

$$t_R = \frac{\Delta R}{D_0} \ T_R \ .$$

(18)

The total time for turning this piece is then

$$T = t_r + t_c = \frac{\Delta R}{D_0} \ T_R + \frac{\Delta R}{D_0} \ \frac{L}{f} \ \frac{2\pi\bar{R}}{v} \ .$$

(19)

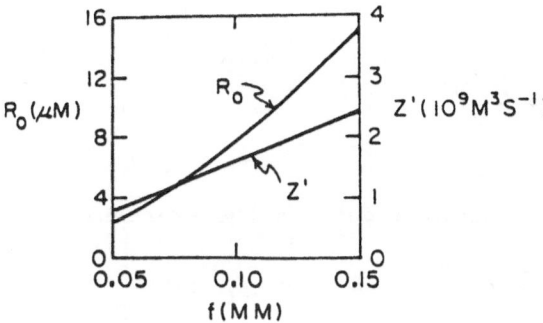

Figure 15. Roughness and effective material removal rate.

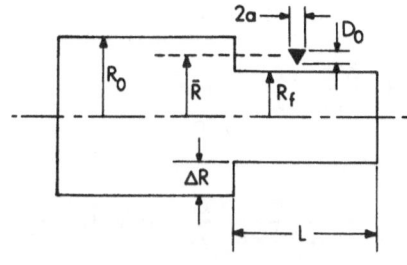

Figure 16

In Fig. 3 data is shown for laser cutting SiC which is fit
very well by the following relationships:

$$D_0 = \frac{m}{v} \quad \text{for} \quad v > v_c$$

$$D_0 = \frac{m'}{v} + b \quad \text{for} \quad v > v_c \tag{20}$$

where

$$v_c = \frac{m-m'}{b} \quad .$$

Inserting this experimental data in the expressions for T gives

$$T = \frac{\Delta R}{m} \; \frac{L}{f} \; 2 \pi \bar{R} \quad \text{for} \; v > v_c \tag{21}$$

$$T = \frac{\Delta R}{m'+vb} \; \frac{L}{f} \; 2 \pi R \quad \text{for} \; v < v_c \quad . \tag{22}$$

It is clear that the time spent repositioning the laser beam makes
a major contribution to the machining time. However, there is no
need to reposition the beam because the focused laser beam is a
cutting tool that can cut equally well in both directions. Instead of repositioning, consider cutting both ways with some care
to ramp the power down at the end of each pass to avoid cutting
too deeply while the beam is turned. Thus the machining time is

$$T = \frac{\Delta R}{m} \; T_r \; v + \frac{\Delta R}{m} \; \frac{L}{f} \; 2 \pi \bar{R} \qquad \text{for} \; v > v_c \tag{23}$$

and

$$T = \frac{\Delta R}{m'+vb} \; T_R \; v + \frac{\Delta R}{m'+vb} \; \frac{L}{f} \; 2 \pi \bar{R} \quad \text{for} \; v < v_c \; , \tag{24}$$

The first of these equations show that for $v > v_c$ the machining
time is a constant determined by the material machined (this
enters through m), the dimensions of the piece to be machined
(ΔR, L and $\bar{R}$) and the feed which one selects in order to achieve
a desired surface roughness.

Since

$$\frac{m}{v} > \frac{m'}{v} + b$$

and

$$m > m' + vb$$

the time required when machining at speeds where Eq. (24) holds is always greater than when machining at higher speeds. This means that for laser machining of ceramics, one can minimize the machining time by taking shallow cuts at high speeds.

Ignoring surface finish requirements, the maximum feed that is permissible is

$$f = 2a \quad .$$

Thus the minimum machining time is

$$T = \frac{\Delta R}{m} \quad \frac{L}{2a} \quad 2 \pi \bar{R} \quad .$$

To achieve a desired finish smoothness, the last pass or two could be done with a smaller feed.

The preceding results can be used to calculate the machining time required to obtain desired surface roughness. The data in Fig. 3 for laser machining SiC in room air shows that $v > v_c$ if the dwell time is less than ~ 1.3 msec. and that $m \cong 0.1$ cm/sec. Under these circumstances a piece with $R_0 = 6$ mm, $R_F = 4$mm, and $L = 10$ mm can be laser machined in the following times:

Roughness	F(mm)	T(sec)
N8	0.060	105
N9	0.082	76
N10	0.130	48

APPLICATIONS

Figure 17 contains two macrophotographs of a ¼" x 20" thread turned on Si_3N_4 by laser machining. In Fig. 17 the rectangular rod from which the threaded section was turned can be seen. Fig. 17a also shows a region in which we demonstrate the ability to produce concave surfaces.

Macrophotographs of laser machined Si_3N_4

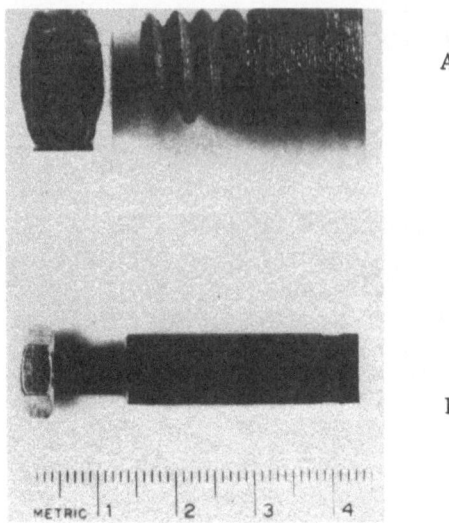

A.

B.

Fig. 17 The rectangular rod was laser
         turned to a 1/4" round and then
         threaded with 20 threads per inch.
         The groove at the other end demo-
         strates the ability of laser ma-
         chining to make concave curved
         surfaces.

     The negatively curved surface was made by focusing the laser
beam so that it struck the center of the rectangular surface at
normal incidence. Thus the corners were out of the focal region
and irradiated at near grazing incidence where the light-to-
material coupling was less. As a result, more material was removed near the center than near the corners and a concave surface
was established. A convex surface (i.e. the round section which
was threaded in Fig. 17) was made by adjusting the focus and
angle of incidence so that the corners were near the focus and
the light near normal incidence at the corners. This makes the
central part of the surface out of the focal region and illuminated at near grazing incidence. Thus more material could be
removed near the corners than near the center when the light was
so focused.

     When laser machining was used to thread Si_3N_4 the focused
beam was at normal incidence on the cylindrical surface. However,
to make a proper thread the position of the focus with repsect
to the previous spiral groove must be adjusted. This is shown in
Fig. 18 where the sequence of focal positions are sketched.

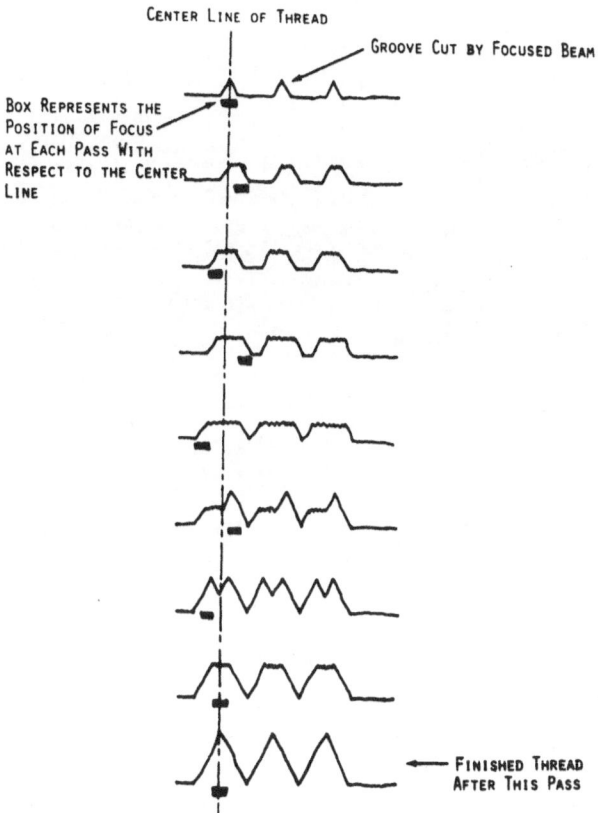

Figure 18   Schematic diagram of the procedure for laser machining
             screw threads.

Figure 19 shows a laser machined piece of SiAlON which originally had the square cross section remaining in the middle of
the piece. The maximum peak to valley nonuniformity of the smooth
surface is 7.5 μm which confirms the potential of laser machining
SiAlON to desired surface quality specifications. Figure 19 also
shows the ½" x 13 screw thread which was laser machined on the
other end of the laser turned SiAlON rod.

Figure 20 shows O-ring grooves cut in alumina using the
laser machining process. The black material is thought to be the
binder used in preparing AV 30.

SUMMARY

Laser machining of ceramics, a form of controlled burning of
materials, has been demonstrated. When the ability of this
technique is properly applied, fine machining of extremely hard
materials is possible.

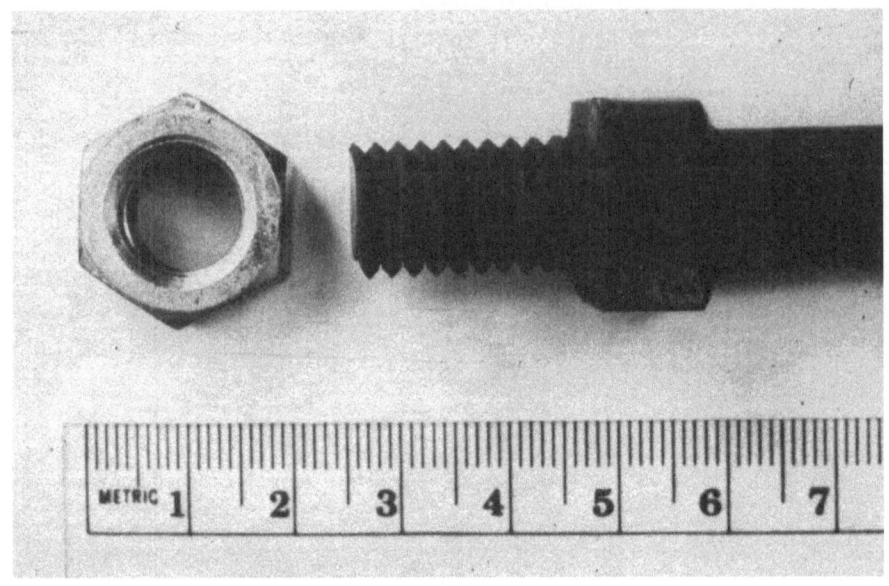

Figure 19

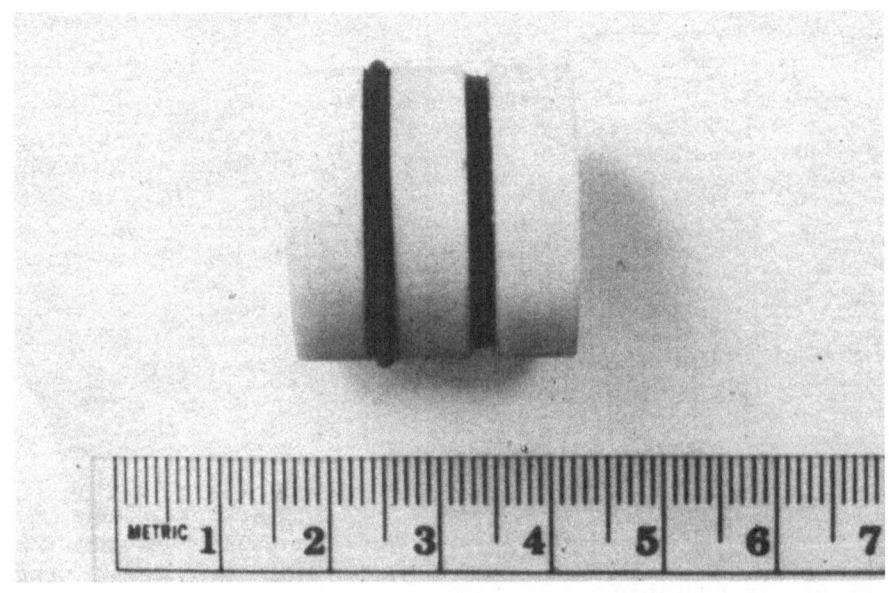

Figure 20

REFERENCES

1. W. W. Duley, "CO_2 Lasers, Effects and Applications," Acad. Press, New York, (1976).
2. G. Boothroyd, "Fundamentals of Metal Machining and Machine Tools", p. 134, McGraw-Hill Book Co., New York, (1975)
3. John B. Kennedy, and Adam M. Neville, "Basic Statistical Methods for Engineers & Scientists", Dun-Donnelley Publishing Co., New York (1976).

LASER-INDUCED STRUCTURAL CHANGES IN SI - METAL THIN FILMS

M. von Allmen

Institute of Applied Physics
University of Bern
CH - 3012 Bern (Switzerland)

This lecture deals with physical processes induced by short (e.g. 50 ns) laser pulses in composite samples. We look at three different kinds of samples:

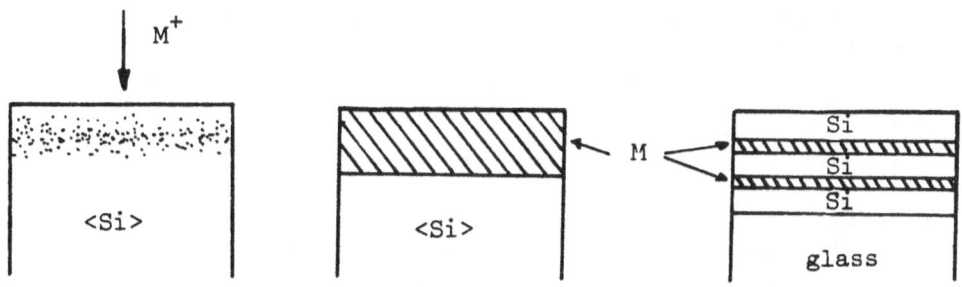

a) ion-implanted <Si>

b) deposited metal film on <Si> substrate

c) metal-Si multi- layers on inert substrate

The heat produced upon absorption of the laser pulse causes melting in a thin surface layer. While the pulse lasts, a melt front penetrates into the solid. After the pulse, heat is conducted away by the solid substrate and the melt front

returns to the surface. This sequence of events can be divided
into 3 parts:

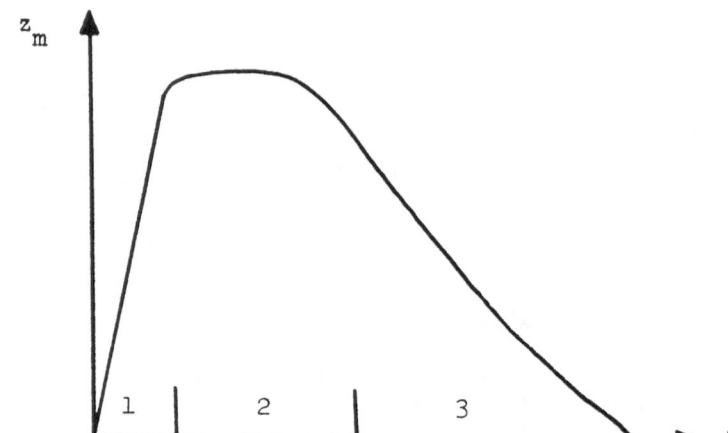

1. Melting

2. Mixing

3. Quenching

Melt front position, z_m, vs. time

MELTING

Temperature $T(z,t)$ in laser-irradiated sample (one-
dimensional heat flow)

$$\rho\, c_p\, \frac{\partial T}{\partial t} = \frac{\partial}{\partial z}\, \left(K\, \frac{\partial T}{\partial z}\, \right) + \frac{I(1-R)}{\rho\, c_p}\, \alpha\, \bar{e}^{\,\alpha z}$$

ρ = density I = laser intensity
c_p = specific heat R = reflectivity
K = thermal conductivity α = absorption coefficient

The transient temperature distributions $T(z,t)$ can be transformed
into a set of <u>isothermal surfaces</u> $T = T_j$, moving at velocities

$$v_j = \left(\frac{\partial T/\partial t}{\partial T/\partial z}\, \right)_{T_j}$$

A special surface is the <u>liquid-solid interface</u> $T = T_i$, at which
latent heat, ΔH, is absorbed or liberated:

$$v_i\, \rho\, \Delta H = K_s\, \left(\frac{\partial T_s}{\partial z}\, \right)_{T_i} - K_\ell\, \left(\frac{\partial T}{\partial z}\, \right)_{T_i}$$

Subscripts s and ℓ denote the solid and liquid phase, respectively. The interface temperature, T_i, is determined by atomic rearrangment dynamics, and is not necessarily a constant.

The above equations also apply for solidification, however with different values of ρ, c_p, K, ΔH, because of redistribution of atoms in the melt.

Further, $T_i > T_m$, for melting, and $T_i < T_m$ for solidification. (T_m = equilibrium melting point).

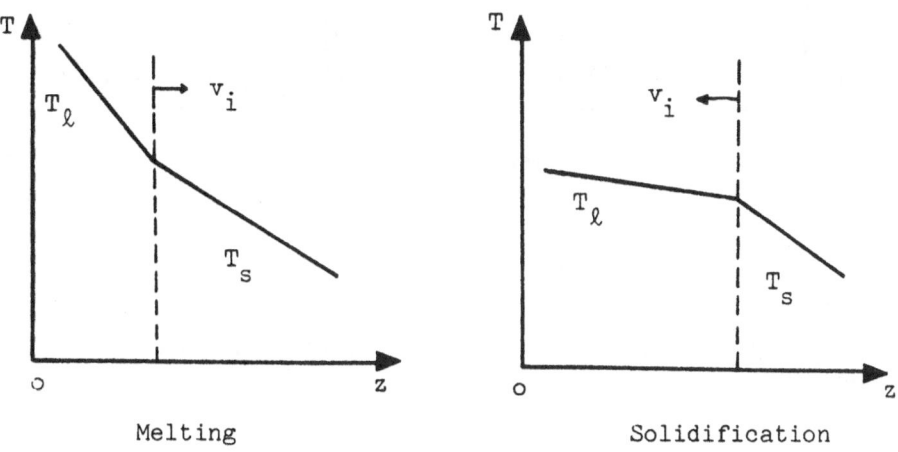

Melting   Solidification

MIXING

      Achieved by melt interdiffusion. Diffusivities D_ℓ in liquid metals are of the order of 10^{-4} cm^2/s.

"Diffusion length" $\delta = 2\sqrt{D_\ell \cdot t}$. (about 60 nm in 100 ns).

If the melt lifetime is taken to be the same everywhere ($v_i = \pm \infty$), the profiles of impurity concentration $c(z,t)$ after time t are obtained from

$$c(z,t) = \frac{1}{\sqrt{\pi}\,\delta} \int_{-\infty}^{\infty} c(z',0) \exp\left\{-\frac{(z-z')^2}{\delta^2}\right\} \cdot dz'$$

For the sample types considered, the following diffusion profiles result:

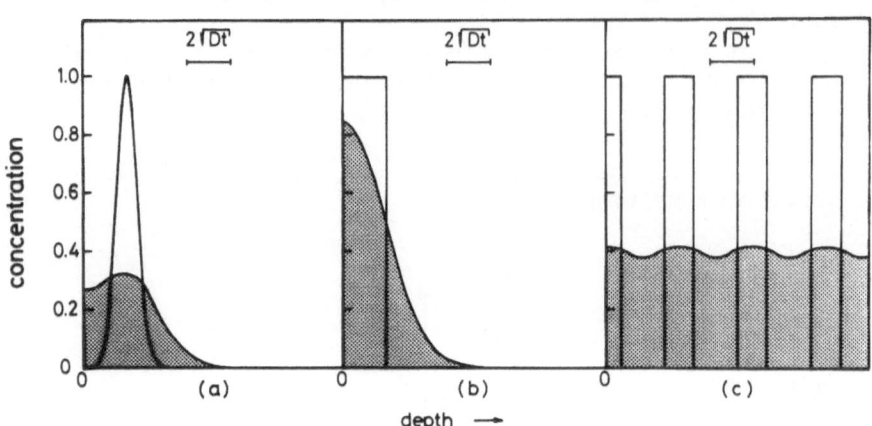

Practically, sample types a - c result in different regimes of
melt composition:

    a) diluted mixture (c = 1 at.%);

    b) inhomogeneous mixture (0 < c(z) < 1);

    c) concentrated mixture (c = 0.1 ... 0.9).

QUENCHING

Crystal growth

        The growth velocity is a function of undercooling ΔT_i =
T_m - T_i and can be written as

$$u = \frac{a\, g_u}{\eta} \left[1 - \exp\left\{ - \frac{\Delta H_m}{k\, T_m} \cdot \frac{\Delta T_i}{T_i} \right\} \right]$$

a = molecular diameter η = melt viscosity
g_u = a constant ΔH_m = heat of melting

g_u, η, ΔH_m, T_m and T_i depend on composition and may vary within
the molten layer.

The <u>viscosity</u> is often fitted to an expression

$$\eta = \eta_0 \ \exp\{Q/k(T-T_\eta)\}$$

η_0, Q, T_η = fitting parameters.

Nucleation

Steady-state homogeneous nucleation rate (Uhlmann):

$$I_n \cong \frac{N_0 \ g_I}{\eta} \ \exp \ \{- \frac{1.024 \ T_m^5}{T_i^3 \ \Delta T_i^2}\}$$

N_0 = density of molecules g_I = a constant
in the melt

If a melt is "instantaneously" undercooled, nucleation
starts only after a "<u>time-lag</u>" $t_n \sim \eta$.

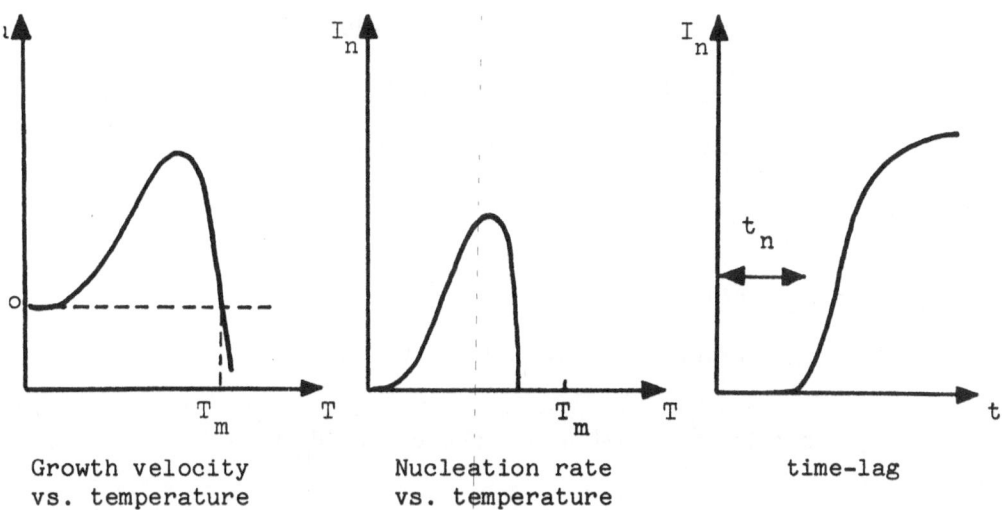

Growth velocity   Nucleation rate   time-lag
vs. temperature   vs. temperature

GLASS FORMATION

How can one predict experimental conditions necessary for
glass formation?

A) <u>Nucleation limited</u>: no crystal in contact with the melt.

B) <u>Growth limited</u>: crystalline substrat acts as a seed.

A) Nucleation theory

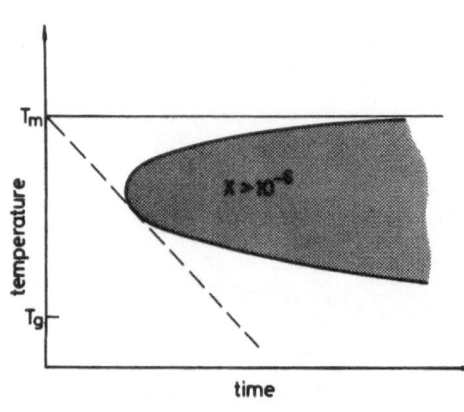

TTT diagram

Crystalline volume fraction:

$$X = \frac{\pi}{3} I_n \cdot u^3 \cdot t^4.$$

I_n = steady-state nucleation rate

T_m = melting point

T_g = glass temperature

(D. R. Uhlmann, J. Non-Cryst. Solids $\underline{7}$, 337 (1972).)

However: – Width of undercooled layer:

$$\Delta z \cong \Delta T_i / (\partial T / \partial z)_i \qquad\qquad \sim 10^{-6} \text{ cm}$$

– Lifetime of undercooled layer:

$$\Delta t \cong \Delta z / v_i = \Delta T_i / (\partial T / \partial t)_i \quad \sim 10^{-8} \text{ s}$$

B) Isothermal velocity

There is a maximum possible growth velocity at $T = T_{max}$:

$$T_{max} = T_m / (1 + \frac{kT_m}{\Delta H_m} \ln 1 + B), \qquad \text{where } B = \Delta H_m / Q$$

$$u(T_{max}) = u_{max} = \frac{a\, g_u}{\eta_o}\, e^{-Q/kT_m}\, \frac{B}{(1+B)^{(\frac{1+B}{B})}}$$

This means, that no crystal can grow if the temperature distribution in the sample is such that

$$v(T_{max}) > u_{max}.$$

SEGREGATION

Equilibrium

Equilibrium segregation
coefficient
$k_o = c_l/c_s$

"zone refining"

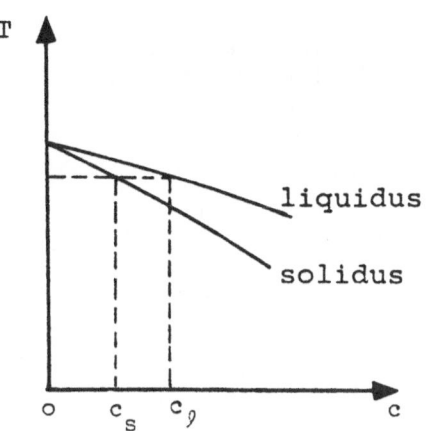

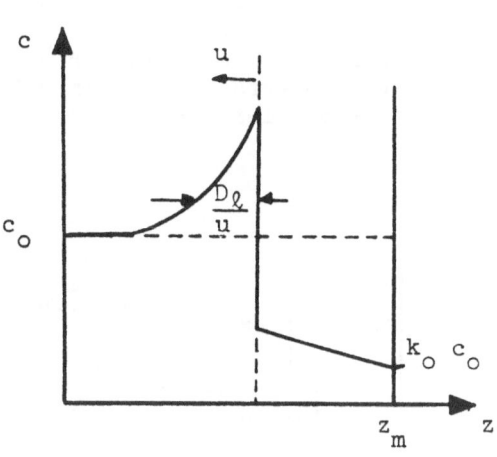

---

k_o: depends on composition and on undercooling, but is practically
constant for small ΔT and $c \ll 1$.

"Non - equilibrium"

l-s interface moves by a

in a time a/u;

impurity atom diffuses by a

in a time a^2/D_i.

Solute trapping $(k \rightarrow 1)$ for

$a/u = a^2/D_i$, i. e. for

$u = D_i/a$.

What is the interfacial

diffusivity D_i?

If: $D_i \cong \sqrt{D_s \cdot D_l} \approx 10^7$ cm^2/s, then $u \lesssim 10$ cm/s.

EFFECTS OF INHOMOGENEOUS MELT COMPOSITION

An inhomogeneous melt composition may result in Constitu-tional Supercooling (CSC).

CSC: Undercooling is larger ahead of the l-s interface than at the interface. This is only possible in the presence of compositional gradients in the melt.

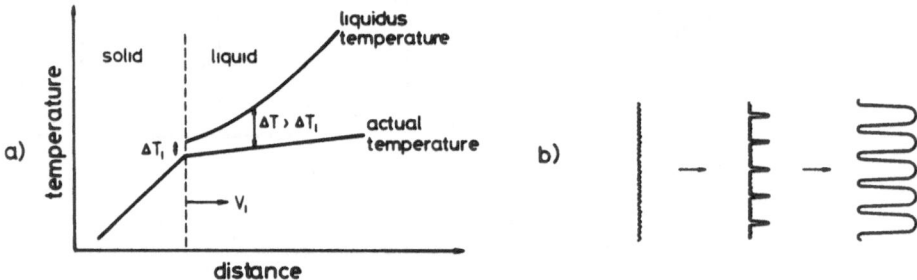

The occurrence of CSC can be predicted from the phase diagram:

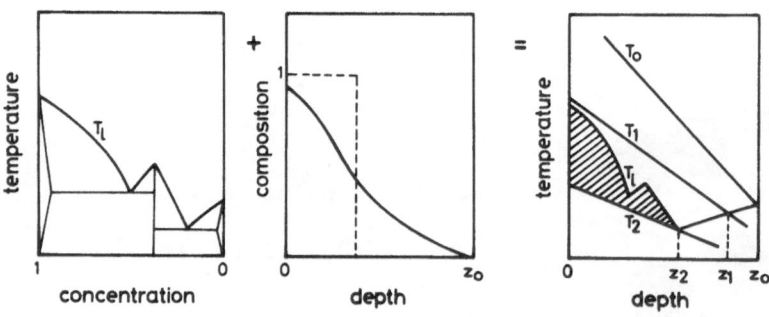

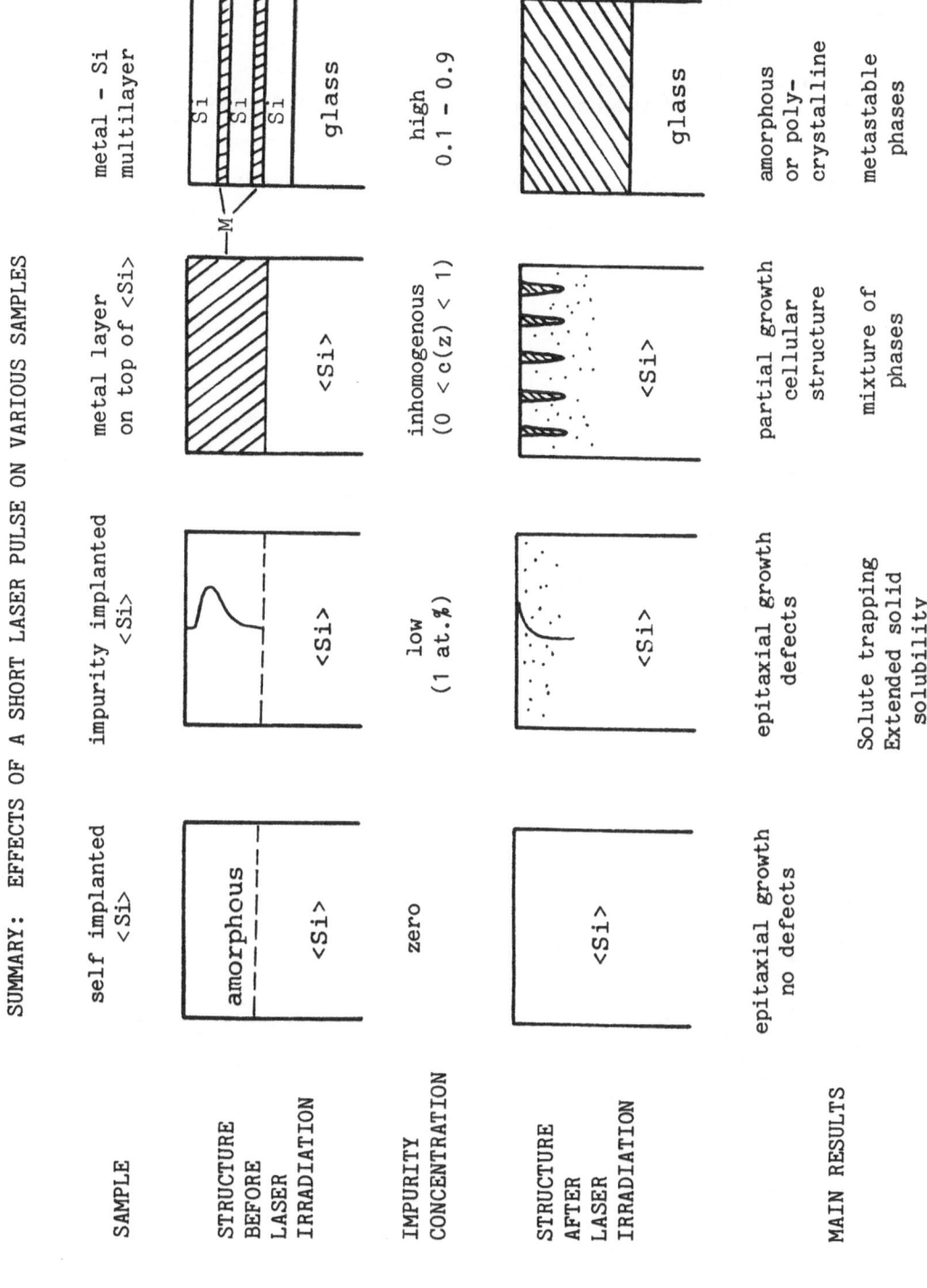

SUMMARY: EFFECTS OF A SHORT LASER PULSE ON VARIOUS SAMPLES

SAMPLE	self implanted <Si>	impurity implanted <Si>	metal layer on top of <Si>	metal – Si multilayer
STRUCTURE BEFORE LASER IRRADIATION	amorphous / <Si>	<Si>	<Si>	Si / Si / Si / glass
IMPURITY CONCENTRATION	zero	low (1 at.%)	inhomogenous (0 < c(z) < 1)	high 0.1 – 0.9
STRUCTURE AFTER LASER IRRADIATION	<Si>	<Si>	<Si>	glass
MAIN RESULTS	epitaxial growth no defects	epitaxial growth defects / Solute trapping Extended solid solubility	partial growth cellular structure / mixture of phases	amorphous or poly-crystalline / metastable phases

Effects of inhomogenous melt composition:

- Growth instabilities because of CSC.
  (See, e.g., B. Chalmers, Principles of Solidification,
  R. E. Krieger Publ. Co. 1977.

- Formation of different phases at different depths.

LASER ORDERING IN ELEMENTAL SEMICONDUCTING FILMS

L.D. Laude, R. Andrew, and L. Baufay

Université de l'Etat à Mons
Faculté des Sciences
B-7000 Mons, Belgium

Laser treatment of semiconducting films which are deposited onto fused silica or, else, NaCl crystal before being flotted off and irradiated, is far from being comparable with the annealing process for ion-implanted layers [1] . In the latter, the single crystal substrate induces an epitaxial growth in the implanted layer which overshadows the first step of the Laser-induced ordering. In free-standing semiconducting films or in semiconducting films which are deposited onto fused silica or any amorphous substrate, more insight is probably achieved in this first step and it is the object of this paper to give some overviews on Laser-induced ordering processes in these films.

Film preparation

A number of preparation parameters must in general be taken into consideration when defining an amorphous elemental semiconducting film. Independently from the actual evaporation technique being used in the preparation, deposition speed, substrate temperature, surface treatment of the substrate, pressure in the vessel in which evaporation is performed are effective parameters which significantly determine the behaviour of the films under thermal annealing in vacuum. This means that thermally induced ordering, although attained by many thermal annealing techniques, is highly sensitive to the set values of the above parameters, in particular regarding the density and character of various defects present in the films at the end of the thermal processes. In the

case of such films being laser irradiated to crystallize,
the situation is, in a way, far more simpler. Concerning the evaporation techniques, a number of these are
currently being used : electron-beam evaporation, rf
sputtering, ion-beam sputtering, Joule effect etc ...

Let us say that the choice of the deposition technique is not essential in performing a laser-induced
crystallization. That choice is essentially governed by
the versatility of the technique, i.e. its ability to
provide reproductible characteristics of the films and
to allow the evaporation of a wide variety of materials
under more or less controled conditions. In the following, examples of laser-induced crystallization will be
given for films which have been electron-beam evaporated
in vacua in the range 10^{-8} to 10^{-10} torr or evaporated
from a heated crucible at 10^{-6} torr, and condensed onto
a fused silica substrate or an air-cleaved NaCl crystal
face. Conclusions should apply to any preparation technique as long as the films are pulsed laser-irradiated
either when they are supported on an amorphous and insulating substrate or when they are self-supported, after
dessolving the NaCl substrate.

Film irradiation

The most striking difference between laser-treatment
and thermal annealing is the fact that the latter must
in any case be performed in controled vacuum whilst the
former may, in general, be performed in air. This is of
significant importance for any development in microelectronics and we will return to this point latter on.
Further, comparing the pulsed laser treatment of the
above films with the one of ion-implanted Si layers,
which constitutes by far the most explored domain of research in the field, the energy required from a laser to
crystallize a film such as described above is much smaller and remains in the range 0.01 to 0.1 J cm^{-2}. The
actual energy depends primarily on the film being either
supported on an amorphous and insulating substrate or
self-supported, the energy being smaller by a factor of
5 to 10 in the latter case. Remarkable is the fact that
such an energy remains the same for either Ge or Si
films, in sharp contrast with thermal annealing procedures. Another interesting fact is that, although not
completely settled, it would seem that the thicker are
the films of one element, the highest would be the power
at nearly identical energies, i.e. the smaller the pulse
duration. Departing from these general conditions of
irradiation, films usually do not show any evidence of

ordering (below 0.01 J/cm^{-2}) or are perforated at too
high energies (above 0.1 J cm^{-2}).

These considerations should be balanced, however,
with laser instabilities (either spatial or temporal).
In general, lasers do not provide for a stable source of
energy. In the pulsed regimes, instabilities of the order of 20 % are present from pulse to pulse ; further,
the pulse intensity is not uniformly distributed across
the beam resulting in inhomogeneities which may attain a
factor of 2 around the average value. These instabilities are responsible for most of the problems encountered in reproducing characteristic crystallization products. Without going into the details of homogeneiza-
tion, we mention here that techniques are available [2]
in this respect which restore homogeneous beams at the
expense of the coherence of such laser beams. Depending
on whether such coherence is needed in nucleating or not
would altere the usefulness of such homogeneizers.

Experimental evidences for ordering in films

Clear evidences of ordering are usually obtained by
X-ray or electron diffraction. Because of a much precise definition of the surface of the film under evaluation, electron diffraction is the prefered technique.
However, it is worth to note that its actual resolution
is limited to, say, 10 or 15 Å in which case short range
ordering would not be evidenced although existing in
amorphous semiconductors. This would urge us to couple
such diffraction analysis with other physical means which
are sensitive to atomic ordering by pointing at either
the electronic structure of the material or its vibrational properties.

Two complementary approaches will be presented here :
transmission electron microscopy and electrical conductivity on identically prepared Ge films. Comparison will
be made with Si films whenever possible.

a. Transmission Electron Microscopy (TEM)

The Ge films which are considered here are 1000 Å
thick, deposited at a rate of 10 Å sec^{-1} in 10^{-6} torr
vacuum, on room temperature, fused silica substrate or
NaCl crystal cleaved in air. For the TEM work, films
are pulled off their silica substrate or floatted off
NaCl and mounted on Cu grids. Laser irradiation is performed either in air when they are still on their substrate or free-standing, in which case in-situ irradiation

is possible. What is usually observed after one pulse of
10^{-6} sec, for instance, is the occurence of several circular crystalline spots [3] on the irradiated film when
laser energy exceeds 10 mJ cm^{-2}. These "stars" are distributed at random and their dimension remains constant
at, say, 10 μdiameter (i.e. 100 times larger than the
film thickness). Increasing the pulse energy does not
change the star dimension but the number of these stars,
although their concentration is not linearly dependent
on the pulse energy. At energies exceeding 50 mJ cm^{-2},
nearly all of the irradiated area of the film is crystallized and is made of an assembly of such stars mixed with
much finer crystallites ($\approx 0.1\mu$). At higher energies
(above 70 mJ cm^{-2}) films start to be punctuated with
holes of increasing importance with energy. Another way
to look at this star formation is to irradiate the same
film with a succession of pulses of nearly the same energy.
Fig. 1. a-h shows a series of micrographs obtained from
a self-supported Ge-Film irradiated with 10^{-6} sec pulses
of 0.01 J/cm^2. In fig. 1a, stars are seen to be isolated within amorphous parts of the film. These progressively reduce and are replaced by stars and polycrystalline material, fig. b to j. Since pulses do not have
systematically the same energy, crystallization products
are seen to vary but, on the average, full crystallization takes place after a train of 10 or 12 pulses.

The star formation may be viewed as the product of
a nucleation event in the film followed by a very rapid
growth. Most of the stars show a fringe of polycrystals
at their upper most border. The size of these polycrystals (≈ 0.1 μ) and their distribution would indicate
that they might be the result of a thermally-induced
crystallization initiated by the latent heat being liberated by the amorphous-crystalline transition within the
larger crystallites belonging to the stars. The radial
distribution of these larger crystallites, on the other
hand, is a clear evidence that their growth is initiated
around a given site in the film, in which site ordering
would develops preferentially due to lacal non-equilibrium conditions affecting the cohesion of the material.
The nature of such a nucleation center is still a con-
jecture but may be approached using arguments presented
elsewhere in this school [4] . Another indication imbed-
ded in this star formation is that their radial growth
could also be favored by a non-homogeneous, possibly
gaussian-like, energy deposition profile. Each star
would then appear to result from a local co-incidence
between film inhomogeneities and laser-beam instabilities.

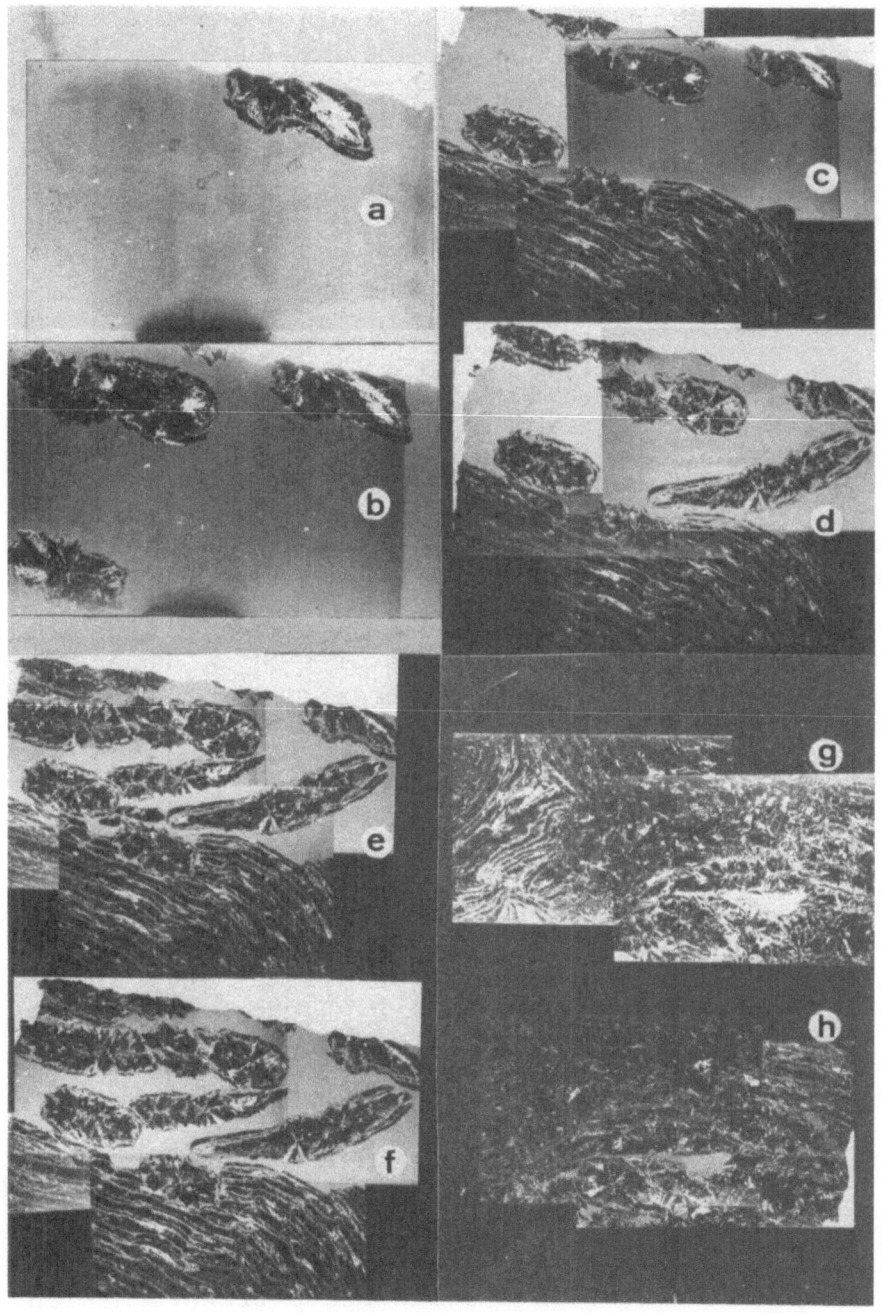

Fig. 1. TEM micrographs of same area of a-Ge film
after one pulse (a) to 8 pulses (h).
Scale: 1 cm is 4 μm.

However, the latter should then be extremely narrow ($\approx$ a few μ), which does not seem to be unconceivable but needs to be evidenced for each particular laser beam actually used.

The star formation described here is not particular to these self-supported Ge films. It also applies to supported Ge films on fused silica and self-supported Si films, with identical stars produced, and therefore seems to be a very general process which may only be slightly affected by film preparation conditions. As an example, Ge films deposited at room temperature onto silica and pulse-irradiated exhibit stars which contain large ($\leqslant$ 200 Å), sometime faceted voids or bubbles [3]. These bubbles are thought to be the product of the coalescence of cavities present in the films before irradiation and crystallization. In films deposited at 150°C, which are still amorphous but with much reduced inhomogeneity (density is then very close to the crystal one), stars have the same characteristics except for the total absence of voids.

This star formation needs to be compared with the scanned, CW laser crystallization of a-GE on fused silica, as reported by J.C.C. Fan et al.[5]. In this experiment, a CW laser beam is focussed to a slit image of 50 μ x 1.5 mm. The a-Ge films are then irradiated through the slit by scanning at a constant speed of 0.2 cm sec^{-1} the films in front of the slit, in a plan parallel to this slit and in a direction perpendicular to the long-axis of the slit. Transmission micrographs of such films show a pattern of successive stripes in the sequence amorphous-fine crystallite-larger crystallites, and again amorphous etc... Each sequence of three extends over 50 μ and the larger crystallites have dimensions 1 x 20 μ. These periodic structural features are observed at low substrate temperature. They disappear at high substrate temperature to leave room for self-sustaining crystallization. The interesting fact is here that the sequence amorphous to crystallites procedes in the same direction as the laser scanning one and not opposite to it, i.e. : in a CW regime heat builds up in the film to progressively reach the crystallization temperature of the film ($\approx$ 700°C), which is however far below the melting point of crystalline Ge (937°C). Therefore, the process is a solid phase transformation, but essentially due to the liberation of the latent heat in the film. That heat progresses rapidly ($\approx$ 200 cm sec^{-1}) along a front parallel to the slit and increases in strength simultaneously ; when the temperature of the

front has reached a sufficient value, then crystallization takes place butthis crystallization <u>precedes</u> by far
the actual displacement of the film in front of the slit.
When the slit image arrives over the crystallized regions,
apparently no further ordering occurs probably because
the laser energy is not large enough to displace the
grain boundaries and increase the size of the once formed crystallites.

Its is only when the slit image covers an amorphous
section of the film that the process resumes. Would this
procedure apply to pulsed laser annealing, then the stars
described above would not exist at random as such (i.e.
large crystals at the center of the stars) or would be
confined to the outskirts of the laser irradiated area
of the film in contrast with observations. This would
tend to prove that pulsed laser annealing of amorphous
elemental semiconducting films is not thermal in essence,
which is effectively the case for a continuous wave
annealing procedure. Indeed, a simple thermal model yields
in the pulsed regime a most probably attained temperature of $\approx$ 350°C at the end of a 10^{-6} sec long pulse of
$\approx$ 10 mJ.cm^{-2} energy. Such conclusions would seem to
find support also in the fact that identical star formations are obtained i) using a ruby laser pulsed at 2×10^{-8}
sec and an energy per pulse reduced to the same value as
above, and ii) for Ge and Si films as well. Finally CW
laser irradiation must be performed under controled Ar
atmosphere to prevent oxydation[5]. Pulsed laser irradiation to the contrary produces the same results under
controled atmosphere or in air. This points at the actual speed at which the crystallization front progresses
through the material in both irradiation procedures. In
the continuous regime, that speed is much too small and
the temperature increase too important to prevent a high
reactivity of the transformed material against ambient
gases.

b. <u>Electrical conductivity</u>

Although more subtle, fluctuations in the density
of the localized electronic states which are associated
with structural inhomogeneities in the irradiated films
are also indicative of configurational short-range ordering which may take place locally, around such discrete
inhomogeneities as voids or cavities and which are related to the existence of unsatisfied (or dangling) bonds.
A model for the nucleation originating at these atomic
sites is developed elsewhere in this school [4] . We would
like here to describe the behaviour of the electronic

conductivity of a-Ge films under low-energy, pulsed laser irradiation and show that actual measurements do provide information on these regions of the film which, although remaining "amorphous" by diffraction means, do evolve under such laser irradiation conditions.

1000 $\overset{\circ}{A}$ thick amorphous Ge films are condensed onto room temperature, fused silica substrate ; gold films, 1 mm apart, are then deposited to allow measurement of the electrical conductivity of a 1 mm^2 section of the a-Ge films. This section of the film is then irradiated through a 1 mm^2 mask with a train of 10^{-6} sec long pulses, at an energy ranging from a few mJ cm^{-2} to 100 m J cm^{-2}. The pulse repetition rate can be varied from 5 to 25 sec^{-1} but is set in general at 10 sec^{-1}. The resistivity of the irradiated film is then recorded on a strip-chart recorder as a function of time during and after irradiation. The resistivity of the film is observed to increase and to saturate after some 10^2 pulses, at all laser energies [6] . The pulse train is then interrupted after 15 sec pulsed irradiation while resistivity is continuously recorded. As mentioned earlier, the supported films which are irradiated at such energies exhibit stars which are not in general inter-connected. The actual proportion of stars against the total volume of the film remains of the order of $\approx$ 5% at, say, 50 mJ cm^{-2}. Therefore, the conductivity (σ) changes which are obtained under low-energy pulsed laser irradiation are essentially due to variations in the transport properties of the still amorphous film. At energies below 10 mJ cm^{-2}, fig.2, the evolution of σ is seen to be completely reversible after interrupting the pulsed irradiation and symmetric about the time t_1 of interruption. At energies between 10 and 18 mJ cm^{-2}, the reversibility of the changes is still obtained but the time constants involved during and after irradiation, respectively, are different. When the energy approaches 18 cm J cm^{-1} the time necessary to return to its original (before irradiation) value σ_o increase dramatically to a few hours. At $18^{\pm}1$mJ cm^{-2}, σ does not return to σ_o and remains at its saturation value σ_s. Above 18 mJcm^{-2}, the process is no longer reversible ; σ reaches a higher σ_s value and relaxes to a final and stable value higher than σ_o. The time constant of this last process remains nearly constant ($\approx$10 min.) when further increasing the laser energy.

The σ behaviour below 10 mJcm^{-2} is easily understood if one assumes that excited electrons decay into the vallence states with a life time of the order of $\sim 10^{-9}$ sec,

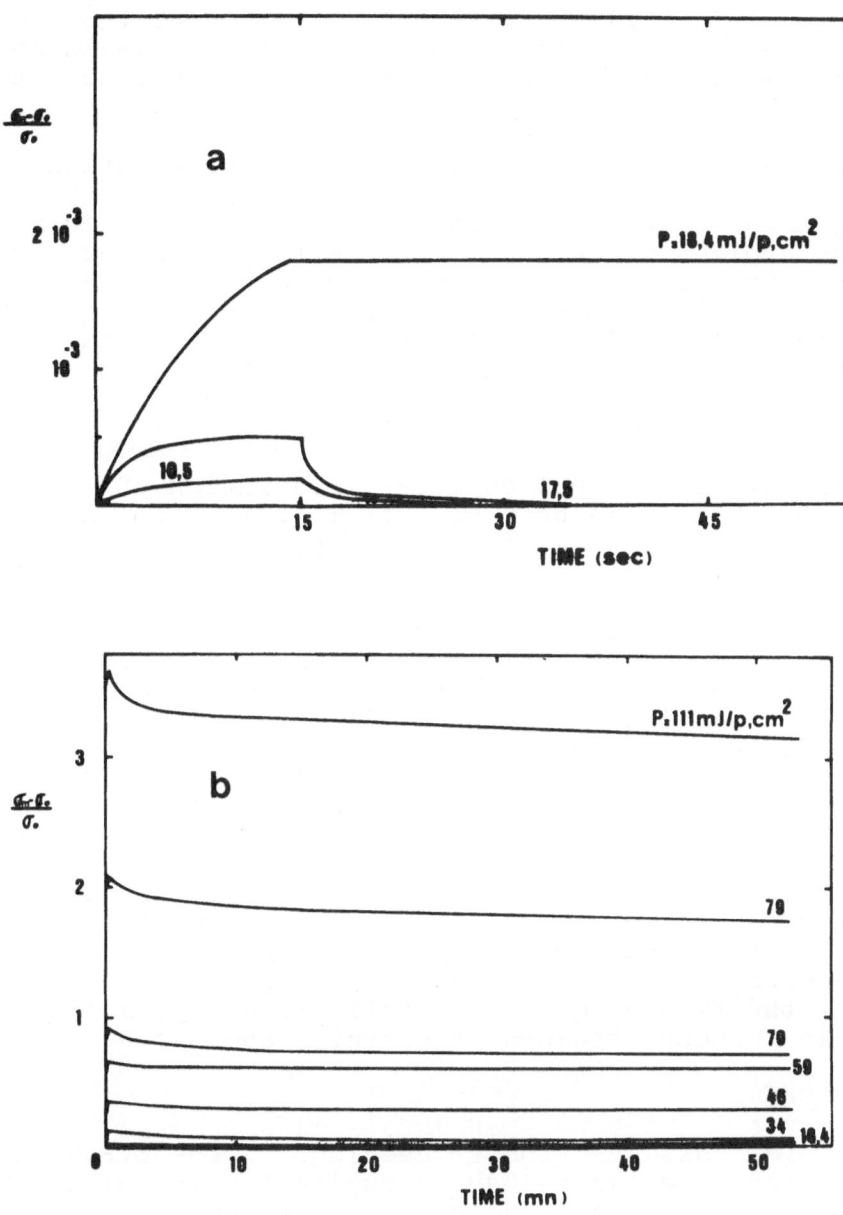

Fig. 2. Conductivity of a-Ge films under pulsed irradiation, at energies 5-18 (a) and above 18 mJ/cm²/
pulse (b).

for instance. The time constant of this relaxation process after t_1 is then simply the one following one single pulse and implies that any fluctuation of the atomic
configuration is elastic. When asymmetry prevails about
t_1, between 10 and 18 mJcm^{-2}, obviously such fluctuation
takes place which is not entirely elastic and is ampli-
fied from pulse to pulse. At 18 mJ cm^{-2} and above, fluctuations are inelastic, i.e. permanent perturbations are
produced in the films. Yet such films appear to be amorphous, except for the few stars. Obviously, their amor-
phicity evolves which does not seem to be related directly with the star formation, but may be considered to represent, at 18 mJ cm^{-2}, the necessary densification stage of the films which should precede any full scale crystallization.

CONCLUSIONS

        The effects of pulsed laser irradiation of amorphous
elemental films have been here examplified to consist of
two processes : i) one macroscopic process which relates
to the formation of large circular crystallization spots,
made of a radial assembly of $\approx$5 µ large crystallites often surrounded by a finely crystallized, 0.5 µ wide region ; ii) one microscopic process which perturbs the
transport properties of the films. Both processes would
seem to be possibly approached via models involving inhomogeneities in both films and laser pulses. A comparison with a continuous wave irradiation of similar films
indicate that the star formation observed in the pulsed
regime would not be of the same nature as the CW regime
whilst the conductivity evolution under a pulsed regime
gives evidence for an irreversible evolution at a critical laser energy of 18 mJ cm^{-2} for 1000 Å Ge films on
fused quartz. That critical energy might be an evidence
for reaching, at that energy, the homogeneous densification of the film prior to any full scale crystallization
which is further attained at energies above $\approx$ 50 mJ cm^{-2}.

REFERENCES

1. See, for instance, the Proceedings of the E.P.S. study
   Conference on Laser-Induced Nucleation in Solids, Mons,
   4-6 Oct. 1979, in J. de Physique C4 - p.1 to 112.
2. C. Hill and D.J. Godfrey in Ref. 1, p. 79.
3. R. Andrew and M. Lovato, J. Appl. Phys. 50, 1142
   (1979) ; and M. Lovato, unpublished.
4. M. Wautelet and M. Failly-Lovato, this school.
5. J.C.C. Fan, H.J. Zeiger, R.P. Gale, R.L. Chapman,

Appl. Phys. Letters, 36, 158 (1979).

6. M. Lovato, M. Wautelet and L.D. Laude, Appl. Phys. Letters, 34, 160 (1979).

7. M. Lovato, R. Andrew, L.D. Laude, M. Wautelet, to be published.

LASER INDUCED STRUCTURAL CHANGES IN THE BULK AND AT DEFECT SITES IN SEMICONDUCTORS

M. Wautelet and M. Failly-Lovato

Faculté des Sciences, Université de l'Etat
·B-7000 Mons, Belgium

INTRODUCTION

When an amorphous semiconducting material is irradiated by a laser beam, some effects may occur depending
on laser photon energy and power. At very low power, the
optical absorption and luminescence properties probe the
electronic structure of the unchanged amorphous phase.
At low power, photostructural changes are observed in
some materials, as evidenced by X-ray diffraction, photoemission, dilatometry, optical absorption,... (Tanaka
1980). More subtile modifications are the origin of persistent photoconductivity observed in some materials
(Sheinkman and Shik, 1976; Lang et al., 1979; Wautelet
et al., 1980). When photo-bleeching occurs, oscillations
in the optical absorption are observed at high power under certain experimantal conditions (Hajto and Apai,
1980). At very high power, nucleation and crystallisation
of the amorphous films occur (see for instance Baeri et
al., 1979; Andrew and Lovato, 1979; Van Vechten et al.,
1979; and other papers in this school).

In this work, attention is paid on the laser power
region where reversible photostructural and photoconductivity changes are observed. Experimental results will
be given and compared with some theoretical models. The
range of validity of the different theories will then be
discussed, with particular emphasis on the case of a-Ge
and structural relaxation at dangling bond sites of tetrahedrally bonded semiconductors.

EXPERIMENTAL RESULTS

   Reversible photostructural changes have been observed on a number of amorphous or glassy materials. A vast
number of works have been performed on the subject during
the last years, so that a complete review is beyond the
scope of this seminar, and only the main results will be
given.

Amorphous Chalcogenides

   Reversible photostructural changes are observed by
X-ray diffraction and thin film dilatometry on a-As_2S_3
and As_2Se_3 and a-GeS_2 and $GeSe_2$ (Hamanaka et al., 1976;
Tanaka, 1980). Thickness changes in the films are of the
order of $5x10^{-3}$ and below. A simultaneous shift of the
optical absorption edge is measured. In a-As_2S_3, photo-
darkening occurs, i.e. the absorption edge is lower in
energy after light exposure. The shift may attain 0.05eV.
In a-GeS_2 and a-$GeSe_2$, one sees photobleaching, i.e. an
increase in energy of the absorption edge after light
exposure. It is important to note that neither photodarkening nor photobleeching are observed in crystalline
materials.

   The electronic structure of amorphous germanium
chalcogenides was also probed by means of the photoemission technique (Takahashi and Harada, 1980). Structures
sensitive to bond angles have been shown to evolve under
light exposure. Photo conductivity and photoluminescence
also change. Typical luminescence is shown in figure 1
for a-As_2Se_3. It is to be noted that the excitation (E)
and luminescence (L) spectra are well separated. This
provides evidence for strong electron-phonon coupling
with a Stoke's shift. This will be discussed in more detail later, in connection with the theoretical models.

   Also intersting is the variation of the electron
spin resonance (ESR) signals under irradiation. To summa-
rise, let us say that ESR gives a measure of the number
of unpaired spins in the studied materials. No ESR signal
is detected in non irradiated a-As_2S_3 and a-As_2Se_3 samples, while complex ones are seen at low temperature after irradiation (Bishop et al., 1975).

   After light illumination, the starting conditions
may be restored either by thermal annealing or by exposure to less than band gap light (Tanaka, 1978).

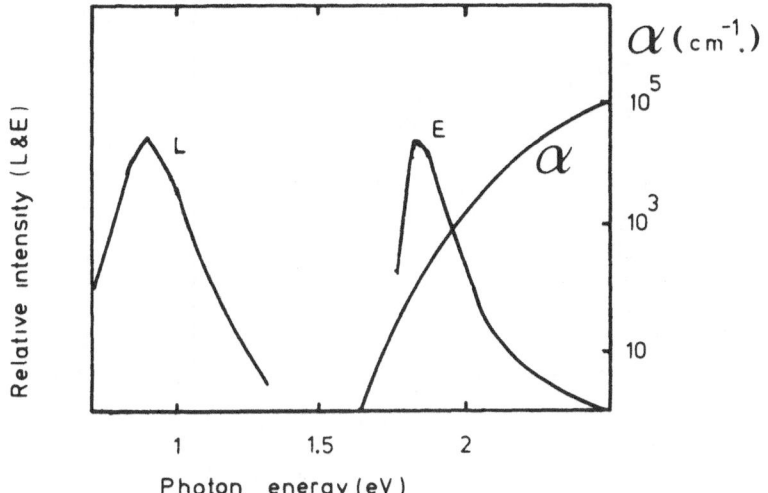

Fig. 1. Typical luminescence (L), excitation (E) and absorption spectra (α), for a-As$_2$Se$_3$ at low temperature.

Amorphous silicon

Neither photostructural changes, photodarkening, nor photobleeching have been observed in a-Ge or a-Si. However, reversible conductivity changes were seen in glow discharged deposited a-Si (Staebler and Wronski, 1977), which is relevant to the present discussion. During light exposure, the photoconductivity increases first by about two orders of magnitude (figure 2), then decreases by nearly a factor of 8 after 4 hours. When the light is turned off, the conductivity decreases nearly by four orders of magnitude from its initial value. This new state is stable at room temperature. The photoconductivity behaviour as a function of the probe intensity, before and after illumination, changes also reflecting a change in the recombination kinetics of the two states. The original state is restored by thermal annealing.

Let us note that luminescence and light-induced ESR have been observed in similarly prepared materials, which contain H bonds (Street and Biegelsen, 1980).

Amorphous germanium

A reversible conductivity change was seen in vacuum-evaporated a-Ge films (Lovato et al., 1979). After laser

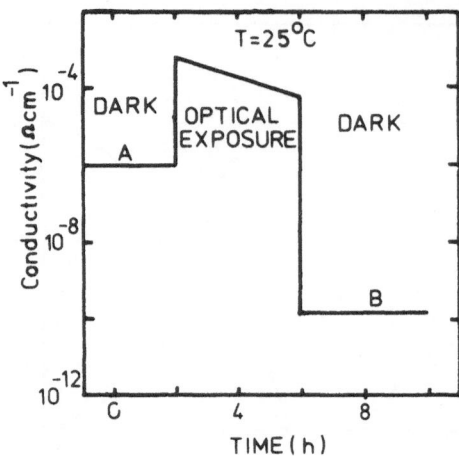

Fig. 2. Conductivity as a function of time before, during
        and after exposure to light in the wavelength
        range 6000-9000 A.

irradiation, the original conductivity is restored after
a time depending on the laser power. This recovering time
extends from seconds to several hours.

THEORETICAL MODELS

     Although apparently different, all these phenomena
seem to be related to similar models of interpretation.
Indeed: 1) they deal with amorphous and covalently bonded materials; 2) they are related with electronic states located in the band gap, since conductivity, luminescence, optical absorption are related to band gap states;
3) they involve probably strong electron-phonon coupling
and metastable states, as seen from luminescence and the
fact that the phenomena are reversible. From these comments, one can say that all these phenomena are related
to localised or defect states. As a first step, let us
see how they are related to gap states.

Bonds and electronic states in amorphous semiconductors

     Let us first consider the case of Si and Ge. These
atoms have four electrons (s and p) outside their closed
shells. They can then hybridise into so-called sp^3 hybrid
states. In this situation, atoms have the tetrahedral

symmetry with four unpaired orbitals. They will combine
with other similar atoms by covalent bonding, resulting
in a diamond-like crystal structure. In the amorphous
phase, it is generally recognised that the fourfold coor-
dination is retained for all atoms, except at cavities,
defects sites, so-called dangling bonds and the surface.
The main difference with the crystalline solid is that
bond angles and bond lengths fluctuate around the crystalline values. These fluctuations induce simultaneous
changes in the electronic structure of the solid.

In order to show this, let us start with a method
of linear combination of atomic orbitals, namely the
bond orbital method (BOM) (Pantelides and Harrison, 1975;
Ciraci et al., 1975). One constructs bond orbitals, b_i,
from sp^3 hybrid ones, located on atoms connected by the
corresponding bond orbitals, i.e. :

$$b_i = (h_1 + h_2)/\sqrt{2} \tag{1}$$

where h_1 and h_2 are hybrid orbitals wavefunctions of
atoms 1 and 2 of the bond. In a regular tetrahedron, the
matrix Hamiltonian is a 4x4 one, given by :

$$\underline{H}_o = \begin{matrix} \alpha & \beta & \beta & \beta \\ \beta & \alpha & \beta & \beta \\ \beta & \beta & \alpha & \beta \\ \beta & \beta & \beta & \alpha \end{matrix} \tag{2}$$

Two eigenvalues are found : $E=\alpha-\beta$ (threefold degenerated)
and $\alpha+3\beta$ (non-degenerated). If a distorsion towards a
threefold symmetry occurs, H_o is no more valid. The matrix Hamiltonian is now :

$$H_1 = \begin{matrix} \alpha_1 & \beta_1 & \beta_1 & \delta \\ \beta_1 & \alpha_1 & \beta_1 & \delta \\ \beta_1 & \beta_1 & \alpha_1 & \delta \\ \delta & \delta & \delta & \alpha_2 \end{matrix} \tag{3}$$

with solutions :
$$E(1,2) = \alpha_1 - \beta_1$$
$$E(3,4) = \{(\alpha_1+\alpha_2+2\beta_1) \pm [(\alpha_1-\alpha_2+2\beta_1)^2 +12\delta^2]^{1/2}\} /2 \tag{4}$$

From the electronic point of view, the trigonal distorsion re-
moves the degeneracy of the threefold degenerated level of the tetrahedron, and splits it into a singlet and a doublet. The total electronic energy is also changed from 4α to $3\alpha_1+\alpha_2$, as calculated from
inter-bond and intra-bond matrix elements. Ciraci et al. (1975)
give :

$$\alpha_1 = 4 \langle b_i(\underline{r})|H|b_i(\underline{r}-\underline{R}_1)\rangle$$

$$\alpha_2 = \langle h_1(\underline{r})|\ H|h_1(\underline{r})\rangle - \langle b_i(\underline{r})|\ H|b_i(\underline{r})\rangle$$

$$\beta_1 = 2\ [\langle b_i(\underline{r})|H|b_j(\underline{r})\rangle + \langle b_i(\underline{r})|H|b_j(\underline{r}-\underline{R}_1)\rangle]$$

$$\delta = \langle h_1(\underline{r})|H|b_j(\underline{r})\rangle + 2\ \langle h_1(\underline{r})|H|b_1(\underline{r}-\underline{R}_1)\rangle\ ,$$

(5)

$|h_1(\underline{r})\rangle$ is the wavefunction associated with the bond parallel to the threefold axis, and $|b_i(\underline{r})\rangle$ correspond to the three equivalent bonds, the so-called back-bonds. $\underline{R}_1$ are the fcc primitive translation vec-

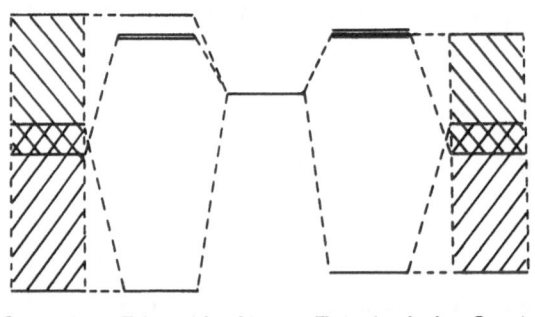

Crystal   Trigonal   Atom   Tetrahedral   Crystal

Fig. 3. Schematic energy level diagram for Si and Ge following
their hybridisation.

tors in the case of crystals. Deformation of the atoms behind the back-bonds are described by the $b_j(\underline{r}-\underline{R}_1)$.

When tetrahedra are only slightly deformed, matrix elements involving bonds on non-adjacent atoms do not vary much, so that α_1 would remain constant. The same occurs for β_1, when matrix elements between back-bonds on the same atom are constant. Then, the solutions $E(1,2)$ of equation (4) are not affected by the slight distorsion of the tetrahedra. The energies of the bonds are easily calculated from atomic s and p orbitals. Let us call u the relative displacement of the atomic nucleus in the direction parallel to the three-fold axis (u=0 for the regular tetrahedron ; u=1 when the three back-bonds are coplanar). After some manipulations, one finds :

$$\alpha_2 = \alpha_o + (u^2-2u)(E_s-E_p)/3$$

$$\delta = \delta_o + (1-u)(1-u^2+2u)^{1/2}(E_s-E_p)/4\sqrt{3}, \qquad (6)$$

where α_o and δ_o are constant. E_s and E_p are the atomic s and p orbitals energies, respectively. Upon distorting slightly the tetrahedron, one obtains that $E(1,2)$ remains constant, while the upper level, $E(3)$, increases and $E(4)$ decreases with increasing u, as shown schematically in figure 3.

It has to be reminded that the previous relations are valid when the tetrahedra are slightly distorted. Although the shape of H_1 is always valid under trigonal distorsion, it is obvious that in the case u=1, α_1 and β_1 as well as α_2 and δ are much modified compared to the tetrahedral case. To prove this, it is sufficient to recall the differences between graphite and diamond.

When bond orbitals are combined to make a solid, one has to make Bloch sums from the bond orbitals wavefunctions. As a result, levels spread into bands as shown in figure 3. From the previous discussion, it comes that energy levels may arise in the gap and that the electronic structure of the amorphous phase may be different from the crystal one, due to bond angle fluctuations (Joannopoulos, 1977).

A similar reasoning may be applied to Se and Te. These atoms have six external electrons. The atomic orbitals are hybridised into (1) two equivalent σ hybrids pointing toward the nearest neighbours, h_1 and h_2, and (2) two equivalent lone-pair hybrids, h_3 and h_4, as shown in figure 4. The σ hybrids of two neighbouring atoms pointing toward each other, interact strongly to form a bonding (B) and an antibonding (AB) orbitals. Te and Se atoms are aligned along chains. When performing calculations, two angles are defined : the bond angle, θ, and the dihedral angle, φ. Due to the structure, electro-nic energy levels have to be calculated by taking into account θ and φ, i.e. more than one atom has to be considered (Hulin, 1963; Chen, 1973; Shevchik, 1974; Joannopoulos et al., 1975; Nizzoli, 1977). Results of electronic structure calculations are schematical-ly indicated in figure 5, together with the associated bands. The lowest band is associated with s states, while the second and third ones correspond to the B and lone pairs respectively. The unoccupied band has AB character. Lone pairs have essentially pure p-like char-acter. From this one can deduce qualitatively the influence of θ and φ on the electronic structure of Se and Te. θ affects mainly

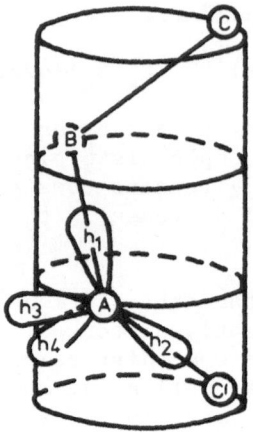

Fig. 4. Configuration of chains in Se and Te, and orbitals described
in text.

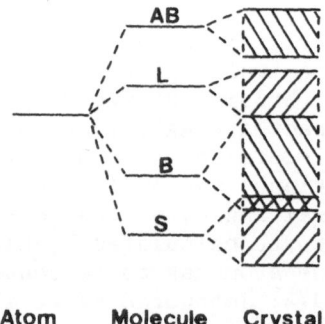

Fig. 5. Schematic energy level diagram for Se and Te.

the hybridisation of the bonds, which have mixed s, p and d character. When $\theta = 90°$, they have pure p character. The s, B and AB levels will then be affected by a variation of θ. As a result, fluctuations in θ may only create gap states in the AB region, i.e. at the top of the fundamental band gap. The dihedral angle has more influence on the lone-pair band, mainly on its shape (Shevchik, 1974).

As is intermediate between Ge and Se in the periodic table of the elements. It has two external s electrons and three p ones. All three orbitals participate to the bonding, so that no lone-pair component is expected. As a result, As is three-fold coordinated.

As_2S_3 and As_2Se_3 structures result from a combination of the bonding characteristics of As and Se, i.e. each As atom is covalently bonded to three S or Se atoms in a triangular pyramidal unit, and each S or Se atom is shared by two As atoms. The resulting crystalline network is shown in figure 6. As and Se atoms are associated in layers. Bonding between layers is Van der Waals like, thus weak, and the crystals may be easily cleaved. Glassy As_2S_3 and As_2Se_3 have the same nearest-neighbour and next-nearest-neighbour coordinations as the crystals. Their electronic structures have been calculated using the tight-binding method (Bishop and Shevchik, 1975). It is interesting to know how structural changes are related to the electronic structure, in order to correlate them with the photodarkening effect. One has to combine As and S or Se orbitals. s bands associated with Se and As are the lowest ones (Figure 7). The As and Se p orbitals are intimately mixed in a bonding p band. This band is slightly separated from the lone-pair band arising from the Se lone orbitals, as in pure Se. Finally, the fundamental semiconducting band gap separates this lone-pair band from the upper-lying antibonding p band. It is to be noted that interactions between layers are neglected in this approach. These inter-layer bondings are probably best seen by lone-pairs, since they are located in the exterior zone of the layers, and directed toward the neighbouring layer. In the tight-binding approach, more interaction between bonding orbitals is known to broaden the corresponding bands. A slight joining between chain is then able to broaden the lone-pair band and to reduce the band gap. A decrease in the anti-bonding energy (i.e. also a reduction of the band gap) may be obtained if the nearest-neighbour distance increases.

The structure of GeS_2 and $GeSe_2$ results also from covalent bonding between the Ge and S or Se atoms. A Ge atom is bonded to four Se atoms, each Se being shared between two tetrahedra. The resulting crystal structure is a layered one, as shown in figure 8. As for As_2S_3, the amorphous structure is quite close to the crystal one. Again a tight-binding approach is helpful. This is summarised in figure 9, following Lannoo and Bensoussan (1977). The lower lying s band is separated from a group of three bands associated with p (Ge and Se) bonding orbitals, lone-pair (Se) and antibonding p

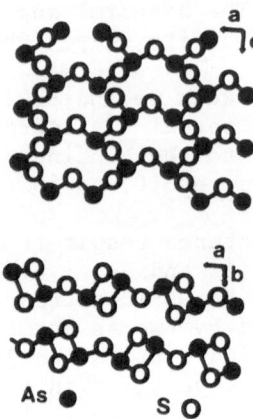

Fig. 6. Atomic structure of As_2S_3.

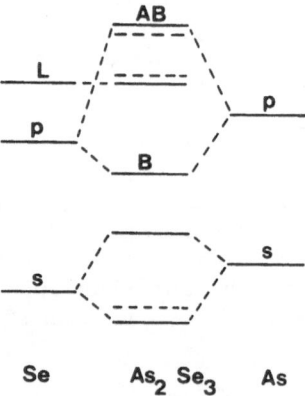

Fig. 7. Schematic energy level diagram of As_2Se_3. Full : crystal
case ; dashed : with a decrease of interlayer distance
and an increase of nearest-neighbour separation.

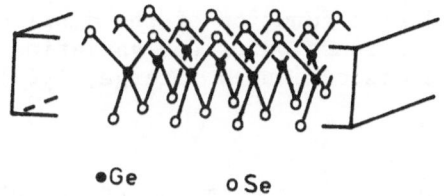

•Ge o Se

Fig. 8. Atomic structure of GeSe$_2$.

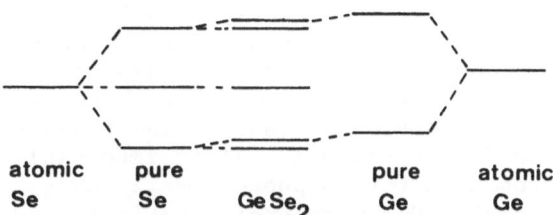

atomic	pure		pure	atomic
Se	Se	GeSe$_2$	Ge	Ge

Fig. 9. Schematic energy level diagram of GeSe$_2$.

orbitals. The Fermi level is located between the lone-pair and anti-
bonding bands.

Total energy and optical excitation

In amorphous materials, orbital geometries vary from atom to
atom, and energy levels in the gap arise from localisation. When an
atom is displaced from its equilibrium value, its associated elastic
energy vary together with the electronic energy. The elastic energy
is known to be a quadratic function of the displacement of the nuclei
more precisely, it is a function of the variation of the configurational coordinate of the distorsion, u (Sturge, 1967), i.e.

$$E_1 = Cu^2 \tag{7}$$

where C is constant for a given atom and takes into account the reaction of the surrounding lattice to the distorsion.

In the case of the distorted tetrahedron, the variation of the
total electronic energy may be calculated from equations (5) and (6):

$$\Delta E_{el} = (u^2 - 2u)(E_s - E_p)/3 \tag{8}$$

During optical absorption, electrons are excited. But the Franck-
Condon principle requires that atomic nuclei do not move during optical excitation. When electrons go from (delocalised) valence to (delocalised) conduction states, the interatomic equilibrium distances of
the non excited and excited solids are the same, i.e. the minima of
the total energy of these states occur for the same u, as illustrated
in figure 10a. If the optical absorption couples defect or impurity
states, the situation may be different. For instance, let us assume
that a transition occurs between a sp^3 fundamental state and sp^2-p_z
excited state. The two hybrid configurations have different energy
versus u curves (Figure 10b). The equilibrium fundamental and excited
states are in A and D, respectively. It is impossible to excite directly the atom from A to D by a photon energy $h\nu = E_2$, due to the
Franck-Condon principle. One must excite optically an electron from
A to B. However, the resulting state, B, is not stable and the system
relaxes upon releasing some of its energy to the solid. It is worth
noting that this vibrational energy is released in the immediate
vicinity of the site which behaves, therefore, as a localised source
of energy. From C, there is some probability to go to A and D. From
D, it is possible to recombine into A by different mechanisms, the
details of which depend on the position of C relative to A and D in
the u-direction. (1) In the situation of figure 10b, it is not possible to go "vertically" from D to A, i.e. no luminescence is expected; (2) If A and D are nearly on the same "vertical" line, luminescence will appear, and the luminescence and absorption spectra will

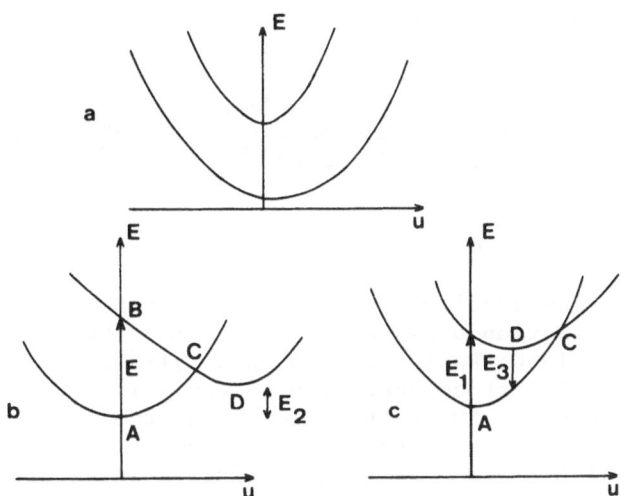

Fig. 10. Configurational coordinate diagram of the total energy of
the fundamental and excited sates corresponding to a)
delocalised states; b) localised states, without luminescence; c) localised states, with luminescence.

coincide (3) If D is between A and C in the u-direction (Figure loc),
vertical transitions are possible paths of recombination, but the
luminescence and absorption spectra will be separated by the so-called Stokes shift : $E_1 - E_3$.

Other "structural" transitions involving three or more configurations are obviously possible and will be discussed separately in
each case. Nonetheless, one can say that all the experimental situations given above may be described by the present model, since they
imply at least the existence of one stable and one metastable states.
In the case described in figure 10b, it is worth noting that one may
attain easily by optical excitation a meatstable state which might

otherwise be obtained with many difficulties by thermal annealing.
Indeed, if E_2 is in the range 1 eV or above, passing through C is
statistically unprobable by only heating the solid, i.e. only a
very small fraction of sites will be excited under normal experimental conditions. Photostructural changes are the most studied examples
dealing with this problem, but it may be easily predicted that new
physical phenomena might be discovered in connection with the subject.

Let us now review the three models given to explain photo-induced changes, namely a dehybridisation of defects, the double well
potential and the self-trapped exciton.

The dehybridisation of defects

This model was developped for the interpretation of the photoconductivity relaxation observed in a-Ge films (Lovato et.al., 1979;
Wautelet et al., 1980; Failly-Lovato et al.,1980). It is based on
the sensitivity of photoconductivity to variations of the electronic
structure in the vicinity of the fundamental band gap. In a-Ge and
Si, it is generally recognised that room temperature conductivity
is associated with states in the gap. Since these are thought to
arise from dangling bonds (DB) (Thomas et al., 1978; Brodsky and
Kaplan, 1979), it seems natural to associate persistent photoconductivity (PP) with DB's properties. The relation between the distorsion of the tetrahedra and their electronic properties was studied
before. The main hypothesis of the model are that (1) there exist
at least two neutral DB configurations : the normal tetrahedral
sp^3 one and the trigonal sp^2-p_z one; and (2) the electronic energy
level of th p_z state of the trigonal configuration lies near the
top of the band gap.

In this case, the effect of laser irradiation with photon energy greater than the band gap is seen as follows : photons excite
electrons from delocalised DB states, either to localised excited
states (Figure 10b) or to delocalised states (Figure 11). The first
process is phonon assisted and occurs with a weak probability. This
seems opposite to experiments. The transition to delocalised states
is probably more realistic. The DB-atom is now in a $(sp^2)^+$ ionic
state, the spatial equilibrium configuration of which is coplanar.
As assumed before, the relaxation to this configuration creates an
empty state in the band gap. The latter may be filled via recombination of an excited electron into the DB site. The system may remain in this excited (neutral) state for a long time, since returning to the original configuration requires a distorsion of the DB
site via C and an activation energy which can be achieved by vibrational activation, but not by optical excitation. Indeed, like the
starting excitation process, this leaves the DB site in a $(sp^2)^+$
planar configuration. This process may also operate with other complex atomic configurations similar to those proposed by Adler (1978).

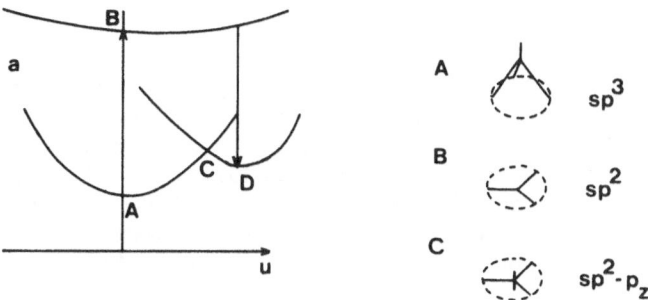

Fig. 11. Configurational coordinate diagram of the total energy at
the dangling bond site in amorphous germanium, and the
corresponding geometries of the dangling bond associated
atoms : A : sp^3 ; B : sp^2 ; C : sp^2-p_z.

In a-Ge, PP is related to this effect since, from state D, electrons
may be excited into a delocalised conduction state via the Fermi-
Dirac distribution law and participate to the electrical conductivity
until they recombine. This is similar to the n-type conductivity in
crystalline semiconductors. The present mechanism may not be important when luminescence is observed, since it is much more rapid.

The present model may also explain the fact that the relaxation
time increases with the light intensity. Until now, it was assumed
that the substrate plays practically no role in the geometry of DB
sites. However, it has to be recognised that after excitation of electrons, the system tends to relax to state D (Figures 10 and 11), but
the bulk reacts to minimise this deformation. This means that the
effective configuration of the DB site is not at D, but between C
and D. If more electrons are excited, it may reasonably be expected
that it will be easier to deform the DB site. In other words, the
excited configuration goes to D and simultaneously the relaxation
time increases since the activation energy increases.

    It is also to be noted that the previous discussion was restric-
ted to one particular DB site. It is obvious that the energy scale
may differ from one DB to any other, since localised electronic
states occupy energy "bands" in the amorphous materials.

The double well potential

    In a-As_2S_3 and GeS_2 type compounds, photostructural changes
are observed together with luminescence at low temperature. As discussed above, the two phenomena cannot be explained by the same model, if one considers only two states (Mollot et al., 1980). In order
to explain photostructural changes, Tanaka (1980) proposed a model
of bistable local bonding geometries, with a corresponding double
well potential. The photo-induced process is as follows : optical

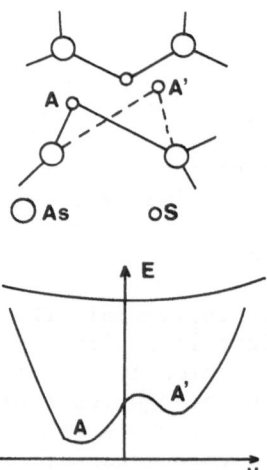

Fig. 12. a) Schematic model of local bonding geometries in As_2S_3.
         b) Corresponding total energy diagram.

absorption results in the excitation of lone-pair (S or Se) electrons
into anti-bonding states. This gives rise to a change in interactions
between lone-pair electrons on two different chalcogens and in bonding between chalcogen and Ge or As atoms. As a result, transition
from A to A' configurations may occur (Figure 12). A' configuration
remains stable after a hole on the chalcogen atom recombines with
an electron, provided the potential barrier between A and A' is sufficient (Figure 12b). As discussed before, photo-darkening in a-As_2S_3
and As_2Se_3 arise from a joining of layers and/or nearest-neighbour
As-chalcogen distance increase. Such an effect would be visible on
the electronic structure, as observed by photoemission. We know of
no photoemission work on these compounds. Fortunately, they exist on
a-GeS_2 and $GeSe_2$ (Takahashi and Harada, 1980), where photo-bleeching
occurs. One observes a decrease in energy of the lone-pair band and
no shift of the p bands. This indicates that no important bond
length variation occurs, but that lone-pair states are affected.

We noted before that luminescence cannot be explained by this
model. This leads to the conclusion that photostructural changes
and luminescence have different origins. The main reason seems to be
that photostructural change are bulk effects, while luminescence
occurs from defect properties. For the bulk effect, this would mean
that the general remark given above, i.e. the equilibrium interatomic
distance is not sensitive to transitions between delocalised states
(Figure 10a) is not valid for the amorphous materials studied here.

The self-trapped exciton

The photo-induced changes in the electrical properties of glow-
discharge deposited a-Si:H observed by Staebler and Wronski (1977)
may be interpreted by a model of charged defect centres or self-
trapped excitons (Elliott, 1979), described by figure 13. Optical
excitation involves a transition from the ground state to an exciton state. Desexcitation may occur by direct return to either the
ground state, or to a metastable state which was shown to be equivalent to the formation of an intimate pair of charged centres,
DB^+-DB^-. The formation energy of this pair is lowered by the Cou-
lomb interaction between them. It is assumed by Elliott (1979) that
in most cases, the excitons will recombine directly but, in a few
cases, an intimate pair of charged defects will form. Once formed,
these defect pairs will be relatively stable and, as discussed above
for our model of a-Ge, can be destroyed only by thermal activation.
This explains the behaviour of the electrical conductivity observed
by Staebler and Wronski (1977). They irradiated undoped and n-type
semiconductors. After optical excitation, the number of charged defects is increased. This decreases the Fermi level (Adler, 1980), as
observed experimentally. More details about this model may be found
in the original works.

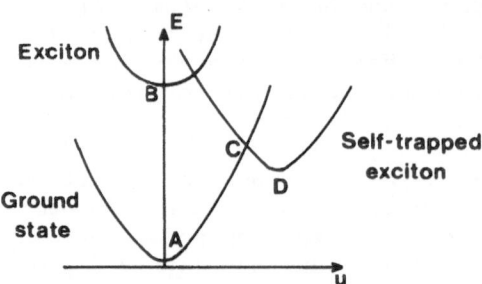

Fig. 13. Configurational coordinate diagram in the self-trapped
exciton model.

CONSEQUENCES OF THE DEFECT RECONSTRUCTION

    The model of dehybridisation of defects is mainly valid for
dangling bond sites in Ge and Si. Apart from amorphous materials,
DBs are also present at vacancies in the crystal, at the surface
and at dislocations. Let us briefly discuss the cases of the Si(111)
surface and dislocations.

Si(111) 2x1 surface

    It is known that three ordered forms exist for the Si(111) sur-
face : 2x1, 7x7 and 1x1. After cleavage in ultra-high vacuum and at
room temperature, the 2x1 structure prevails. This is a metastable
phase which transforms irreversibly into the 7x7 one by heating. It
is generally admitted that the Si(111) 2x1 surface is characterised
by alternative rows of lowered and raised atoms (Haneman, 1961).
These raised and lowered atoms are thought to be negatively and positively charged, respectively (Appelbaum and Hamann, 1976). This
is similar to the situation discussed above. There exist two energy
configurational coordinate curves corresponding to each kind of
surface atoms. During laser irradiation, the relative populations
of the electronic states of the raised and lowered atoms are modified, i.e. their electronic energy changes. The electron population
of the once raised atoms decreases since valence electrons are excited, which will produce the electronic energy of these raised atoms

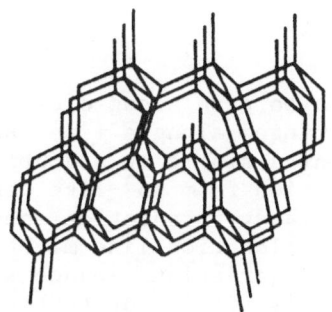

Fig. 14. Example of dislocation in Si.

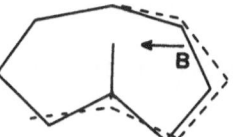

Fig. 15. Two-dimensional effect of dangling bond reconstruction
in an amorphous material.

to be lowered. As a result, their total energy is modified, leading
to a different deformation of the original tetrahedra. In contrast,
the energy of the initially lowered atoms will be increased. This
should modify the atomic geometry of the surface in the direction
of a 2x1 to 1x1 transformation. It is expected that this will have
consequences on surface phenomena, more particularly on crystal
growth.

The dislocation

Dangling bonds are also seen along dislocations (Figure 14).
They are periodically arranged along a line. However, it is not
known whether they present a single or a double periodicity along
the line. If a transformation from sp^3-like configuration to an
sp^2-p_z one occurs during laser irradiation, one may easily see that
the displacement of the dislocation line might be enhanced. As seen
from figure 15, bond angle variations induce strains in the cavity
and it might be that this is sufficient to break another bond (B).
This process modifies the bonding energy of B, hence decreasing the
migration energy of the dislocation. Calculations are needed to prove
this point quantitatively, since dislocations displacements are
important in nucleation and crystal growth processes.

CONCLUSIONS

Laser-induced structural changes may then be explained using
simple models involving electron-phonon couplings. In particular,
a reconstruction at dangling bond sites in amorphous Ge and Si occurs
probably, and would have important consequences about laser-induced
processes. The fact that long-lived metastable states may be obtained
by light irradiation, but not by thermal annealing, is an intersting
subject for future research, and might lead to applications in the
domain of semiconductor technology.

This work was supported by project IRIS of the Belgian Ministry
for Science Policy.

REFERENCES

Adler, D., 1978, Phys. Rev. Lett., 41:1755.
Adler, D., 1980, J. Non-Cryst. Solids, 35-36:819.
Andrew, R., and Lovato, M., 1979, J. Appl. Phys., 50:1142.
Appelbaum, J.A., and Hamann, D.R., 1976, Rev. Mod. Phys., 48:479.
Baeri, P., Campisano, S.U., Foti, G., and Rimini, E., 1979,
    J. Appl. Phys., 50:788.
Bishop, S.G., and Shevchik, N.J., 1975, Phys. Rev. B, 12:1567.

Bishop, S.G., Strom, U., and Taylor, P.C., Phys. Rev. Lett.,
     34:1346.
Brodsky, M.H., and Kaplan, D., 1979, J. Non-Cryst. Solids, 32:431.
Chen, I., 1973, Phys. Rev. B, 7:3672.
Ciraci, S., Batra, I.P., and Tiller, W.A., 1975, Phys. Rev. B, 12:
     5811.
Elliott, S.R., 1979, Phil. Mag. B, 39:349.
Failly-Lovato, M., Andrew, R., Laude, L.D., and Wautelet, M., 1980,
     to be published.
Hajto, J., and Apai, P., 1980, J. Non-Cryst. Solids, 35-36:1085.
Hamanaka, H., Tanaka, K., and Iizima, S., 1976, Sol. St. Comm., 19:
     499.
Haneman, D., 1961, Phys. Rev., 121:1093.
Hulin, M., 1963, Ann. Phys.(Paris), 8:647.
Joannopoulos, J.D., Sclüter, M., and Cohen, M.L., 1975, Phys. Rev. B,
     11:2186.
Joannopoulos, J.D., 1977, Phys. Rev. B, 16:2764.
Lang, D.V., Logan, R.A., and Jaros, M., 1979, Phys. Rev. B, 19:1015.
Lannoo, M., and Bensoussan, M., 1977, Phys. Rev. B, 16: 3546.
Lovato, M., Wautelet, M., and Laude, L.D., 1979, Appl. Phys. Lett.,
     34:160.
Mollot, F., Cernogora, J., and Benoit à la Guillaume, C., 1980,
     J. Non-Cryst. Solids, 35-36:939.
Nizzoli, F., 1977, Il Nuovo Cimento, 39B:135.
Pantelides, S.T., and Harrison, W.A., 1975, Phys. Rev. B, 11:3006.
Sheinkman, M.K., and Shik, A.Ya, 1976, Sov. Phys. Semicond., 10:128.
Shevchik, N.J., 1974, Phys. Rev. Lett., 33:1572.
Staebler, D.L., and Wronski, C.R., 1977, Appl. Phys. Lett., 31:292.
Street, R.A., 1978, Phys. Rev. B, 17:3984.
Street, R.A., and Biegelsen, D.K., J. Non-Cryst. Solids, 35-36:651.
Sturge, M.D., 1967, Solid State Physics, 20:92.
Takahashi, T., and Harada, Y., 1980, J. Non-Cryst. Solids, 35-36:1041
Tanaka, K., 1978, Sol. St. Comm., 28:541.
Tanaka, K., 1980, J. Non-Cryst. Solids, 35-36:1023.
Thomas, P.A., Brodsky, M.H., Kaplan, D., and Lepine, D., 1978, Phys.
     Rev. B, 18:3059.
Van Vechten, J.A., Tsu, R., Saris, F.W., and Hoonhout, D., 1979,
     Phys. Lett., 74A:417.
Wautelet, M., Laude, L.D., and Andrew, R., 1980, Phys. Lett., 77A:274.

INTENSITY DEPENDENT ABSORPTION IN SEMICONDUCTORS*

Alan F. Stewart   and   Michael Bass

Center for Laser Studies
University of Southern California
University Park, Los Angeles, California 90007·

Laser calorimetry has been used to study intensity dependent absorption process in semiconductors at 1.06 and 1.318 µm. New measurements of two photon absorption and two photon excited free carriers are reported. Saturation of the one photon absorption has been observed and evidence is presented for a pulse width dependence of the two photon absorption.

The technique of laser calorimetry has been shown to be extremely useful in investigation of two photon absorption (TPA) in semiconductors.[1] The calorimeter measures the total absorption due to all processes and by varying the laser intensity we have been able to measure a variety of intensity dependent phenomena. We report in this letter new measurements of TPA, observations of an apparent saturation of one photon absorbing transitions, two photon induced free carrier absorption, and an indication of a pulse-width dependence of TPA in semiconductors. Additional experiments have confirmed our earlier predictions concerning the role of internal reflections in a calorimetric measurement of TPA. The results of the experiments are summarized in Table I where the 1.06 µm TPA coefficients for two different pulse durations and those at 1.318 µm are listed. Also included are measurements of free carrier absorption cross sections and observed changes in the low intensity absorption due to saturation. Crystals of the same material but with different defect and impurity concentrations were found to have extremely different absorption properties.

---

*A version of this paper appears in the December 15, 1980 issue of Applied Physics Letters.

The experimental procedure was as described in Ref. 1. The total absorption of each room temperature sample was measured as a function of the intensity and the data was evaluated taking into account reflections at the rear surface of the sample. Studies were conducted using a Q-switched Nd:YAG laser operating in the TEM_{00} mode at 1.06 and 1.318 μm. Oscillation was confined to a single longitudinal mode by forming a intracavity etalon with gain[2]. The laser was made to operate with pulse durations of ∿11.4 and ∿26.4 nsec. defined as the full width at 1/e of the peak intensity. The pulse duration was varied by adjusting the aperture size and dye concentration in the passive Q-switch. Operation at 1.318 μm was obtained through the use of a Pockel's cell Q-switch and dispersive optics with coatings designed for minimum reflectivity at 1.06 μm. The pulse duration at 1.318 μm was 80 nsec. precluding a direct comparison with measurements at 1.06 μm with the same pulse duration.

When measuring TPA in high index semiconductors it is essential to know if field enhancement due to internal reflections occurs and to what extent it affects the results. If a single internal reflection is allowed at the exit surface of the crystal Eq. 1a and b describe the propagation of the incident and reflected beams[3]:

$$\frac{dI}{dZ} = -[\alpha + \beta(I + 2I_R)]\ I \qquad\qquad (1a)$$

$$\frac{dI_R}{dZ} = [\alpha + \beta(I_R + 2I)]\ I_R\ . \qquad\qquad (1b)$$

Here α is the linear absorption coefficient in cm^{-1} and β is the TPA coefficient in cm/MW. Equation 1a and b are valid only under the assumption of a constant phase relationship between the two beams in the form of a standing wave in the crystal. The equations presented in Ref. 1 differ in that the coefficient of 2 for the coupling terms was given as 1. In Ref. 1 we treated the case where there is a random phase relationship between the two beams and intensities, not fields, add as in calculations involving incoherent sources[4].

The effect of internal reflections on the measured total absorption was determined using samples of CdTe, CdSe and GaAs with the appropriate antireflection coatings on one surface only. In this way the total absorption as a function of intensity could be measured with no internal reflections or exactly one internal reflection simply by turning the sample around. The TPA coefficient, β, was obtained from data taken with the antireflection

coating on the exit surface using Eq. 1a with $I_R = 0$. Once the
value of β was known the data obtained with the antireflection
coating on the entrance surface was analyzed using Eq. 1a and b
with the coefficient of the cross terms as the fitting parameter.
As a final check, uncoated but otherwise identical samples of
these materials were measured and again the data was evaluated
using Eq. 1a and b. In all cases a cross term coefficient of
unity resulted in an excellent fit to the data and gave the same
value of β for the uncoated sample as had been obtained for the
sample with the AR coating on the exit surface. This proved that
the AR coating did not contribute to the measured TPA. As a result
of the preceding tests we concluded that the contribution of in-
ternally reflected beams to TPA could be adequately analyzed using
an addition of the intensities in such uncoated samples as were
to be used in the remaining experiments. It should be noted that
if the true coefficient of the cross terms in Eq. 1a and b had
some other value between 0 and 2, assuming a value of unity leads
to an estimated uncertainty in β which is at most 20%.

Figure 1 shows the results of the measurement obtained using
a 1.06 µm laser incident on an uncoated crystal of CdSe approximately 2 mm thick. The theoretical fit (solid line) assumes a
coupling coefficient of unity in Eq. 1a and b. In fitting these
equations to the data, numerical integrations must be performed
over the Gaussian spatial and temporal profiles of the incident
pulses. The linear absorption coefficient indicated by " ◄ "
on the ordinate axis was determined both by using a cw laser and
from the ordinate intercept for the fit drawn through the high
intensity data points.

Occasionally, the linear absorption coefficient determined
using a cw laser was significantly larger than the low intensity
measurements made using a pulsed laser. This effect was most ob-
servable in crystals of CdTe, ZnTe, and $CdSe_xSe_{1-x}$ in which
$\alpha \geq 0.1$ cm^{-1}. Fig. 2 shows data taken for two different crystals
of CdTe at 1.06 µm (a) and at 1.318 µm (b). The low intensity
1.06 µm data taken on crystal #1 with a pulsed laser in Fig. 2(a)
indicates a total absorption almost 40% smaller than the corresponding cw measurement. In the upper plot of Fig. 2b, data for
for the same crystal of CdTe shows a minimum of the total absorption at intensities near 1MW/cm^2. At larger intensities, TPA
becomes dominant and total absorption coefficient increases almost linearly. Similar data for CdTe crystal #2 appears in the
lower plot of Fig. 2b but in this case the line drawn through
the high intensity points has a negative ordinate intercept.

There is abundant evidence for the existence of impurity and
defect states with energy levels in the band gaps of these crystals both in the literature[5] and in spectrophotometer data taken

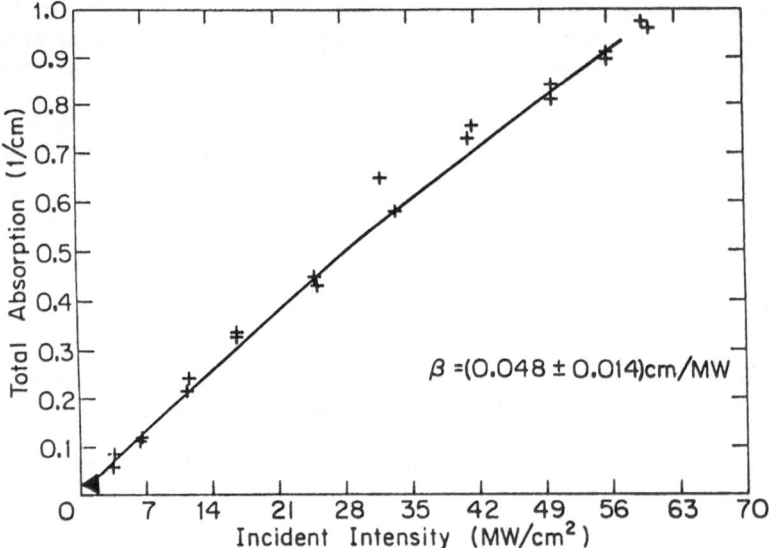

Figure 1.   Total Absorption versus Intensity for CdSe (Sample #1)
             Showing Two Photon Absorption at 1.06 μm. In this
             plot two photon absorption is indicated by $\alpha_{TOT} \propto I$.
             The linear absorption coefficient from cw measurements
             is indicated by the black triangle. The solid curve
             is a theoretical fit derived from Eq. 1a and b with
             a cross term coefficient of one.

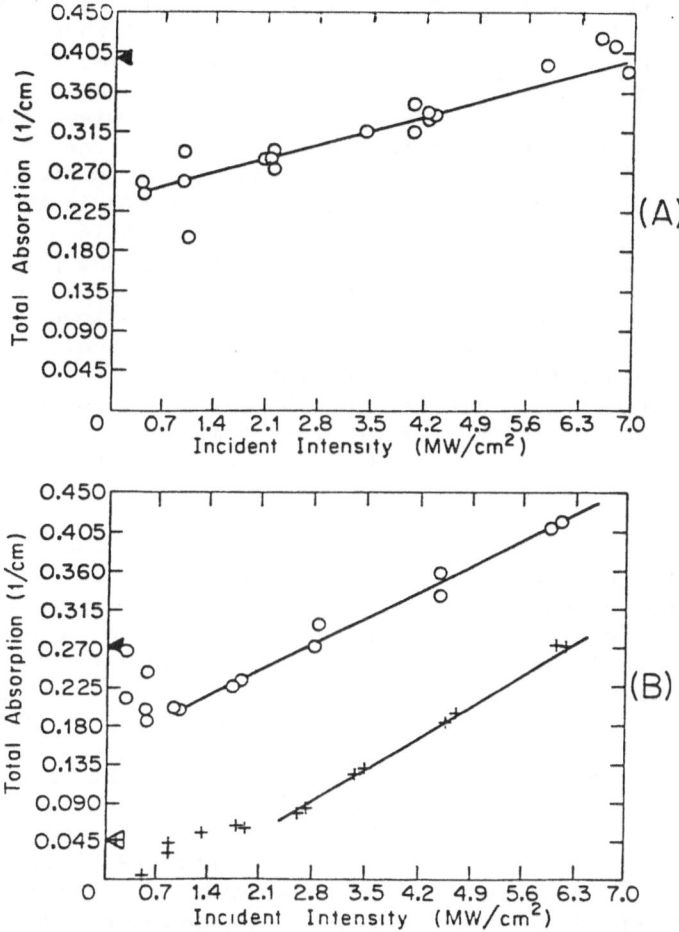

Figure 2.   Total Absorption versus Intensity for CdTe Showing
            Saturation of the Absorption and TPA.   The data in
            (s) is for CdTe Sample No. 1 at 1.06 μm.   In (b) the
            0's indicate data for this sample and the +'s are for
            CdTe Sample No. 2 and both are at 1.318 μm.   The
            linear absorption coefficient from cw measurements are
            indicated by blackened triangles for Sample No. 1 and
            an open triangle for Sample No. 2.

for the samples in this study[6]. We believe that the observed
changes in the low intensity absorption in Fig. 2 are the result
of saturation of single photon absorbing transitions involving
impurity and defect states in the band gap of these crystals.
Just as for a two level system, saturation results in lower absorption. However, it is important to realize that if these deep
levels are populated by absorption of the laser beam, then there
is an increased probability of absorption of a second photon in a
transition to the conduction band. This is a stepwise, resonant
TPA process[8] and results in a rate of TPA which increases rapidly
at low intensities as the one photon transition into these levels
become saturated. Consequently, for low intensities β will depend on the intensity. A model for the absorption in a crystal
which includes saturation of transition involving one photon as
well as TPA as outlined above can help in interpreting the data
in Figures 2(a) and (b). However, the data in the lower plot in
Fig. 2B requires closer examination. Simple graphical analysis
shows that in order for the ordinate intercept of such a plot to
be negative, the TPA coefficient must be a function of the intensity or some other higher order process must contribute to the
total absorption near $1MW/cm^2$. Based on the data obtained at
higher intensities (see Table I) it can be shown that in this
range free carrier absorption does not contribute significantly.
We suggest that the intensity dependence of β described above is
responsible for the observed behavior of these crystals. In
addition, the presence of resonant TPA involving defect and
impurity states can help to explain the existence of a pulse
width dependence in TPA (see Table I) and the disparity between
measurements reported in the literature[7].

At higher intensities and in other materials we have observed quadratic terms in the data for the total absorption
versus intensity. This represents an absorption process dependent on the intensity cubed. In Fig. 3 we present data obtained
for Si at 1.318 μm. The low intensity fit assumes TPA only while
the full range of data is fitted with a quadratic polynomial.
The observed behavior may be caused by absorption by two photon
excited free carriers[7]. If so, the coefficient of the quadratic
term in the polynomial may be used to deduce the free carrier
absorption cross section. In the limit of a recombination time
long compared to the laser pulse duration the free carrier
absorption cross section may be accurately estimated by[7]

$$\sigma \sim \frac{c4h\omega}{\beta\tau}$$

where τ is the 1/e of the intensity pulse duration and
c is $3\sqrt{3}$ times the coefficient of the quadratic term in the

polynomial fit. In Si, the measured carrier absorption cross
section was found to be $(3.6 \pm 1.3) \times 10^{-20}$ cm^2 which is smaller
than previous estimates by two orders of magnitude[8].

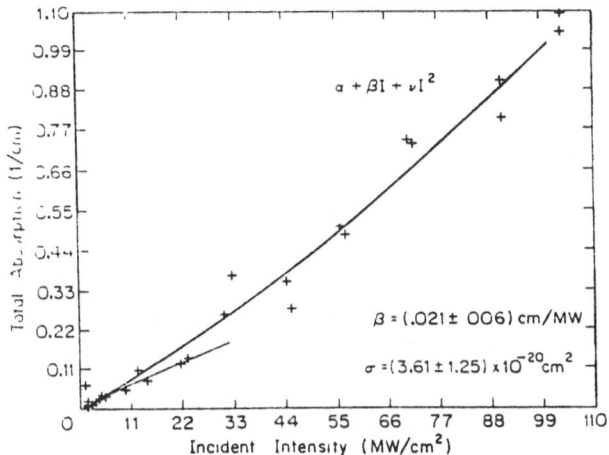

Figure 3.   Total Absorption versus Intensity for Si Showing
            Contributions from Higher Order Processes. This data
            was obtained at 1.318 μm. The cw absorbtivity of this
            crystal at 1.318 μm was 0.006 cm^{-1}. A quadratic term
            in the polynomal fit corresponds to a three photon
            process.

        The results of these experiments are summarized in Table 1.
Our measurements of TPA coefficients are found to agree very well
with previous state-of-the-art transmission measurements[7]. The
primary source of error in these and all experiments designed to
measure nonlinear phenomena is uncertainty in determining the absolute intensities used. The absolute error in the TPA coefficients
reported here is due to uncertainty in laser spot size measurement,
temporal profile measurement, the determination of the absolute
energy of each pulse, and the analysis of the heating and cooling
curves.
        In conclusion we have reported on three previously unobserved
intensity dependent absorption processes in various semiconductors
including a pulse width dependence to TPA. The effect of internal
reflections on a measurement of TPA in a high index material can
be adequately treated by the addition of the intensities in most
cases. In addition to TPA we have observed changes in the low

Table I. Measured Optical Properties of the Materials Studied

Material	λ(μm)	τ(nsec)	α(cm⁻¹)	β(cm/MW)	Δα(cm⁻¹)	σ(cm²)	Orientation
CdTe#1	1.06	11.4	0.0398±0.04	0.053±0.016	0.15	–	$\vec{k} \parallel [110]$
		37.8		0.078±0.023	0.16		$\vec{E} \parallel [00\bar{1}]$
	1.318	79.2	0.272±0.027	0.120±0.036	0.09		
CdTe#2	1.06	11.4	0.049±0.005	0.037±0.011	–	1.77 ± 0.062 $\times 10^{-18}$	
		26.4		0.046±0.014			
	1.318	79.2	0.048±0.005	0.135±0.041			
ZnTe	1.06	27.6	2.06±0.2		.73		
GaAs	1.06	11.4	1.50±0.15	0.030±0.009			
CdSe#1	1.06	27.6	0.018±0.002	0.048±0.014			$\vec{k} \parallel \vec{e}$
CdSe#2	1.06	11.4	0.013±0.001	0.025±0.008		2.63 ± 0.92 $\times 10^{-18}$	
		26.4		0.038±0.011			
	1.318	79.2		0.067±0.030			
CdS$_{0.25}$Se$_{0.75}$	1.06	11.4	0.090±0.009	0.075±0.020	0.026		
		26.4		0.21±0.063	0.071		
CdS$_{0.5}$Se$_{0.5}$	1.06	11.4	0.022±0.002	0.032±0.01			
		26.4		0.135±0.041			
KRS-V	1.06	37.8	0.0047±0.0005	0.0016±0.005			
Si	1.318	79.2	0.0061±0.0006	0.021±0.0006		3.61 ± 1.25 $\times 10^{-20}$	$\vec{k} \parallel [111]$

Legend

λ = laser wavelength
τ = laser pulse duration
α = linear absorption coefficient
    measured cw

β = two photon absorption coefficient
Δα = difference between α measured cw and obtained from
     pulsed laser measurements
σ = free carrier absorption cross section

intensity absorption which can be attributed to saturation of absorbing transitions as in a two level system. The existence of defect and impurity states in the band gap provides a reasonable explanation for the observed saturation of absorption. A higher order process dependent on the intensity cubed has been observed. We presume this to be linear absorption by two photon excited free carriers and have estimated the absorption cross section for free carriers in these materials near one micron.

We wish to acknowledge useful discussions with Drs. R. T. Swimm, R. Quimby and N. Koumvakalis. This work was supported by National Science Foundation Grant No. ENG. 7820470.

REFERENCES

1. M. Bass, E. W. Van Stryland, and A. F. Stewart., Appl. Phys. Lett., 34(2):142 (1979).
2. D. Bua, D. Fradin, M. Bass, IEEE J. Quant. Electron., QE8:916 (1972).
3. V. V. Arsenev, V. S. Dneprovskii, D. N. Klyshke, V. S. Fokin, V. U. Khattatov, Sov. Phys. JETP, 36:407 (1978).
4. M. Born and E. Wolf,"Principles of Optics", Pergamon, Oxford, p. 256 (1975).
5. R. H. Bube,"Photoconductivity and Related Phenomena", J. Mort and D. M. Pai, eds., Elsevier Scientific, New York, p. 118 (1976).
6. A. F. Stewart, Ph.D. thesis, University of Southern California, 1980, (unpublished).
7. J. H. Bechtel and W. L. Smith, Phys. Rev. B, 13:3515 (1976).
9. J. F. Reintjes and J. C. McGraddy, Phys. Rev. Lett., 30:901 (1973). These authors considered the possibility of a resonant TPA process.

THEORETICAL ASPECTS OF INTERACTION OF LASER RADIATION WITH HOT ELECTRONS IN SOLIDS

A.S. Epifanov

FIAN Academy of Sciences of the USSR Lebedev Physical

Inst.Leninsky Prospekt 53 MOSCOW (USSR)

INTRODUCTION

Hot electric current carriers in semiconductors and solid dielectrics, that is electrons and holes, attract attention of physi-
cists beginning from the 4th decade of our century. The first theoretical studies applied to the properties of the broad band gap insulators and were stimulated by the phenomenon of electric breakdown. It should be noted that at the time the scant knowledge of
band structure and electric properties of solids prevented scientists from through understanding of electron processes in the conduction band caused by external electric field.

Appreciable progress was achieved as a result of the investigations undertaken on semiconductors. In this connection I should
like to mention the starting works by Reider and Schockley [1]. The
advantage of semiconductors as subjects of investigations lies in
the fact that there are free carriers in the conduction band even
in weak electric fields as well as in no field at all.
Experimental methods were devised which made it possible to distinguish the hot carrier effects from those due to the change in the
free electrons (or holes) concentration. The rapid progress in the
theory of solids stimulated further experimental works.
Intensive studies on the one hand made it possible to explain numerous pecularities of the transport phenomena in strong electric
fields, on the other hand resulted in discoveries of new effects,
sometimes surprising, caused by the heating up of carriers in strong
electric fields. Such phenomena as Gunn effect, negative differential
conductivity, all kind of instabilities etc. are the basis of a
large number of electronic device and instruments. All that testi-
fies of the great practical importance of the hot carrier properties investigations.

A new impetus has been given recently to researches by the
creation of high power laser systems, in the first place because
of a new practical problem that arrises during constructions of
such systems. That is the problem of damage to optical materials
caused by the high-power laser radiation. The main points of this
problem are as follows.

Let a laser beam of wave-length λ pass through a sample of
some optical material. This material is transparent at the wavelength λ , that is the absorption coefficient α is so small that
can be neglected. As the radiation intensity I is increased but
still remains less than some value I_{cr} nothing is changed including the absorption coefficient. When I becomes equal to I_{cr} a damage to the sample is caused. This "optical breakdown" is usually
accompained by a bright luminescence in the form of a spark. The
critical intensity I_{cr} is generally called "breakdown threshold."
If we estimate the energy ε_{ab} which would have been absorbed during the laser pulse with starting value for the absorption coefficient and $I \leq I_{cr}$ we can easily see that this energy is by several orders of magnitude less than the energy ε_d necessary to produce irreversible changes in crystal structure. Thus one arrives at
the conclusion that a mechanism must exist whereby a great nonlinear absorption is induced, a mechanism that becomes noticeable
only when the laser intensity is nearly equal to the breakdown
threshold.

A lot of possible explanations of the laser breakdown phenomenon have been proposed. Here we confine ourselves to the only mechanism that is connected with the hot electron problem. This mechanism, which is usually called "electron avalanche ionization,"
can be visualized in the following manner[2]. At first, some processes (f.e. multiphoton ionization of the valence or impurity electrons into the conduction band) give rise to a certain initial number of "starting" electrons. These electrons become accelerated in
the field of the electromagnetic wave and reach an energy exceeding the impact ionization energy generating new free electrons.
The processes of the electron heating up and of the impact ionization are repeated and thus initiate the cascade process of multiplication. Damage takes place if the electron-number density reaches, during the time of action of the laser pulse, the critical value at which the energy absorbed by the conduction band electrons
and transferred to the lattice is sufficient for the development
of irreversible processes. This concise description already shows
that the first and the main problem we meet considering the electron avalanche mechanism of nonlinear absorption, concerns the
interaction of the electromagnetic field of a laser beam with the
hot electrons in solids.

There are several possible approaches to the analysis of the
hot carriers interaction with electromagnetic field. Firstly, the

so called method of a "test electron". Secondly, the direct solution of the kinetic equation for the conduction band electrons in
the presence of the high-power laser radiation. And finally, the
calculation of the probability that some "lucky" electrons would
reach an energy exceeding the impact ionization potential undergoing
such collisions with phonon that change the direction of the electron momentum to the opposite nearly simultaneously with the change
of the sign of the electric field.

    The "test electron" method consists in calculating the rate
at which an electron gains and loses energy in collisions with
phonons in the presence of electromagnetic field. Such an approach makes it possible, in a number of cases, to estimate the mean
value of the electron energy and kinetic coefficients for transport phenomena. The analysis of the conditions for a stationary
distribution to the possible provides an estimation for the critical
field. The main shortcomings of this approach are caused by the fact
that the presence of an instability is stated, butdynamics of the
process is left out of consideration. In connection with the breakdown problem in electric fields this led to a great variety of
"breakdown criteria" proposed to fit experimental data. The argument
was stopped only by the Keldysh's work[3] in which a consistent analysis of kinetic equations had been done. By now the method of kinetic equation can be considered as the only reliable way of solution of different problems concerning the hot electron effects. The
only question that remains when we use this method is that of
"lucky" electrons mentioned above. The analysis of the paper[4] has
shown that the usual method of deduction of Boltzmann equation,
which we shall discuss later in more detail, is inapplicable to the
case of the intermediate field strength because of the important
role of the electrons whose free path is much greater than the mean
one. The momentum distribution over directions of such electrons
is obviously anisotropic and the first terms in the Legendre polynomial expansions of the basic equation are not sufficient. One
can say also that all this is due to the strong dependence of the
electron energy distribution on the free path length which must be
observed when

$$\Omega < \nu$$

where Ω is the frequency of the electromagnetic radiation, ν is
the frequency of electron-phonon collisions. And even though the
number of the electrons with the free path exceeding considerably
the mean value is small enough, their contribution of the total distribution at high energies turns out to be important. As it will
be clear later, in the limit of large frequencies

$$\Omega > \nu$$

the distribution function is no more dependent on free path length
so that the role of "lucky" electrons becomes less important.

Estimations show that in the region of laser wave-length

$$\lambda = 0,1 - 20\mu,$$

the usual kinetic equation remains applicable to the hot electron problem if the field strength E is of order of magnitude or greater than 10^5V/cm, that is when the laser beam intensity

$$I \gtrsim 10^8 \ W/cm^2.$$

Before considering the above mentioned kinetic equations we should make some remarks.

1. In order to make all calculations as simple as possible and obvious enough and to be able to arrive at perceptible results all our considerations will be performed in the parabolic conduction band approximation, that is the electron energy dependence on the quasimomentum will be chosen in the form

$$\varepsilon = \frac{p^2}{2m} \ ,$$

where m is effective mass of electron, assumed to be constant. We shall obtain correct results, at least qualitatively as regards the dependencies of observed values on the parameters of experiment. Numerical calculations taking into account the real band structure, that have been carried out recently, show that quantitatively our results will not also much differ from the exact values. All our equations and methods of solution can be easily generalized, but the final results either become very complicated or can't be obtained at all without numerical evaluations.

2. To be specific we shall speak only about hot electrons. In those cases when holes paly an important role their properties can be readily taken into account.

3. It so happened that the breadwon problem gave rise to the investigations of the properties of hot carriers: electric breakdown in static fields, optical damage in laser fields. That is why we shall choose the problem of electron avalanche ionization induced in solids by the high-power laser irradiation as an example of the general use of equations and methods of their solution. It should be noted that this is one of the most difficult problems since it demands a solution in the quasistationary (but not stationary) approximation. That means we can assume that the distribution of electrons over energies corresponding to the given radiation intensity is established "instantaneously" (stationarity with respect to the time scale $\sim \tau_\varepsilon$ which is the characteristic relaxation time of the electrons along the energy axis which is usually longer by one or two orders of magnitude than the relaxation time of the electron momentum over the directions in collision with phonons). Whereas a non-zero electron flux along the energy axis is induced

by the impact ionization processes and gives rise to changes in electron distribution during times much longer than τ_ε (non stationary with respect to the time scale $\sim \gamma-1$, where γ is the electron-avalanche developement rate) the concentration N of the conduction band electrons increases according to the simple law

$$N = N_0 e^{\gamma t}$$

where N_0 is the initial concentration of "starting" electrons. It is obvious that a large number of stationary problems can be solved by the methods we shall be considering here just putting $\gamma = 0$.

4. We will put aside all the effects due to the anisotropic character of the electron distribution function in the quasimomentum space.

The main part of our condiserations will be based on the quantum kinetic equation, but to make the picture complete we begin with the classical Boltzmann equation.

CLASSICAL BOLTZMANN EQUATION IN DIFFUSION APPROXIMATION

We should observe that the applicability of the Boltzmann equation to the case of strong fields is not obvious beforehand. However, Keldysh has proved that if the field can be assumed to be quasi-classical and the electron-phonon interaction is weak enough:

$$h\nu \ll <\varepsilon> \ ,$$

where $<\varepsilon>$ is the mean energy of electrons, it can be used also in the case of strong electric fields.

The usual approach to the analysis of the Boltzmann equation consists in expansion of distribution functions $f(\varepsilon,\chi)$ (ε is the electron energy, χ is the cosine of the angle between the electron momentum and the polarization vector of the field) over the Legendre polynomials:

$$f(\varepsilon,\chi)= \sum_m f_m(\varepsilon)P_m(\chi) \ ; \quad f_m(\varepsilon) = \frac{2m+1}{2} \int_{-1}^{+1} f(\varepsilon,\chi)P_m(\chi) \ d\chi \ .$$

Let us assume now that the length of the electron mean free path and the field wave-length satisfy the inequality

$$l \ll \lambda \ ,$$

that allows us to ignore the distribution function dependence on the spatial coordinates. Let the field strength be $Re \ E^*$

$$E^* = E\,e^{\,i\Omega t}\,,\ E = |E|\,.$$

The procedure is completely analogous to that in static fields which is described in books, so we can omit it here. Let us note only that the distribution function is seeked in the form

$$f_1^* = f_1 \exp\ \{i\,(\,\Omega t - \phi)\,\}\,,$$

where

$$0 \leqslant \phi \leqslant \frac{\pi}{2},\quad \frac{df_1}{dt} \sim \gamma f_1 \ll f_1/\tau_1$$

Naturally we have to assume that the relaxation time τ_1 exists. It can be shown that

$$\cos\ \phi = 1\,/\,(1 + \Omega^2 \tau_1^2),$$

and we arrive at the following equation

$$\frac{1}{3}\ \frac{e^2 E^2}{m\sqrt{\varepsilon}}\ \frac{\partial}{\partial\varepsilon}\ (-\frac{\varepsilon^{3/2}\,\tau_1}{1+\Omega^2\tau_1^2}\ \frac{\partial f_0}{\partial\varepsilon})+(\frac{\partial f_0}{\partial t})_{coll} = \frac{\partial f_0}{\partial t} \tag{1}$$

(e is the electron charge)

wh ere $(\frac{\partial f}{\partial t})_{coll}$ is the collisions integral which can be developped in the usual way.

We shall see later that the diffusion approximation for the quantum kinetic equation leads to the same result and in that way it will be proved that the use of this approximation is equivalent to the assumption of the classical character of the hot electrons interaction with electromagnetic fields.

We shall see also that when the radiation quanta are large enough, specific quantum effects become apparent and the diffusion approximation which is very pleasant to work with, is no more adequate for the problem.

One more remark concerning the region of applicability of the above written equation. The justification for the transformation of the Boltzmann equation to the form (1) lays in the assumption that

$$e\,El \ll\ <\ \varepsilon\ >$$

which, generally speaking, may not be good enough even in fairly weak fields. If this inequality does not take place the distribu-

tion function becomes essentially anisotropic. But a number of investigations show (see f.e. [5]) that also in the case when the energy gain of an electron on the free path is much greater that the
phonon energy, the distribution function once more becomes nearly
isotropic and so one can use again the diffusion equation if

$$e \; El \gg \hbar \, \omega ,$$

where $\hbar \, \omega$ us the phonon energy (a mean value).

QUANTUM KINETIC EQUATION

The quantum kinetic equation for the conduction band electrons
in the presence of a strong electromagnetic field was obtained by
Melnikov and Epstein[6].

The Hamiltonian of the system is

$$H = \frac{1}{2m} \sum_p \left[p - \frac{e}{c} A(t) \right]^2 a_p^+ a_p + \sum_k \hbar \omega_k B_k^+ B_k + \sum_{p,k} B_k a_{p+k}^+ a_p (B_k + B_k^+).$$

Here m and e are respectively the electron effective mass and
charge, a^+ and a are creation and annihilation operators for the
electron with the canonical momentum p, B_k^+ and B_k are the operators for the phonon with the wave vector $k/\hbar$, $\hbar\omega_x$. being the
energy of this phonon, B_k is the matrix element of the electron-
phonon interaction, A (t) is the vector potential of the electromagnetic field:

$$E\,(t) = - \frac{1}{c} \frac{\partial A}{\partial t} = E \cos \Omega t ,$$

c being the light velocity.

Using the equation for the density matrix ρ of the system

$$\frac{\partial \rho}{\partial t} = - \frac{i}{\hbar} \left[\rho \; H \right] ,$$

under the conditions that will be discussed a little later, one can
obtain the following kinetic equation

$$\frac{\partial n_p}{\partial t} = \frac{2\pi}{\hbar} \sum_k |B_k|^2 \sum_{l=-\infty}^{+\infty} I_l^2 \left(\frac{eEk}{m\,\Omega^2} \right) \{ \left[n_{p+k}(N_k+1) - n_p N_k \right] \times$$

$$\times \; \delta(\varepsilon_{p+k} - \varepsilon_p - \hbar\omega_k - l\hbar\Omega \,) +$$

$$+ \left[n_{p+k} N_k - n_p (N_k + 1) \right] \; \delta(\varepsilon_{p+k} - \varepsilon_p + \hbar\omega_k - l\hbar\Omega) \Big\} \qquad\qquad (2)$$

where

$$n_p = \langle a_p^+ a_p \rangle \; ,$$
$$N_k = \langle B_k^+ B_k \rangle \; ,$$
$$\langle A \rangle = \mathrm{Tr} \; (A\rho) \; .$$

The above mentioned conditions are:

1. The electron-phonon interaction must be weak enough,

$$\hbar / \tau \ll \varepsilon \quad (\; \frac{1}{\tau} = \nu),$$

This condition on the one hand makes it possible to decouple correlation functions and on the other hand is necessary to make the distribution function equation local in time.

2. The electron free path has to be small compared to the wavelength

$$l \ll \lambda \; .$$

Owing to this fact we can consider all processes to be local in space.

3. The condition

$$\frac{v}{c} \ll 1 \; ,$$

v being the electron velocity,
connected with the previous one and permitting us to neglect the effects of the magnetic field of the wave.

4. The frequency of the electron-phonon collisions must be less than the frequency of the field

$$\nu \ll \Omega$$

This condition is not quite necessary and when the contrary inequality takes place a slight amendment is to be made to equation (2). Still it is interesting to note that when the quantum effects are important (that is when the energy of photons is large enough) this condition is usually satisfied.

The kinetic equation (2) can be easily interpreted from the physical point of view. It describes the change in the distribution function due to jumps of electrons when they absorb or emin ρ quanta of electromagnetic field, photons, with simultaneous absorption or emission of a phonon. However, obvious as it is, its direct application to various problems is almost impossible. That is why we are going to derive from it some approximate equations which have permitted in the last years to obtain a lot of interesting and important results on the behaviour of hot electrons in solids in the electromagnetic wave of laser radiation. Certainely, the use of approximate equations is very like the attempt to get out from a quagmire - if the head is pulled out then the tail is stuck in.

Obtaining comparatively simple equations we lose a part of the information that the initial one contains, in particular we can no more describe the effects due to the anisotropic character of the electron distribution. Besides that we have to make numerous assumptions and a good deal of them can't be easily justified or verified, therefore final results are rather more qualitative than we would like them to be. But as a sort of excuse, I should note that they were just the qualitative theoretical dependences of the breakdown threshold on such parameters as initial lattice temperature, wave frequency, duration of the radiation pulse that gave rise to the purposeful experimental investigations aimed at determining the role of the electron avalanche in laser-induced breakdown of real materials - dielectrics and semiconductors. As to the quantitative values, sometimes theoretical results fit the experiment much too well, taking into account the great number of assumptions made by theorists as well as the real degree of experimental accuracy. We could recall situations when this agreement of results even led to premature conclusions that later turned out to be either uncorrect or inexact.

In general the solid body is such a complicated subject for investigations that it happens to be very seldom that one may really expect a good quantitative agreement of theoretical values with experimental. And the more is said in a paper about "excellent agreement" the more cautious a reader should be.

Now we set about our procedure to derive thus criticized approximate equations and the greatest simplification can be achieved by use of the diffusion approximation.

4. DIFFUSION KINETICAL EQUATION

Now our task is to obtain the diffusion equation of the Fokker – Planck type:

$$- \frac{\partial}{\partial \varepsilon} \tilde{S} (\varepsilon,t) = \tilde{g} (\varepsilon) \frac{\partial f(\varepsilon,t)}{\partial t} + \tilde{R} (\varepsilon,t; f(\varepsilon)), \quad (3)$$

where

$$\tilde{S} (\varepsilon,t) = - \tilde{g} (\varepsilon) \left[\tilde{D} (\varepsilon) \frac{\partial f(\varepsilon, t)}{\partial \varepsilon} + \tilde{Q} (\varepsilon) f (\varepsilon,t) \right],$$

is the flux of the electrons through an isoenergetic surface ε,

$\tilde{D} (\varepsilon)$ is the diffusion coefficient along the energy axis,

$Q (\varepsilon)$ is the rate of energy loss due to the spontaneous emission of phonons,

$\tilde{g} (\varepsilon)$ is the density of the number of states, and $\tilde{R} (\varepsilon,t; f(\varepsilon))$ is the term describing the outflow or inflow of electrons as a result of ionization.

We have to make some assumptions of course. The anisotropic part of the distribution function we consider to be small as compared to that isotropic $f(\varepsilon,t)$. We also substitute isotropic collisions for those anisotropic:

$$\mathcal{J}_n^{\ell} (\frac{e \mathbf{E \cdot K}}{m \Omega^2}) \longrightarrow \int_0^1 \mathcal{J}_{\ell}^2 (\frac{e E K}{m \Omega^2} x) \, dx .$$

Then we expand the distribution function up to the second term:

$$\tilde{f} (\varepsilon') \approx \tilde{f} (\varepsilon) + \Delta\varepsilon \, \tilde{f}' (\varepsilon) + \tfrac{1}{2} (\Delta\varepsilon)^2 \tilde{f}'' (\varepsilon),$$

where

$$\Delta\varepsilon = \pm 1 \, \hbar\Omega \pm \hbar \, \omega_k .$$

This expansion can be valid only if

$$| \Delta\varepsilon \frac{\partial \tilde{f}}{\partial \varepsilon} | << \tilde{f} (\varepsilon) .$$

It is clear that this inequality can't be satisfied for all electron energies and numbers of quanta 1 . To make the mistake due to this assumption insignificant we should demand the condition

$$1_{max} \, \hbar \, \Omega| \, \frac{\partial \tilde{f}(\varepsilon)}{\partial \varepsilon} \, | << \, \tilde{f}(\varepsilon)$$

to be satisfied. Here 1_{max} is the maximum number of photons that can be effectively absorbed (or emitted) so to change considerably the distribution function. The last condition can be analyzed on the grounds of our further results. Then it is found that it can be rewritten in the form

$$e \, E1 \, / \sqrt{1 + \Omega^2 \, \tau^2} \, >> \, \frac{\hbar\omega}{2N_{ph} + 1} \, , \qquad (4)$$

where $\hbar\omega$ is the mean energy of those phonons that play an important role in the process of absorption of the electromagnetic field energy by electrons, N_{ph} is their number.

Estimations show that this condition is satisfied in the case of dielectrics if the laser radiation intensity is near to the breakdown threshold. As regards the semiconductors even at such intensities the validity of the last assumption should be verified on each occasion.

If we let $\Omega \to 0$ the inequality (4) is transformed to the well known condition of applicability of the diffusion equation in the case of the static electrical fields.

After some straightforward calculations we obtain the following expression for the coefficients of our equation

$$\tilde{D} \, (\varepsilon) = \tilde{D}_E(\varepsilon) + \tilde{D}_o \, (\varepsilon); \qquad \tilde{Q} \, (\varepsilon) = \tilde{Q}_o \, (\varepsilon)$$

$$\tilde{D}_E(\varepsilon) = \frac{e^2 E^2 1 \, p(\varepsilon)}{6 \, m \, \Omega^2 \, \tau^2} \quad ;$$

where $\tilde{D}_o$ and $\tilde{Q}_o$ are the respective coefficients in the absence of laser field.

The obtained equation is completely equivalent to the classical diffusion approximation presented above. To make sure of that the only thing we have to do is to expand the collision integral in (1) in the same way as we have done for equation (2).

Thus we have shown that the diffusion approximation for the quantum kinetic equation does not extend the sphere of applicability of the classical approach and so the above mentioned assump-

tions are in fact the conditions under which it can be used. I'd like
to emphasize the principal condition for the classical approach

$$l_{max} \, \hbar\Omega << <\varepsilon>$$

It leads to (4) if we take into account that $n_{max} \sim eE \, \dfrac{1}{\hbar\Omega \, \sqrt{1+ \Omega^2 \tau^2}}$

(Which is the argument of Bessel functions if we do not cofine ourselves to the case $\Omega\tau >> 1$, a more general expression for D_E is

$$D_E = \frac{e^2 \, E^2 \, pl}{6m(1+ \Omega^2 \tau^2_{\, 1})} \quad) \ .$$

I t is interesting to note that the relaxation time τ_1 remains
the same with the electromagnetic field on. But this is not true for
the general quantum case in which one has to write an integral equation similar to (2) for the value of τ_1.

Another approximation is somewhat disharmoniously called.....

5. DIFFERENTIAL-DIFFERENCE QUANTUM KINETIC EQUATION

One can obtain this equation restricting the expansion of distribution function to be

$$\tilde{f} \, (\varepsilon\pm \, l\hbar\Omega \pm \hbar\omega \,) \approx f(\varepsilon\pm \, l\hbar \, \Omega \,)\pm \hbar\omega \, f'_\varepsilon \, (\varepsilon\pm l\hbar\Omega)+ \frac{1}{2} \cdot (\hbar\omega)^2 f''(\varepsilon\pm l\hbar\Omega \,).$$

As the general result takes a lot of place, we write here the
equation that allows only for one-photon processes. It has the form

$$\sqrt{\varepsilon}\frac{\partial f(\varepsilon,t)}{\partial t} = \frac{\partial}{\partial\varepsilon}\{ \, \sqrt{\varepsilon} \, \Big[\, D_o(\varepsilon)f'_\varepsilon \, (\varepsilon,t) + Q_o(\varepsilon) \, f(\varepsilon,t)\Big]\} \, +$$

$$+ \, W_1 f \, (\varepsilon+ \hbar\Omega \,) \, + W_2 f(\varepsilon-\hbar\Omega \,)-(W_1+W_2)f(\varepsilon)+ W_{11}f'_\varepsilon \, (\varepsilon+ \hbar\Omega \,) \, +$$

$$+ \, W_{21}f'_\varepsilon \, (\varepsilon- \hbar\Omega \,) + W_{12}f'' \, (\varepsilon+ \hbar\Omega \,) + W_{22} \, f'' \, (\varepsilon- \hbar\Omega \,). \qquad (5)$$

The part of this equation in the brace describes the process
of diffusion due to electron-phonon collisions only, all other

terms depend on the field strength and correspond to electron-phonon-photon collisions. Terms containing f'_ε $(\varepsilon \pm \hbar\Omega)$ and f''_ε $(\varepsilon \pm \hbar\Omega)$ are of some special interest. Their presence is caused by the fact that as a result of an electron-phonon-photon collision an electron jumps to the energy ε starting from $\varepsilon \pm \hbar\Omega \pm \hbar\omega$ but not from $\varepsilon \pm \hbar\Omega$ (otherwise only terms $f(\varepsilon \pm \hbar\Omega)$ would have been present).

Certainly, the equation (5) is much more complicated than differential equation (3), and to solve it one has to invent special methods for any particular problem. One of this methods we shall consider later while speaking of the electron avalanche ionization in solids.

We are finishing the installation of our theoretical machinery. Our next task is to see how it works.

6. THE SOLUTION OF ELECTRON-AVALANCH PROBLEM IN DIFFUSION APPROXIMATION

The electron avalanche problem in connection with the laser breakdown in solids is formulated as follows.

We'll call "the interaction region" the part of the volume of a sample where the heating up of free carriers is possible under the influence of laser radiation. For instance, if a laser beam is focused inside the sample we can consider the caustic volume to be the interaction region. To make our case more definite and simple we assume the intensity of laser radiation to be constant all over the interaction region. This restriction is removed by some tedious but straightforward calculations. For the same reason we assume that the intensity does not vary through the laser pulse (the pulse is of a "rectangular form"; we'll consider mostly laser pulses of duration from 10^{-11} to 10^{-7} sec). We are to determine the field strength in the wave at which enough energy to produce irreversible processes can be absorbed in the interaction region. This absorption is due to free electrons multiplying in the avalanche ionization process.

While solving this problem one meets 3 stages.

Firstly, it is necessary to estimate the energy that would be sufficient to cause damage. This energy is influenced by particular features of the damage development mechanism. The last can be formation of a fused region, or cracking , or anything else. This is an interesting question but we can leave it alone. The thing is (as we shall see a little later) that the increase of temperature of the lattice is explosive-like which permits us to do without the exact knowledge of the breakdown energy.

Secondly, considering the process of lattice heating up we have to determine at what rate the avalanche ionization must be

developping to induce the absorption sufficient for breakdown.
This question poses a relation between the pulse duration t_p and
the avalanche development rate γ.

And finally, we have to determine the dependence of γ on the
laser radiation intensity. We begin with this part of the problem.

The form of Eq.(3) is significantly different in the regions
$\varepsilon \leqslant \mathfrak{T}$ and $\varepsilon \geqslant \mathfrak{T}$ [1] ; unasmuch as at $\varepsilon \geqslant \mathfrak{T}$ the term R is important (it is
essential to take into account the outflow of the electrons as a re-
sult of ionization, it is precisely this process which governs primarily the distribution function), whereas in the region $\varepsilon < \mathfrak{T}$ it
is important only in the vicinity of the point $\varepsilon = 0$, and can be
taken into account in the boundary conditions.

It will be convenient subsequently to change over to the dimen-
sionless variable $x = \varepsilon/\mathfrak{T}$ ($\mathfrak{T}$ being the effective ionization potential) and use all the functions of ε , introduced above, as functions
of x without a tilde, so that the coefficient $D(x)$ and $Q(x)$ will
have the dimension sec^{-1} after division by the corresponding power
of I.

From the solution of Eq.(3) in the region $\varepsilon \geqslant \mathfrak{T}$ we can determine the quantity

$$\sigma = S(1)/f(1) \, Q_o(1). \tag{6}$$

If the value of σ is known, the last relation (6) is in fact the
first boundary condition for the region $\varepsilon \leqslant \mathfrak{T}$.

It can be seen that $\sigma \to \infty$ when $f(1) \to 0$. We have usually the
condition $\sigma \gg 1$ well satisfied (which is related to the fact that
the mean time necessary for electrons with energy $\varepsilon \geqslant \mathfrak{T}$ to produce
an impact ionization is about $10^{-15} \dots 10^{-16}$ Sec, that is much
shorter than all other characteristic times), so that we can often
assume

$$f(1) = 0 \tag{6a}$$

to be the first boundary condition.

Now we have to find the second boundary condition. Since it describes the number of electrons taking part in ionization process,
we can write

$$2S(1) = S(x_1) - S \, (0 \div x_1)$$

---

1) $\mathfrak{T}$ is the effective potential for impact ionization.

where the term $S(0 \div x_1)$ is the outflow of the electrons in the region $x \leqslant x_1$ due, for example, to recombination. Region $x \leqslant x_1$ is that where the electrons come after impact ionization and is narrow enough, so that the exact form of the distribution function in this region does not play any role in the determination of the avalanche rate and we can rewrite the second boundary condition in the form (in the absence of recombination processes which can be allowed for in the breakdown criterion)

$$2S(1) = S(0). \tag{7}$$

This is the usual "flux doubbling" condition currently used in various multiplication problems.

We seek the solution of Eq.(3) with boundary conditions (6,7) in the form

$$f(x,t) = e^{\gamma t} f(x).$$

It can be shown that the eigenvalue spectrum for γ contains parallel with negative values (which would be essential for us if times shorter than 10^{-11} Sec were of interest to us; negative eigenvalues are necessary to describe the coming of the system to the equilibrium state) one and only one positive value which gives us the electron-avalanche rate:

$$N = N_0 \, e^{\gamma t} , \tag{8}$$

where

$$N = \int g(x) \, f(x) \, dx .$$

We shall see later that the breakdown criterion can be approximately used in the form

$$\gamma t_p \simeq 15,$$

where t_p is duration of the laser pulse. If we choose our pulse-duration region to be nanoseconds we can see that the conditions are satisfied with large margin, so that a good approximation for the rate γ of the cascade ionization can be obtained from the following successive-approximation procedure.

In Eq.(3), after substituting (8), we put $\gamma = 0$ and obtain the zero-order approximation for the distribution function $^0f(x)$; with the aid of the boundary conditions we determine the first approximation for the rate $^1\gamma$ and obtain $^1f(x)$, substituting $^1\gamma \, ^0f(x)$ in equation (3), etc.

In accordance with the foregoing, we rewrite (3) in the form

$$\frac{\partial}{\partial x} \{g(x)[D(x)(1-g(x)^{o}f'(x) + Q_{o}(x) \ ^{o}f(x)]\} = 0 ,$$

so that (dropping the signe ^{o})

$$f'(x) + \frac{Q_{o}(x)}{D(x)(1+g(x))} \ f(x) = - \frac{S(1) \ g(1)}{g(x)D(x)(1+g(x))} ,$$

and, taking the boundary condition (6) into account, we obtain

$$f(x)=f(1)\exp\{-F(x)\}\{1- \sigma \int_{1}^{x} dy\exp\{F(y)\} \frac{g(1)Q^{o}(1)}{g(y)D(y)(1+g(y))} \}$$

where

$$F(x) = \int_{1}^{x} \frac{Q^{o}(y)dy}{D(y)(1+g(y))} .$$

We next carry out the procedure of finding the first approximation for γ :

$$S(0)-S(1) = \gamma \int_{0}^{1} g(x) \ f(x) \ dx ,$$

or

$$\gamma \simeq Q_{o}(1) \ \sigma/ \int_{0}^{1} g(x) \ \frac{f(x)}{f(1)} \ dx. \tag{9}$$

Now we shall apply this solution to the analysis of dependences of the breakdown threshold on the laser radiation frequency and the initial lattice temperature. Nowdays the results for all kind of electron-phonon scattering are known. We confine ourselves to the case of deformation-potential electron scattering on low frequency phonons in the high-temperature approximation. That last demands the condition

$$KT \gg \sqrt{2m \ V_{s}^{2} \ \mathbb{I}}$$

to be satisfied (V_{s} is the sound velocity, $\hbar\omega_{g} = V_{s}g$).

For the coefficients of the diffusion equation we have then

$$D_{o}(x) = \frac{4V_{s}^{2} \ m \ x^{3/2}}{\sqrt{2m\mathbb{I}} \ l_{ac}} ,$$

$$Q_o(x) = \frac{2V_s^2 \sqrt{2mI}}{kT \, l_{ac}} \, x^{3/2},$$

$$l_{ac} = \frac{\pi\rho\hbar^4 \, V_s^2}{m^2 \mathcal{E}_1^2 \, kT},$$

and we obtain in accordance with the foregoing

$$f(x) = -f(1) \frac{\sigma}{\delta} \exp\{-F(x)\} \int_1^x dy \, \exp\{F(y)\} \frac{1+\eta y}{y^2(1+\eta y + g)},$$

where

$$F(x) = \frac{1}{\delta} \left[x - \frac{g}{\eta} \ln\left(1 + \frac{\eta x}{1+g}\right) \right],$$

$$g = \frac{e^2 E^2}{6m^2 V_s^2 \Omega^2}, \quad \delta = \frac{kT}{I} \; ; \quad \eta = \frac{2I}{m\Omega^2 l_{ac}^2}.$$

We should get used to the designations q and δ because they will be used for dimensionless intensity and temperature respectively.

For the avalanche rate γ we have

$$\gamma = Q(1)\left\{ \int_0^1 \left(\sqrt{x} \, e^{-F(x)} \int_x^1 \frac{1+\eta y}{\delta y^2(1+\eta y+g)} \, e^{F(x)} \right) dx \right\}^{-1}.$$

The inner integral can be estimated by means of the saddle-point method:

$$\gamma \approx \Theta^{-1} \exp\left\{- \frac{1}{\delta} + \frac{g}{\eta\delta} \ln\left(1 + \frac{\eta}{1+g}\right)\right\} \approx \Theta^{-1} \exp\left\{- \frac{1}{(g+1)\delta} - \frac{\eta g}{2\delta(1+g)^2}\right\},$$

where

$$\Theta = Q^{-1}(1) \left(\frac{\eta}{g+1}\right)^{-3/2} \Gamma\left(\frac{3}{2}\right) \Psi\left(\frac{3}{2}, \frac{5}{2} + \frac{g}{\eta\delta} \; ; \; \frac{g+1}{\eta\delta}\right).$$

All over th e reasonable range of parameters the quantity Θ does not change more than by 2 orders of magnitude.

For analysis and estimations it is convenient to use the following expression for the critical field

$$E^2_{cr} = \Lambda \, \frac{Im^2 v^2_s}{2kT \, e^2} \; (\, \Omega^2 + \frac{I}{m \, l^2_{ac}}),$$ (10)

where

$$\Lambda^{-1} = \frac{1}{12} \, \ln \, (\frac{t_p}{15\theta}) \sim 1 .$$

We have usually $0{,}6 < \Lambda < 1{,}3$.

The character of the threshold dependence on the laser beam frequency is quite obvious from this expression. Another important dependence is that on the initial lattice temperature. We speak of the initial, starting lattice temperature; but since the temperature in the interaction region increases during a laser pulse, there is a question, of course, what sense such a dependence can have.

It turns out that considerable change in lattice temperature occurs only when the concentration of free electrons becomes as great as $10^{17} - 10^{18}$ cm^{-3}, but that means that the conditions for the general outburst are satisfied the avalanche process being at top development. Before that, during the laser pulse, the absorption coefficient in the interaction region is small and the heating up of the crystal lattice is negligible. All this we shall consider in more detail a bit later when discussing the breakdown criterion.

Taking into account that the mean free path of the electrons in the high-temperature approximation is inversely proportional to the lattice temperature

$$l_{ac} \sim 1/T ,$$

it is sufficient to take a glance at expression (10) for the critical field in order to understand that its dependence on the temperature turns out to be quite different in the cases when the electron-phonon frequency is greater or lower than the wave frequency. Namely, if $\Omega\tau << 1$ then $E_{cr} \sim \sqrt{T}$, but if $\Omega\tau >> 1$ then $E_{cr} \sim 1/\sqrt{T}$.

It is evident that in the low temperature approximation (zero-oscillations scattering) the critical field doesn't depend on the lattice temperature while its dependence on the field frequency remains the same. It is easy to obtain the temperature T_t at which the threshold begins to depend noticeably on the initial lattice temperature

$$kT_t = -\frac{1}{2} V_s \sqrt{2m\bar{I}} \ .$$

The quantity T_t is usually within the range 200-400°K.

The experimental studies undertaken in the last years with the purpose to find out the role of the electron avalanche mechanism in laser-induced breakdown of real crystals were mainly based on the foregoing results. They succeded in determining the conditions under which this mechanism is predominating. However, these studies and the comparison between theoretical and experimental results have showed that the diffusion approximation is not always valid, especially in the case of high frequency fields when $\hbar\Omega$ is not much less than $\bar{I}$ (we should expect it certainly), as well as in the case of picosecond pulses when the condition $\gamma \ll Q$ ins't satisfied with large enough margin.

The last fault can be easily remedied; some other methods of solution of the diffusion equation are developped. A more accurate consideration leads, for example, to the following transcendental equation that relates γ and E_{cr}:

$$\frac{1}{\gamma_0 \alpha} \ (2- \frac{I_2\left[4(\gamma_0\alpha)^{1/2}\right]}{2} = F(\alpha)- \frac{4}{3}, \ (\Omega\tau >>1)$$

where I_2 is the modified Bessel function, $F(\alpha)$ is specified by the series

$$F(\alpha) = \sum_{m=0}^{\infty} \frac{2^{m+2}}{(2m+3)!!(2m+1)} \ ,$$

and admits   the following integral representation

$$F(\alpha) = \frac{1}{\alpha^{1/2}} \int_0^\alpha \xi^{-2}\gamma(\frac{3}{2} \ , \ \xi) \ e^\xi \ d\xi \ ,$$

where $\gamma (\frac{3}{2} \ , \ \xi)$ is the incomplete Gamma-function,

$$\gamma_0 = \gamma/Q(1), \qquad \alpha = \left[(1+g)\delta\right]^{-1} \ .$$

But the most remarkable fact is that a fairly good approxima-

tion can be obtained if we simply replace all the coefficients dependent on energy by their values at $\varepsilon = \mathfrak{J}$ (considering them to
be constant). After such a procedure the solution of the boundary
problem is trivial and we arrive at the following equation determining the dependence of γ on E

$$2 \, \Delta e^{-\alpha/2} = \Delta \mathrm{ch} \, \frac{\Delta}{2} - \alpha \mathrm{sh} \, \frac{\Delta}{2} , \qquad (11)$$

where

$$\Delta = \sqrt{\alpha^2 + 4 \, \gamma_0 \alpha} .$$

This fact has also a much more general character. Calculations similar to those made above were performed for other types of
kinetic coefficients $Q(x)$ and $D(x)$, and in particular for the case
of polar and nonpolar scattering by high-frequency ("optical")
phonons. The results allow us to state that not only the character
of the function γ (g) but also the num erical values depend little
on the concrete form and on the behavior of the kinetic coefficients as $x \to 0$ i.e., finally on the concrete dispersion laws and on
the energy dependence of the frequency of the electron-phonon collisions. A peculiar exception occurs in cases when electron
"runaway" is possible, but consideration of this effect is outside
the scope of the present discussion.

7. AVALANCHE IONIZATION PRODUCED BY LARGE RADIATION QUANTA[9]

We start from the differential-difference kinetic equation and
making use of the last comment of Sec.6 assume all coefficients to
be constant. Also we omit here the terms that contain the deriva-
tives of the distribution function in the displaced points
$x' = x \pm x^{(+)}$ as well as the terms that describe multiphoton intra-
band transitions with participation of phonons, since these terms
are not essential at the electromagnetic-radiation frequencies and
intensities of interest to us. Thus we have the following equation

$$\gamma_0 \delta^1 f(x) = f''(x) + \delta^{-1} f'(x) - 2sf(x) + sf(x + x_0) + sf(x - x_0) , \qquad (12)$$

where
$$S = g / x_0^2 .$$

(+) $x = \hbar \Omega /$

At first we consider only cases when $\mathfrak{I}/\hbar\Omega = n$ is an integer. Introducing the notation

$$f_k(x) = f\left[(n-k)x_0 + x\right] , \quad 0 \leq x \leq x_0, \quad k = 1,2,\ldots n,$$

we obtain the following system of n linear second order differential equations

$$\delta^{-1} \gamma_0 f_1 = f''_1 + \delta^{-1} f'_1 + Sf_2 - 2Sf_1$$

$$\cdots\cdots\cdots\cdots\cdots\cdots\cdots\cdots\cdots\cdots\cdots\cdots$$

$$\delta^{-1} \gamma_0 f_k = f''_k + \delta^{-1} f'_k + Sf_{k+1} + Sf_{k-1} - 2Sf_k \qquad (13)$$

$$\cdots\cdots\cdots\cdots\cdots\cdots\cdots\cdots\cdots\cdots\cdots\cdots\cdots$$

$$\delta^{-1} \gamma_0 f_n = f''_n + \delta^{-1} f'_n + Sf_{n-1} - Sf_n \quad .$$

The solution of this system is of the form

$$f_k(x) = \sum_{p=1}^{2n} A_k^p \, e^{\lambda_p x}$$

the values of λ_p being obtained in the usual manner from the characteristic equation

$$\Delta_n = 0 \qquad (14)$$

where Δ_n are the corresponding determinants of order 2n, for which the following recurrence relation holds :

$$\Delta_n = \left[\lambda(\lambda + \delta^{-1}) - (2S + r)\right] \Delta_{n-1} - S^2 \Delta_{n-2} , \qquad (15)$$

with

$$\Delta_0 = 1, \quad \Delta_1 = \lambda(\lambda + \delta^{-1}) - S - r \qquad r = \gamma_0 \delta^{-1} \quad .$$

Thus,

$$\Delta_n = \frac{1}{y_1 - y_2} \{ (y_1 + S) y_1^n - (y_2 + S) y_2^n \} ,$$

where y_1 and y_2 are the roots of the characteristic equation for the difference relation (15). Solving (14), we get

$$\lambda (\lambda + \delta^{-1})^{(p)} = \gamma_0 \delta^{-1} + 4S \cos^2 \frac{\pi p}{2n+1} , \quad p = 1, 2, \dots n,$$

and the coefficients of $\exp (\lambda_p x)$ in the distribution functions are

$$A_k^p = G_k^p A_0^p ,$$

where

$$G_k^p = (-1)^{k+1} \sin \frac{2 \pi p k}{2n+1} / \sin \frac{2 \pi p}{2n+1} .$$

The sough value of the avalanche rate γ_0 is then obtained from the solution of the equation

$$/ U_{1m} / = 0 .$$

In the determinant $/U_{1m}/$, 2n-1 rows are formed as a result of the conditions that the distribution function and the flux be continuous at the points $x_k = kx_0$, $k = 1, 2, \dots n-1$, while the remaining two rows correspond to boundary conditions simila to (6,7), but with allowance for the fact that x_0 is finite:

$$f_1 (x_0) = 0$$

$$2 \{ f_1'(x_0) + \delta^{-1} f_1(x_0) - S \int_0^x f_1(x) dx \} - f_n'(0) - \delta^{-1} f_n(0) = 0 .$$

The elements of our determinant are thus

$$U_{1,p} = e^{\lambda_p x_0}$$

$$U_{2,p} = G_n^P(\lambda_p + \delta^{-1}) - 2\lambda_p e^{\lambda_p x_0} + \frac{2S}{\lambda_p}(e^{\lambda_p x_0} - 1);$$

$$U_{2+k,p} = G_{k+1}^P e^{\lambda_p x_0} - G_k^P, \quad k = 1, \ldots n-1;$$

$$U_{n+1+k,p} = \lambda_p U_{2+k,p}, \quad k = 1, \ldots n-1; \quad p = 1, 2, \ldots 2n.$$

In the frequency region of interest to us we have

$$x_0 / \delta \gg 1 \qquad (\hbar\Omega \gg kT) \qquad\qquad\qquad (16)$$

and the terms containing exponentials of $\lambda_i x_0$ with $\lambda_i < 0$ can be neglected in the calculation. We consider first the case when

$$S\delta x_0 \ll 1, \qquad\qquad\qquad\qquad (17)$$

and according to (16)

$$S\delta^2 \ll 1.$$

Then the solution in the lowest orders is of the form

$$\gamma_{0,n} \frac{n^{n+2}}{n!} (q\delta)^{n+1} + C_{2n-1}^{n-1} \frac{n^{2n}}{\delta} (q\delta^2)^n.$$

Thus, in contrast to the diffusion solution, in the limit (17), and when $\gamma_0 \ll 1$ the rate γ_0 is a power-law function of the field

$$\gamma \backsim E^{2(n+1)}$$

In the expansion in powers of $S\delta x_0$, the coefficients of some of the terms are of order n^k, so that if the inequality (17) is not satisfied rigorously enough this expansion cannot be used: with increasing n, the region where $\gamma_0(E)$ is given by a power law

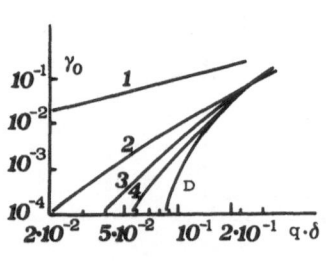

Fig. 1

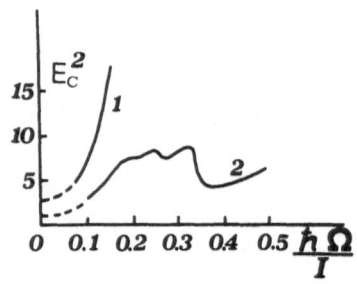

Fig. 2

such as above shifts into the region of pulses of longer duration.
This explains how the transition to the diffusion solution takes
place. More general $\gamma_0(q)$ dependences can be also calculated
subjecte to the rather weak restriction (16), which is satisfied
at n = 2...6 all the way to $\gamma_0 \simeq 0.1$. ($\gamma_0 = S\delta$ at n = 1).

   The results are given in Fig.1, which shows for comparison
also the plot of γ_0 (qδ) obtained in the diffusion approximation.
It is seen that at n >5 the diffusion solution describes quite well
the function $\gamma_0(q\delta)$ in the entire pulse-duration region of interest to us.

   So far we confined ourselves to the case $I/\hbar\Omega$= n with n
integer. The described method can be generalized also to the case

$$I/\hbar\Omega = n + p/m,$$

where n,p and m are integers. We must subdivide I into mn+p parts,
after which we determine f_1 in the region

$$1 - \frac{p}{mn+p} \le x \le 1, \text{ in the region } 1 - \frac{m}{mn+p} \le x \le 1 - \frac{p}{mn+p} , \text{ and}$$

so forth, and obtain thus a system of equations for 2n+1 functions
in which only functions with even (odd) indices are coupled one
with the other in the kinetic equations. The expansion is

$$\gamma_\circ \simeq S \; \delta\{ \; \frac{(S \; \delta Z_2)^n}{n!} + \frac{(S\delta \; Z_1)^{n+1}}{(n+1)!} + \eta_n \; (S \; \delta Z_2)^{n+1} + \text{cross terms} \; \}$$

where Z_1 and Z_2 are respectively the length of the even and odd segments into which the unit interval was broken down.

Fig. 2 shows the calculated plots of the critical field againts the frequency of the laser radiation for two values of avalanche rate : 1 - $\gamma_\circ$ = 10^{-5} sec^{-1}, 2 - $\gamma_\circ$ = 3.10^{-2} sec^{-1} (picosecond and nanosecond pulses respectively). It is seen that the character of the frequency dependence of the critical field (plotted in arbitrary units) is determined essentially by the duration of the laser pulse. At nanosecond durations the break-down threshold can decrease with increasing frequency of the field even at $x_\circ$ >0.3 and $E_{cr}(\Omega)$ is an oscillating function. The oscillations are due to the presence of sharp maxima of the distribution function f (x). Allowance for the exact boundary condition can lead to the appearance of additional shallow maxima and to a certain smoothing of the frequency dependence of the critical field.

LATTICE HEATING AND BREAKDOWN CRITERION

We have obtained expressions giving the field dependence of the avalance rate γ . But this is insufficient to carry out th e comparison between experimental and theoretical results. This because in a real experiment the value of γ cannot be measured directly, and we usually know only such quantities as the duration of the laser pulse, its temporal form, etc. That is why we have to determine relationship between these parameters and the rate γ . This relationship is usually called "breakdown criterion."

When the field strength is great enough, the electronic density may increase by several orders of magnitude even during a very short time interval. Certainly, the breakdown is not caused by this mere fact, but by the great increase in absorption coefficient which is due to the increase in concentration of the conduction band electrons . The energy absorbed by electrons is transferred to the lattice through the electron-phonon collisions .

There are a lot of processes that can be then initiated in the lattice, such as thermoelastic waves, local fusion, and so on, that lead to breakdown. Most of these processes have been investigated in great detail, but we shall not discuss them here because for such a nonlinear process as the electron avalanche it is quite unnecessary to know the exact amount of the critical energy.

Now we are going to obtain the value of the whole power loss
of the electrons due to the electron-phonon interaction

$$\tilde{Q}_{loss} = I \int dx g(x) f(x) \left\{ Q(x) - \frac{1}{g(x)} \quad \frac{\partial}{\partial x} \left[g(x) D_o(x) \right] \right\} \simeq \frac{4}{\sqrt{\pi}} Q(1)$$
$$(q \delta)^{3/2} IN \quad (\Omega\tau \gg 1)$$

The equations that describe the cascade generation of electrons and the heating of the lattice can be written in the form

$$\frac{dN}{dt} = \gamma^\circ \phi (\theta) N, \quad \frac{d\theta}{dt} \beta \theta^x N, \tag{18}$$

where $\theta = T/T_o$, T_o is the initial lattice temperature; if scattering by acoustic phonons predominates, the parameters are:

$$\beta = 4Q(1)(q \delta_o)^{3/2} I / \pi^{1/2} c\rho T_o, \quad \delta_o = k T_o / \mathcal{J} \quad , \quad x = 3/2$$

Here C is the specific heat of the lattice, ρ is the density,

$$\gamma = \gamma^\circ \phi (\theta), \quad \gamma^\circ = \gamma (T_o).$$

The   solution of the system (18) is

$$\gamma\overset{\circ}{t} = \int_1^\theta \frac{d\theta'}{\theta'^{3/2} A(\theta')}, \quad A(\theta') = \frac{\beta N_o}{\gamma^\circ} + \int_1^{\theta'} d\theta'' \theta''^{-3/2} \phi(\theta'')$$

$$\tag{19}$$

(No is the initial electron density) and this is also the desired
breakdown criterion. Since the integral in (19) always converges,
the function $\theta(t)$ has a vertical asymptote, i.e. the temperature
rise has an explosive character. That is why it is unnecessary to
know the final temperature (or the breakdown energy); it suffices
to calculate the first integral in (19) as $\theta \rightarrow \infty$. It can be esti-

mated also that in a time exceeding $0.9t_p$ the heat rise is less than 10° (in a breakdown field). This is precisely the circumstance that allows us to disregard the temperature dependence of the avalanche rate γ and to obtain the simple criterion.

So the solution of our problem of the electron avalanche ionization in solids is completed, at least to some extent. We had an opportunity to see how our general methods can be used. Certainly this is only one of numerous problems we meet when dealing with hot carriers in semiconductors and isolators. But even inside this problem there are some questions left. I just am going to enumerate them.

1. The question of the role of the electron spatial diffusion. Nowadays this question is answered but we have no time to discuss it here.

2. The influence of all kind of defects and impurities on the electron-avalanche process. It is interesting to note that in principle impurities might as well slow down this process as accelerate it.

3. The question of the role and origin of initial electrons starting the cascade multiplication process.

4. The development of avalanch ionization when a short-time recombination is allowed.

We always spoke of the "effective" impact ionization potential. This is because the electron energy required to produce a new electron in the conduction band can be both greater and less than the energy band gap ε_{gap}. It is greater when the conservation laws cannot be satisfied if the electron energy is near to $\mathfrak{I}$. We often meet this situation in semiconductors. As to the broad gap dielectrics, we should remember of a number of conduction subbands, and when the radiation quanta are large enough the direct transition between them must be taken into account. Thus an electron has to gain a comparatively small energy by means of electron-phonon-photon collisions and after that it jumps quickly to the ionization energy. These direct processes can make the effective potential to appear to be less than ε_{gap}.

And this is the last remark I wanted to make.

REFERENCES

1. Shockley W., Bell System Tech. Journ. 30 , 990; 1951

2. Prokhorov A.M., Fund.and Appl.Laser Physics, Proc.Quant.
    Electr. Conf., Isfagan, Wiley Intersc.Publ., 1971, p.51

3. Keldysh L.V. Sov. Phys. - JETP 10, 509, 1960

4. Keldysh L.V., Zh. Eksp. Teor.Fiz. ,48, 1962

5. Baraff G.A., Phys.Rev., A 26, 133, 1964

6. Epshtein E.M., Sov.Phys. - Solid State 11, 2213, 1970

7. Epifanov A.S., Sov.Phys. - JETP, 40, 897, 1975

8   Epifanov A.S., Manenkov A.A., Prokhorov A.M., Sov.Phys.JETP
    43, 377, 1976

9. Gorshkov B.G.,Epifanov A.S.,Manenkov A.A., Sov.Phys.JETP 49
    309, 1979.

IMPURITY CLOUDS IN SILICON AND GERMANIUM

V.P. Kalinushkin, V.V. Voronkov, G.I. Voronkova,
V.N. Golovina, B.V. Zubov, T.M. Murina, and
A.M. Prokhorov

The Physical Institute of the
Academy of Sciences of the USSR
117333, Moscow, USSR

INTRODUCTION

Local inhomogeneities in semiconductors (i.e. structural defects and regions of enhanced impurity concentration) can influence crystal properties and device parameters. The influence of structural defects is probably caused by their action as impurity s egregation sites [1,2]. Thus the impurity inhomogeneities are the most interesting object of investigation. Most methods such as electron-probe microanalysis, transmission electron microscopy (TEM), etc. detect only regions of high impurity concentration, i.e. inclusions (particles captured during crystal growth), precipitates and their colonies. On the other hand, probe resistivity measurement reveals only regions of large size. However the inhomogeneities of small size (say 10 μm) and of relatively low concentration can have a crucial effect on crystal properties.For instance they control the quality of germanium detectors [3,4]. Such impurity inhomogeneities can be detected and studied by low-angle scattering of IR laser beams [5,6].

THE METHOD OF LOW-ANGLE LIGHT SCATTERING

The impurity inhomogeneity causes local deviations of the dielectric constant. The relative deviation (denoted here by $\tilde{\varepsilon}$) is likely to satisfy the Rayleigh-Gans condition [7], i.e. $| \tilde{\varepsilon} | \ll \lambda/a$ where a is a linear size of the inhomogeneity and λ is the wavelength (inside the crystal). Then the scattered light intensity I is proportional to $\varepsilon_m^2 \, a^6 \, \lambda^{-4} C$ where $\tilde{\varepsilon}_m$ is the maximum value of $\tilde{\varepsilon}$ and C is the concentration of scattering inhomogeneities.

The dependence of I on the scattering angle θ, is controlled by
the profile function $\overset{\sim}{\varepsilon} / \overset{\sim}{\varepsilon}_m = f(r/a)$, r being the distance from
the centre of the inhomogeneity (spherical symmetry is assumed). If
$a > \lambda$ then most scattering takes place in the region of low angles,
$\theta \leqslant \lambda /2 \pi a$, that is I decreases rapidly at these small angles.
If $a << \lambda$ (Rayleigh scattering) then the intensity I is constant
at low angles. The latter condition of small size is usually satisfied by inclusions, precipitates, and structural microdefects.

Consequently if the measured I (θ) - function has a pronounced
fall at low angles, one can reconstruct the profile function
f(r/a) and, particularly, find the characteristic size a. After
that, one can calculate the combination $C \overset{\sim}{\varepsilon}_m^2$ using the absolute
value of I. If independent information about concentration C is
available, then it is possible to evaluate $\overset{\sim}{\varepsilon}_m$.

Let us assume that the scattering inhomogeneity is a small
crystal region with enhanced concentration of dissolved impurity
atoms. These atoms can be electrically charged. Then the inhomogeneity is a region of enhanced free carrier (electron or hole)
concentration. The region is electrically neutral as its size a
exceeds the Debye screening length. The electric field of the light
wave causes vibration of free carriers and corresponding strong
polarisation of the crystal region (as compared to the polarisation due to impurity atoms themselves). In other words the free
carrier contribution to $\overset{\sim}{\varepsilon}$ is much greater than that of impurity
atoms of the same concentration [5] . Free carrier contribution to
$\overset{\sim}{\varepsilon}$ is proportional to λ^2 [7], so the intensity I does not depend
on λ. For the inhomogeneities of constant $\overset{\sim}{\varepsilon}$ the intensity de-
creas es as λ^{-4}. Thus for sufficiently large λ the scattering
by free carrier inhomogeneities will dominate.

The main difficulty of the described method is the small intensity of scattered light; usually it is less than the sensitivity of photo-detectors for the middle IR - range ($\sim 10^{-5} \div 10^{-6}$
watt/ster.). In the present work this difficulty was overcome by
using a coherent light source (CO_2 - laser) and optical hetero-
dyning, which raised the sensitivity up to $(10^{-8} \div 10^{-10})$watt/
ster.

SCATTERING DIAGRAMS FOR Si AND Ge

Floated-zone silicon and Czochralski-grown germanium crystals
of high purity were mainly dislocationless; several crystals with
dislocations were also investigated. Fig.1 shows a typical scattering diagram I (θ). The region up to $\theta \sim 1.5°$ is caused by the
divergence of the incident laser beam. The fast fall of I(θ) in
the interval from 1.5 to 4° can be approximated by a profile
function $f = (1 + r^2/a_1^2)^{-2}$ where the size is $a_1 \sim 20$ µm. The slower
fall in the interval from 4 to 11° corresponds to optical inho-

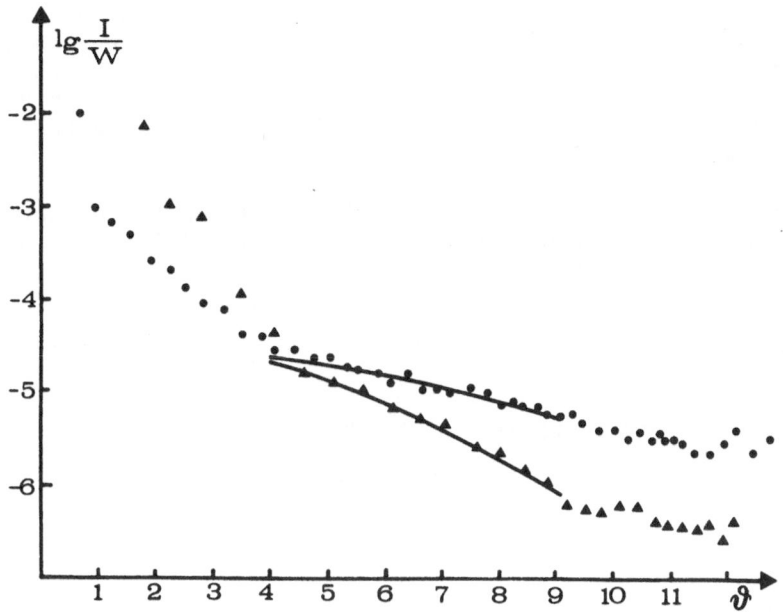

Fig. 1 Behaviour of normalized scattered intensity
as a function of θ

mogeneities of Gaussian profile $f = \exp(-r^2/a_2^2)$, the characteristic size being $a_2 \sim 6$ μm. At $\theta° > 11°$ the intensity I reaches
the plateau value corresponding to scattering by small – scale
inhomogeneities, $a < 2$μm.

To clarify the nature of the observed optical onhomogeneities
one should investigate the influence of different sample treatments on the scattering diagram. We shall discuss only the inhomogeneities of Gaussian type that have been studied in more detail
5,6,8,9.

1. The bulk photo-excitation of germanium samples [6] leads to a
strong increase of the scattered intensity I (θ) in the angular
interval of slow fall (the size a keeps the initial value 6 μm).
This shows that the Gaussian inhomogeneities are the crystal

regions of enchanced free carrier concentration.

2. Heat treatment of dislocationless p–Si samples at T ≥ 1000°C
followed by rapid quenching leads to the disappearance of the
slow fall portion of the scattering diagram [5]. However, the initial
scattering diagram was restored after approximately a month at
room temperature or after several hours at 200°C. If the high-
temperature heat treatment was followed by slow cooling then the
initial diagram was retained.

3. Increasing of heat treatment duration t, leads to an increase
of the inhomogeneity size a (as compared to its initial value a_o)
according to the simple diffusion law [8]

$$a_t^2 = a_o^2 + 4\ Dt. \tag{2}$$

This dependence is shown in fig.2. The coefficient D(T) in (2)

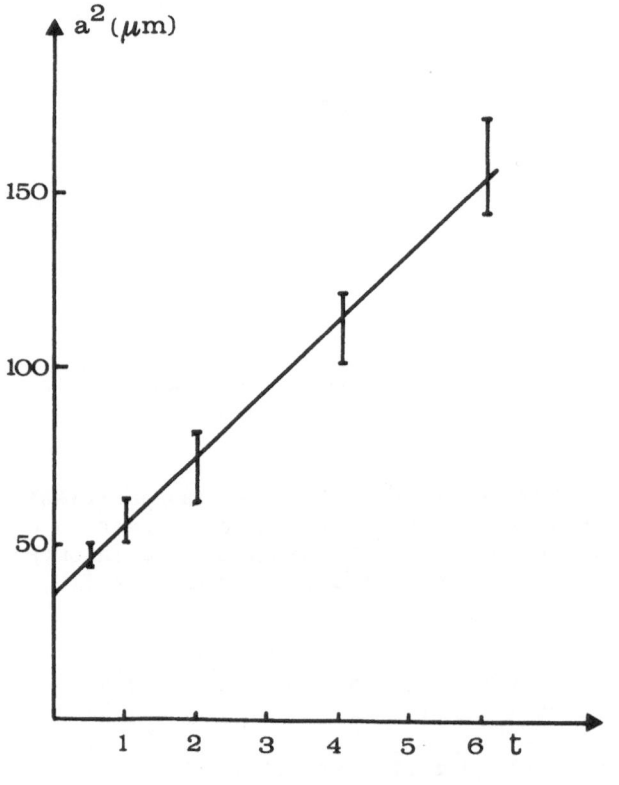

Fig.2   Dependence
         of the inho-
         mogeneity
         size a as a
         function of
         the heat
         treatment
         duration t

coincides with the diffusion coefficient of oxygen in silicon [10]
for the three temperatures T that were used (1000, 1050 and
1100°C). For instance $D = 2.10^{-11}$ cm^2/sec at T = 1000°C. Heat treatment of germanium samples [9] at 750°C leads to a similar increase of
a according to equation (2), and D again coincides with the diffusion coefficient of oxygen in germanium.

4. Impurity clouds

The above data enable us to derive some conclusion on the nature
of the optical Gaussian inhomogeneities. The nucleus of each inhomogeneity is a local crystal region with enchanced concentration
of dissolved oxygen atoms: hereafter it is called "oxygen cloud".
Oxygen in Si and Ge is an electrically neutral impurity:therefore
s cattering by oxygen clouds is weak. However, at sufficiently low
temperature, oxygen can form electrically active complexes with
s ome fast-diffusing impurities. As a result the initial oxygen
cloud turns into a more complex impurity cloud that is a region
w ith a high concentration of charged impurity centres (and with
h igh free carrier concentration). We shall call this process
"activation of clouds". Activated clouds contribute significantly
to scattering of light. Heat treatment leads to dissociation of
oxygen-impurity complexes. The fqst-diffusing impurities leave the
clouds and distribute homogeniously over the crystal. On quenching
the reactivation has no time to proceed which corresponds to
disappearance of scattering. Diffusion of fast-diffusing impurities at room temperature requires about one month to activate the
oxygen clouds again.

The Gaussian profile of the oxygen cloud (which retains after
heat treatment) clearly points out that the clouds are the result
of free diffusion of oxygen atoms that were instantaneously liberated from a point source (i.e. the source of oxygen had a size
much smaller than a_o). Using the value of a_o and the cooling rate
of the crystal under growth conditions, one can calculate the
temperature T_f at which oxygen started to diffuse, i.e. the liberation of oxygen from the sources took place [8,9]. It turned out
that T_f is 120°C below the melting point. The most probable oxygen
sources are the oxide particles captured from the melt by growing
the crystal. Dissolution rate of oxide inclusions strongly depends
on the interstitial supersaturation as the dissolution process
involves absorption of silicon interstitials. At some critical supersaturation the dissolution rate can increase sharply. It means
that the inclusion works as an instantaneous source of oxygen
atoms.

5. Relation with swirls, clouds and microdefects

The observed intensity of scattering by impurity clouds was

higher for dislocationless crystals of Si and Ge that exhibit
large etch pits usually forming (in the case of Si) the so called
swirl-pattern. We shall call these etch pits simply "swirls".
Crystals without swirls show no scattering by clouds [5] . According to the described model, this fact does not necessarily mean
that clouds are absent but can merely mean that they are not
activated (crystals without swirls seem to contain a high concentration of very small defects [11,12] that may serve as precipitation sites for fast-diffusing impurities). Indeed, after short
heat treatment at 1000°C (followed by quenching and exposure at
200° C) scattering by gaussian inhomogeneities (a 6 μm) appeared.
Etching of the treated samples showed the appearance of swirl-
p atterns.

It should be noted that heat-treated and quenched samples of
silicon show no swirls even if the swirl-pattern was present in
as-grown crystals. After a month (or more) at room temperature T_r
the swirl-pattern in treated samples was restored, together with
scattering. The correlation between swirls and scattering clouds
was checked by additional experiments. Two samples of silicon
(w ith swirls in as-grown state) were heat-treated at 1000°C and
quenched: the first sample, as usual, down to T_r and the second
one down to 200°C (and then slowly cooled to T_r). The first sample
as usual, showed no scattering clouds and no swirls, while the
second one exhibited strong scattering by clouds and typical
swirl-patterns.

It follows from these observations that etch swirls are
actually caus ed by activated impurity clouds. The etching effect
is probably due to strong local deviations of resistivity at the
clouds. Up to now etch swirls were attributed to structural microdefects - dislocation loops of intrinsic type [13,14]. The individual loop (or loop cluster) is found by TEM [13-15] under the etch
h illock (some etchants produce hillocks instead of pits). It is
i mportant that not every hillock corresponds to a loop [15].
Furthermore silicon crystals studied in [16] showed the typical
swirl-pattern but did not contain loops.

Now it is possible to outline the general picture of inhomogeneity formation in silicon. Oxide inclusions give rise to oxygen clouds at temperature T_f (120°C below the melting point). Later
the loops are formed at the same crystal sites (but not at every
one). Finally, at low temperature the activation of clouds takes
place. Some clouds contain loops (the size of a loop is usually
less than a cloud size a). The swirl-pattern revealed by etching
is caused not by microdefects (loops) but by activated clouds.

It should be stressed that the last assertion has a limited
validity. It is of course possible that etching does reveal micro-

defects in some cases. Such methods as copper or lithium decora-
tion evidently reveal microdefects and not clouds.

Finally we wish to underline that impurity clouds represent
a new type of impurity inhomogeneities. Oxygen cloud is formed not
by accumulation of impurities as some crystal region but, on the
contrary, by diffusional expansion from the initial centre (oxide
inclusion). This type of impurity inhomogeneities probably has a
universal character and can exist in all melt-grown crystals.

REFERENCES

1. A.J.R. de Kock, Philips Res. Repts 1 (1973) 1.

2. G.A.Rozgonyi and C.W.Pearce, Appl.Phys.Le-t. 32 (1978) 747.

3. V.K.Eremin, N.B.Strokan and N.I.Tisnek, Fiz.techn.polupr.
    (Soviet Physics - Semiconductors) 9 (1975) 1575.

4. V.K.Eremin, N.B.Strokan, N.I.Tisnek and A.Sh.Shamagdiev, Fis.
    techn.polupr. (Soviet Physics - Semiconductors) 12 (1978) 718.

5. V.V.Voronkov, G.I.Voronkova, B.V.Zubov, V.P.Kalinushkin,
    B.B.Krynetskiy, T.M.Murina and A.M.Prokhorov, Fis.tverd. tela
    (Soviet Physics - Solid State) 19 (1977) 1784.

6. V.V.Voronkov, G.I.Voronkova, B.V.Zubov, V.P.Kalinushkin,
    B .B.Krynetskiy, T.M.Murina and A.M.Prokhorov, Fiz.tverd.tela
    (Soviet Physics - Solid State) 20 (1978) 1365.

7. H .C.van de Hults, Light scattering by small particles
    (Wiley, N.Y.Chapman and Hall, L. 1957).

8. V.V.Voronkov, G.I.Voronkova, B.V.Zubov, V.P.Kalinushkin,
    E.A.Klimanov, T.M.Murina and A.M.Prokhorov, Fiz.techn.polupr.
    (Soviet Physics - Semiconductors) 13 (1979) 846.

9. V.V.Voronkov, G.I.Voronkova, B.V.Zubov, V.P.Kalinushkin,
    T.M.Murina, E.A.Petrova, A.M.Prokhorov and I.M.Tiginjanu, Fiz.
    techn.polupr. (Soviet Physics-Semiconductors) 13 (1979) 1137.

10. C.Haas, J.Phys.Chem.Solids 15 (1960) 108.

11. A.J.R. de Kock, Acta Electron. 16 (1973) 303.

12. N.V.Veselovskaja, E.G.Sheihet, K.N.Neymark and E.S.Falkevich,
    in : Rost i legir. polupr. krist. i plen. (Growth and Doping
    of Semicond.Crystals and Films, Proc. All-Union Symp. on
    Growth and Synthesis of Semicond.Crystals and Films) 4th
    1975 (Publ.1977) part 2, pp. 284-286 (Russian).

13. H.Foll and B.O.Kolbensen, Appl.Phys. 8 (1975) 319.

14. P.M.Petroff and A.J.R. de Koch, J.Crystal Growth 30 (1975) 117.

15. L.I.Bernewitz, B.O.Kolbesen, K.R.Mayer and G.E.Schuh, Appl.
    Phys. Lett. 25 (1974) 277.

16. E.Nes, Phys.Stat.Sol. (a) 33 (1976) K5.

AMORPHOUS SILICON SOLAR CELLS

J.I.B. Wilson

Department of Physics, Heriot-Watt University

Edinburgh EH14 4AS

1. WHY AMORPHOUS SILICON ?

The first reports of amorphous silicon photovoltaic diodes appeared in 1976[1], and since then several other device applications have been suggested[2,3,4,5], but it is the promise of cheap solar cells with efficiencies greater than the present 5-6% which excites most attention. Whilst crystalline Si p-n solar cells do work effectively, they are too expensive for terrestrial applications, unless their ability to operate for long periods without maintenance can offset the initial capital investment, at a site where there is no electrical power line. Present efforts to reduce the cost of solar cell electricity either attempt to use high efficiency cells (based on GaAs and $Ga_xAl_{1-x}As$) with cheap optical concentrators, or attempt to use large-area thin film cells of modest efficiency but cheap construction. In locations where there is a large amount of diffuse, scattered sunlight, optical concentrators are useless, and in any location they require a solar tracking mechanism with its attendant maintenance problems. Thin film cells are not commercially available, although those based on Cu_xS/CdS have been on the verge of success for some years despite lifetime problems[6]. As a half-way stage, and to retain the attractions of familiar silicon technology, there have been several ingenious techniques of producing planar forms of crystalline silicon without the necessity for growing and slicing ingots[7], and this does result in some reduction of cost. Merely by automating the production stages (diffusion of dopants, deposition of contacts, annealing), and perhaps using high energy laser and electron beam processing[8,9], it is possible to make further economies. In order to achieve the projected costs of the

413

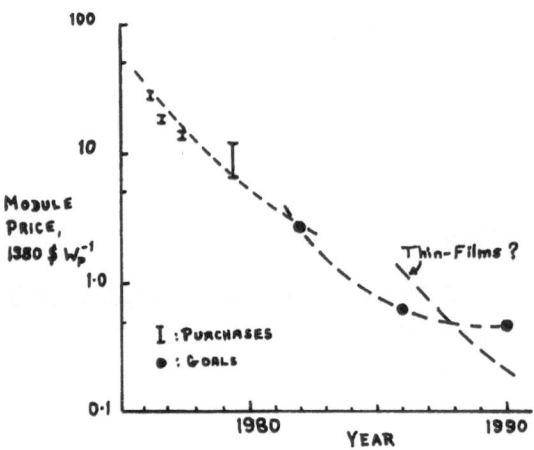

Fig. 1   US Department of Energy cost targets, and purchases
          of solar cell modules, in 1980   per watt of generating
          capacity in 1 kW m^{-2} sunlight (after Forney, 1979 [10]).

US Department of Energy10 it is essential to develop thin film
solar cells with efficiencies of at least 8%, and amorphous
silicon, based as it is on an accepted semiconductor, is therefore very attractive. Even with the present limited performance
there are commercial possibilities, and two Japanese companies
(Sanyo and Fuji) have announced forthcoming production of dome-
stic electronics with liquid crystal displays powered by
amorphous silicon photovoltaic batteries from indoor lighting.

2. GROWTH OF AMORPHOUS SILICON

The amorphous silicon literature is particularly widespread,
for not only are device researchers interested, but there are
many groups who are interested in the intrinsic properties of the
material itself as an example of structural disorder. The
proceedings of the most recent Amorphous and Liquid Semiconductor
Conference (1979) give a detailed picture of this activity[11], but
a more ordered account of most aspects is to be found in a
review volume edited by Brodsky[12]. Since the manner of producing
amorphous silicon decides whether its properties are controllable

or not we shall describe the most common methods of growing the particular material of interest for devices.

Amorphous silicon deposited from the elemental vapour on to cold substrates cannot have its electrical conductivity controlled by doping, for the Fermi level is controlled by the high density of localized energy levels introduced into the bandgap by defects such as unsatisfied bonds. If some hydrogen is included in the correct way then the mid-gap state density is reduced from over 10^{20} eV^{-1} cm^{-3} to less than 10^{18} eV^{-1}cm^{-3}, for hydrogen

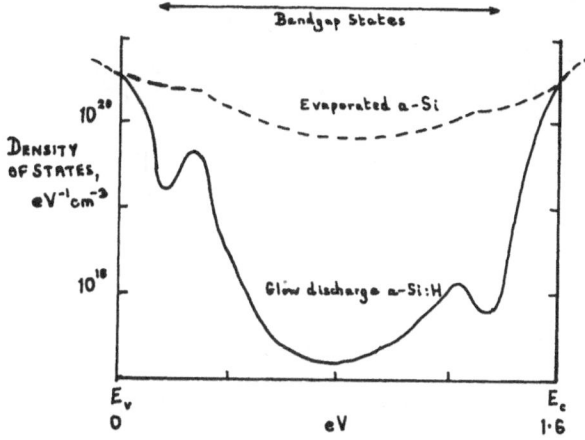

Fig. 2   The density of states in the bandgap of amorphous silicon. The conduction and valence bands have tails of localised states, which extend into the gap states produced by bonding defects (after Spear, 1977 [13]).

appears to remove dangling bonds, and consequently effective doping can take place.

The first successes with this approach used the plasma decomposition of silane (SiH$_4$), with phosphorus or boron doping by adding traces of phosphine or diborane to the gas[14,15]. Hydrogenated amorphous silicon (a-Si:H) was found to be a good semiconductor with different transport and optical properties from crystalline or pure amorphous silicon. Since then hydrogen has been incorporated in sputtered amorphous silicon[16], in silicon from thermally-decomposed silane[17], and in polycrystalline silicon[18], with varying degrees of success in reducing their

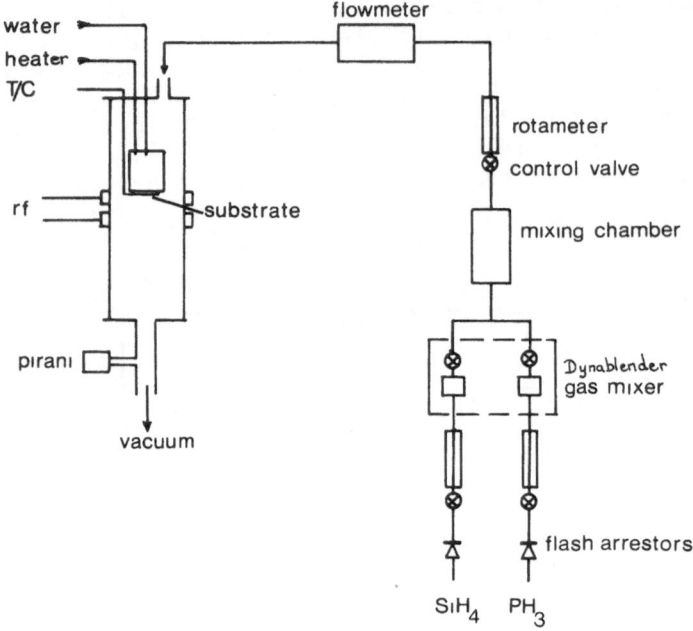

Fig. 3   Apparatus for the rf glow discharge deposition of doped
amorphous silicon from silane and phosphine.

defects state densities. The glow discharge decomposition technique still produces the best films, and the exact hydrogen content
and structure depend not only on the gas flow, temperature, and
plasma energy, but also on the geometry of the reaction vessel.
The optimum conditions are low rf or dc powers (a few watts
cm^{-2}), low pressures (~ 0.1 torr), high gas flow rates, and
substrate temperatures of 200–300°C, resulting in a deposition
rate of 1-2 $\mu m\ hr^{-1}$. Some laboratories use silane diluted with
argon or hydrogen, but it is more common to use pure silane, and
a dopant mixutre. Other related compounds – the fluorinated and
chlorinated silanes – have also been decomposed in this way, and
yield similar films of rather more complex structure[19].

There are also similarities between this method and the laser decomposition of SiH_4[20,21] or $SiCl_4$[22], which is not entirely a pyrolitic process.

3. THE STRUCTURE OF A-Si:H

The tetrahedral bonding of crystalline silicon is retained in the amorphous form with the same bond length, but there is a variation in the bond angles which leads to a breakdown of long-range order. When a network of this type is built up some bonds will be at orientations which cannot be satisfied by other silicon

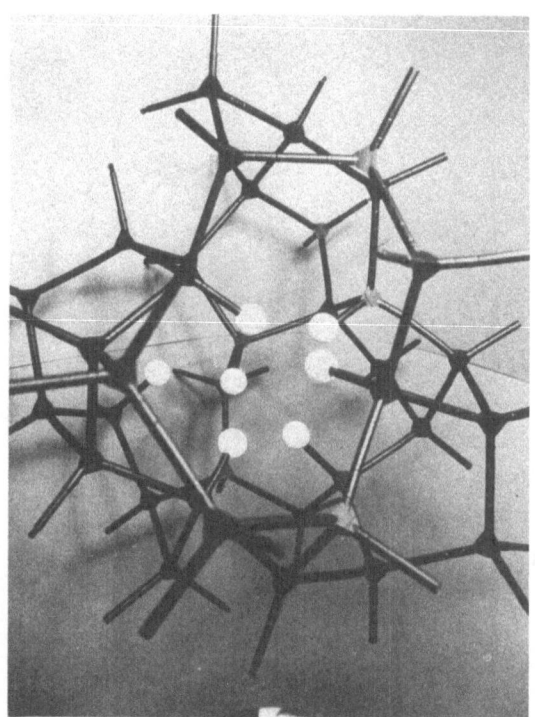

Fig. 4   A ball-and-stick model of a continuous random network. The white balls represent hydrogen, attached to dangling bonds from silicon atoms.

atoms, and it is these which are passivated by hydrogen atoms[23]. If the deposition conditions are far from ideal (a cold substrate, high silane pressure, or high rf power) then instead of ~ 5 atomic % of hydrogen, up to 30 atomic % may be included, and

there will be some silicon atoms with more than one hydrogen atom
attached. In the limit, a polysilane chain will form, and such
deposits are useless for solid state devices. There is the possibility that high hydrogen content films may be composed of
silicon-rich regions separated by hydrogen-rich regions separated
by hydrogen-rich filaments perpendicular to the substrate[24]

At temperature in excess of 700°C the films recrystallise,
but at lower temperatures than this, hydrogen is already being
evolved, with consequent changes in film properties. There are two
stages of hydrogen loss, one at ∿ 350°C and one at >500°C, and it

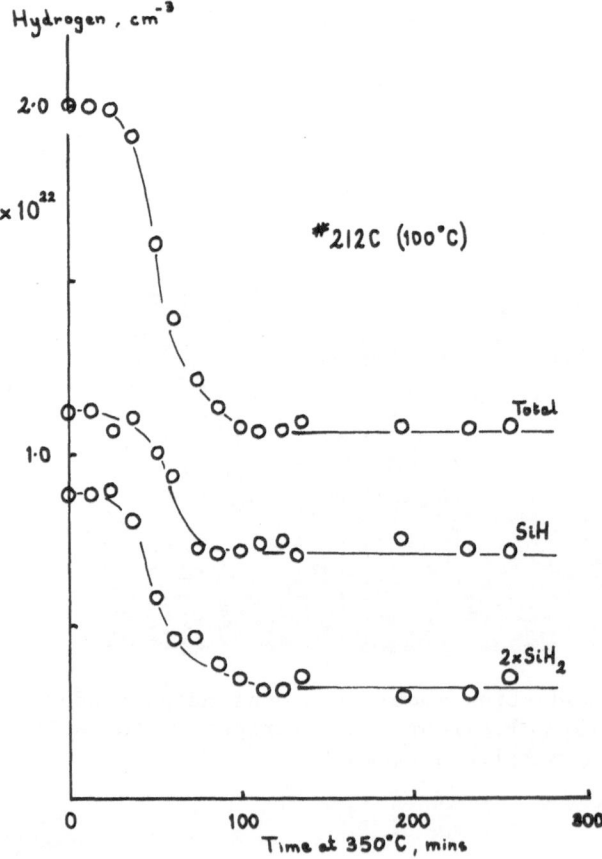

Fig. 5 The thermal effusion of hydrogen from a low substrate
        temperature a-Si:H film, held at 350°C, from in situ
        infrared absorption measurements[25].

now appears that hydrogen is simultaneously lost from both-SiH-
and -SiH$_2$-groupings[25]. The secondary-bonded hydrogen in particular
is undesirable, but cannot be preferentially removed by normal
heating. The hydrogen content may be determined in several ways,
including thermal evolution in a vacuum[26], resonant nuclear reactions[27], α-particle elastic scattering[28], secondary ion mass spectroscopy[29], infrared absorption[30-33]. Whilst infrared spectroscopy
does not allow the hydrogen content with depth to be determined,
it is a non-destructive method which gives a rapid determination
of the SiH and SiH$_2$ groups, as well as SiO.

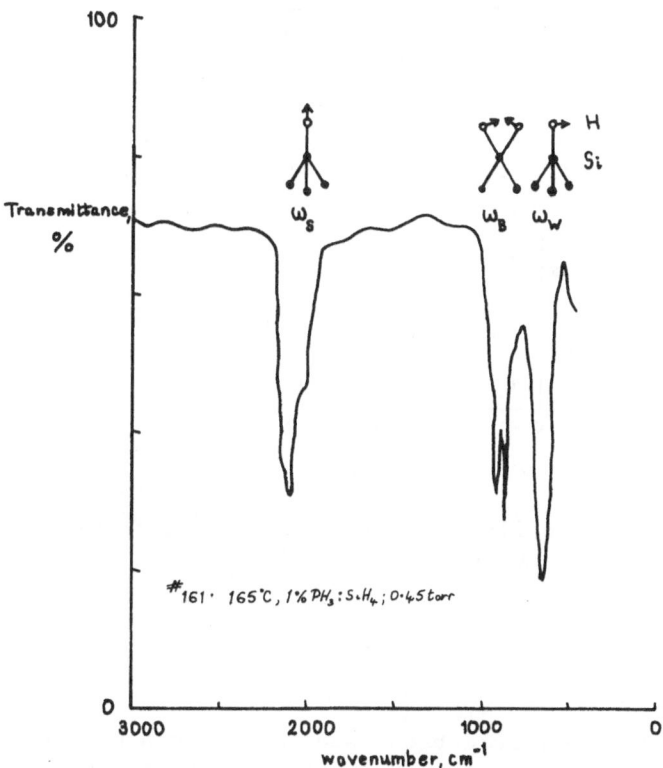

Fig. 6   The infrared absorption of a low substrate temperature,
doped a -Si:H film with representations of some of the
stretching, bending, and wagging vibration modes.

Films grown at > 250°C do not usually contain SiH_2, and so do not absorb in the bending-mode region of 840-900 cm^{-1}. There is still some dispute over the exact assignments of all absorption peaks, with arguments over the relative importance of adjacent-SiH_2-groups to the 840 cm^{-1} band.

A recent result of great interest is that phosphorus-doped films contain more hydrogen than the corresponding undoped films, and boron-doped films contain less[34]. Nuclear resonance hydrogen

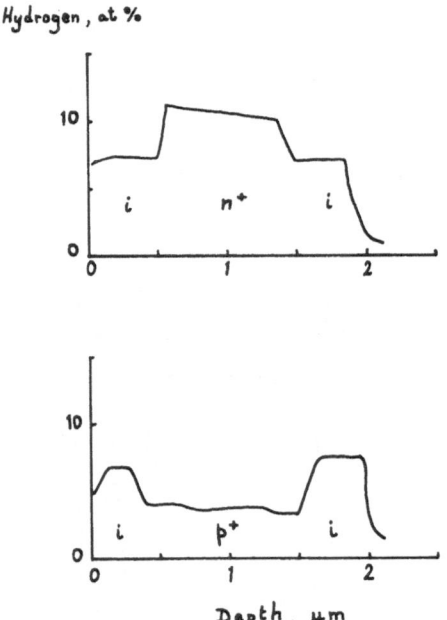

Fig. 7   Hydrogen profiles of i-n$^+$-i and i-p$^+$-i films obtained by resonant nuclear reaction[34].

determination has therefore been used as a method of dopant profiling stepped and graded p-n junctions. The cause of this disparity is not known, but is most likely an indication of the species into which the plasma splits the gases, SiH_4, PH_3, B_2H_6.

4. ELECTRICAL AND OPTICAL PROPERTIES OF A-Si:H

The investigations covered by this section are so extensive that only a few points of direct relevance to device operation will be considered.

Since the films <u>are</u> amorphous their mobility is low, and there is a considerable difference between the Hall mobility and drift mobility. Carriers can move either through the localised band-tail states (by hopping), or through the usual extended states further up in the respective bands. Hole transport is worse than electron transport, and even at room temperature is dominated by localised states, thus p-type material is a less-useful semiconductor than n-type, which itself is poorer than undoped a-Si:H. The room temperature effective electron drift mobility is ~ 1 cm^{-2} V^{-1} s^{-1}, which is slightly lower than that in the extended states alone, due to trapping and release of electrons by localised states. The conductivity of doped a-Si:H can be adjusted between $\sim 10^{-9}$ (Ω cm)$^{-1}$ and 10^{-2} (Ω cm)$^{-1}$ for phosphorus doping, with a minimum

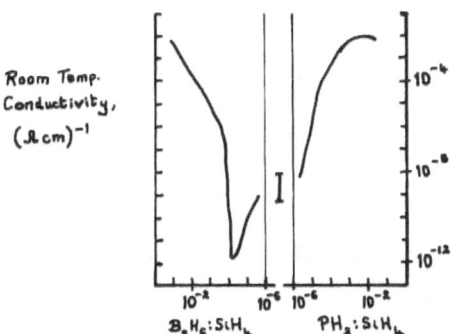

Fig. 8   The room temperature conductivity of rf glow discharge a-Si:H films doped with phosphine or diborane (after Spear and Le Comber[35]). Undoped films have the value indicated in the centre of the figure.

of $\sim 10^{-12}$ (Ω cm)$^{-1}$ for slight p-type doping and $\sim 10^{-2}$ (Ω cm)$^{-1}$ for heavy p-type doping: undoped material is slightly n-type. The Fermi level cannot easily be moved closer than ~ 0.2 eV to the band edges (defined by the "mobility gap", or separation between the extended state boundaries). The donor band lies ~ 0.2 eV

below the conduction band extended states. The actual distribution of gap-states must be included in any calculation involving
space-charge width at junctions, and it has a large effect on
the photoconductive and photoluminscent properties of the material.

The optical gap of a-Si:H varies with hydrogen content,
between ∿ 1.5 eV and ∿ 2.2 eV. Because of the lattice disorder,
E(k) selection rules are released and the material behaves as
though it had a direct bandgap. Thus, although the refractive

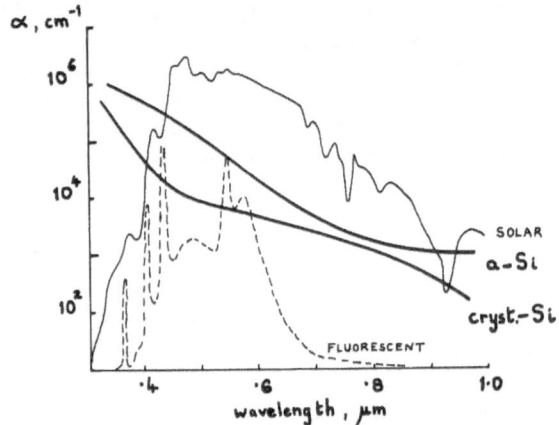

Fig. 9   Typical optical absorption coefficients of a-Si:H and
         crystalline silicon, superimposed on the spectra of AM1
         sunlight and fluorescent lighting.

index of a-Si:H is similar, it has a different absorption spectrum
from that of crystalline silicon, and so is in fact a more effective solar absorber. This is fortunate, for it compensates for
the virtual absence of any diffusion current in the amorphous silicon solar cell (which is the dominant component in conventional
crystalline cells) due to the short minority carrier diffusion
length (∿ 400 Å). It has also been proposed that free electrons
and holes are only produced by optical excitation if there is an
electric field present to dissociate the excitons which result
from optical absorption[36]. Thus an efficient solar cell based
on a-Si:H must have a wide space-charge region (requiring undoped
material) covering the depth of the device where optical
absorption occurs.

5. SOLAR CELLS

Several different diode structures have been tested as solar cells, from simple Schottky diodes to variously-doped p-i-n diodes, the best overall efficiency being between five and six percent

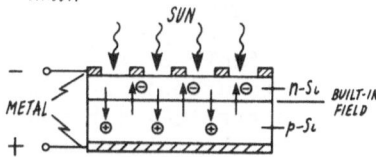

1) SEMICONDUCTOR ABSORBS SOLAR PHOTONS,
2) PHOTON ENERGY RELEASES MOBILE ELECTRICAL
*   CHARGES WITHIN THE SEMICONDUCTOR,*
3) THE +ve AND −ve CHARGES ARE
*   SEPARATED, BEFORE THEY CAN RECOMBINE*
*   (GIVING HEAT), BY AN ELECTRIC FIELD,*
4) METAL CONTACTS TRANSFER ELECTRICAL
*   CHARGES TO A LOAD IN AN EXTERNAL*
*   CIRCUIT*

Fig. 10   The three features required of a solar cell, with a schematic cross-section of a crystalline silicon p-n cell.

(for small) Schottky-type cells). Before describing some of these devices in more detail, the standard parameters used to compare performances will be explained[37].

There are some differences in behaviour between crystalline and amorphous cells, but the current-voltage characteristics of all cells are described well by the same equation:

$$I = I_0 (\exp (eV/nkT) - 1 - I_L \qquad (1)$$

That is, the net current is given by the superposition of the usual dark diode current and a photocurrent, I_L. An ideality factor, n, is included as an engineering parameter to allow all sources of leakage current, I_0, to be added together, whereas different paths dominate at different bias ranges (diffusion, recombination in the space-charge region, tunneling, etc.). Thus

The photovoltaic characteristic is the same shape as the dark
characteristic, but shifted downwards by $|I_L|$, unless there
is an additional photoconductive effect which increases the slope
of the highly-conducting branch of the curve. In amorphous cells,
both I_o and n have different values when illuminated and so these
parameters must be measured under normal operating conditions. The
best way is to use the relation between the open circuit voltage,
V_{oc}, and the short circuit current, I_{sc}, making measurements at
different irradiances:

$$V_{oc} = \frac{nkT}{e} \ln \left(\frac{I_L}{I_o} + 1 \right) \qquad (2)$$

$$I_L = I_{sc} \qquad (3)$$

These equations neglect the effects of series resistance, R_s
(which mainly reduces the photocurrent) and shunt resistance, R_{sh}
(which mainly reduces photovoltage), which reduce the maximum
power obtainable from the cell, but these terms may easily be
added if necessary. Amorphous silicon cells which use underlined undoped
material fortunately have their series resistances reduced by
photoconductivity in normal operating conditions, and so R_s can
be neglected. The maximum power point lies close to the knee of
the characteristic and gives the matched load for the maximum
power conversion efficiency. The ratio of this power to the
product of V_{oc} and I_{sc} is the "fill factor", FF, which is a
measure of the squareness of the characteristic and which is
reduced by poor series and shunt resistances.

The photocurrent, I_L, may be calculated if the optical
absorption coefficient, photon flux, reflection loss, and
transport parameters are known. It increases linearly with illumination intensity and therefore V_{oc} increases logarithmically
with intensity, up to a saturation value at which the band-bending
at the junction is flattened.

An increase in temperature will decrease V_{oc}, because of the
increase in leakage current, I_o, but will have little effect on
I_{sc}. Below 0°C, I_{sc} will decrease as R_s increases, and V_{oc}
eventually saturates as tunneling dominates I_o. A plot of V_{oc}
againts temperature enables the barrier height, $e\phi_B$, of a
Schottky cell to be determined from the intercept of the extra-
polated linear plot.

$$I_o = A\, T^2 \exp (e\phi_B/kT) \qquad (4)$$

(Substitute equation (4) into equation (2) to obtain this result).
There is some argument as to whether this thermionic emission

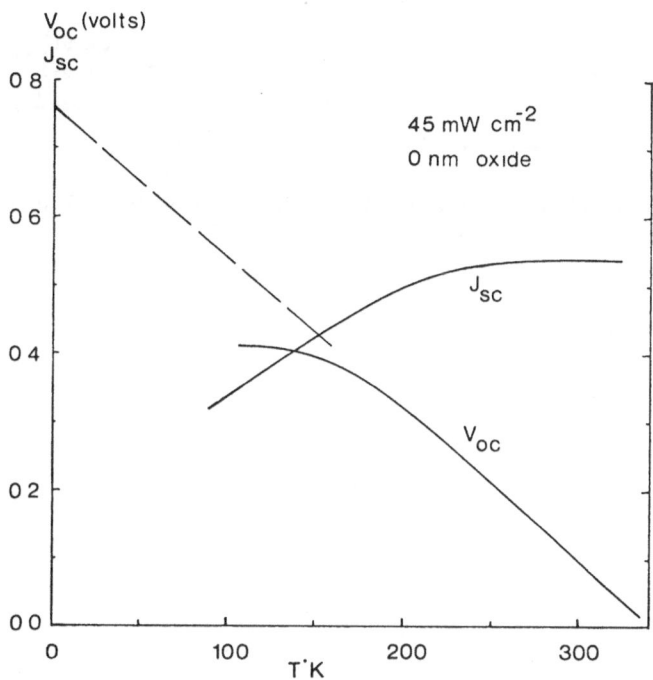

Fig. 11   The effect of temperature on open circuit voltage and
short circuit current of an a-Si:H Schottky cell. The
extrapolation indicated gives an intercept equal to be
product of n and ϕ_B.

theory is applicable to low mobility a-Si:H, but diffusion theory
gives similar results for the barrier height as it is not clear
what values should be used for the pre-exponential parameters.

In order to get a high barrier in a Schottky junction it is
necessary to use a high work-function metal with an n-type semiconductor, but these expensive high-temperature metals may be
avoided by adding a thin (1-3 nm) interfacial insulating layer
to a lower work-function contact, to suppress the dark current.
This is our own approach[38], and in practice many "Schottky" cells
are in reality MIS cells. The explanation of this effect is that

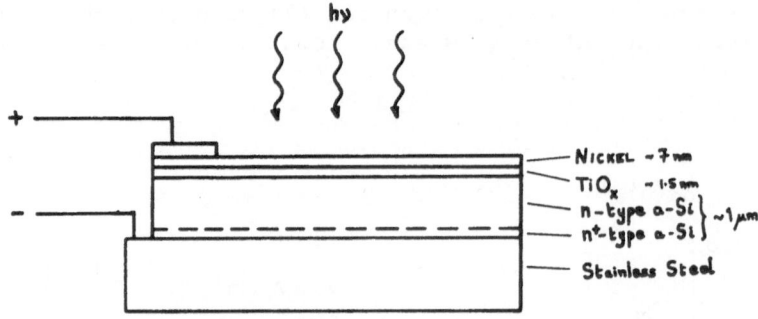

Fig.12   A schematic cross-section of an MIS a-Si:H cell[38]

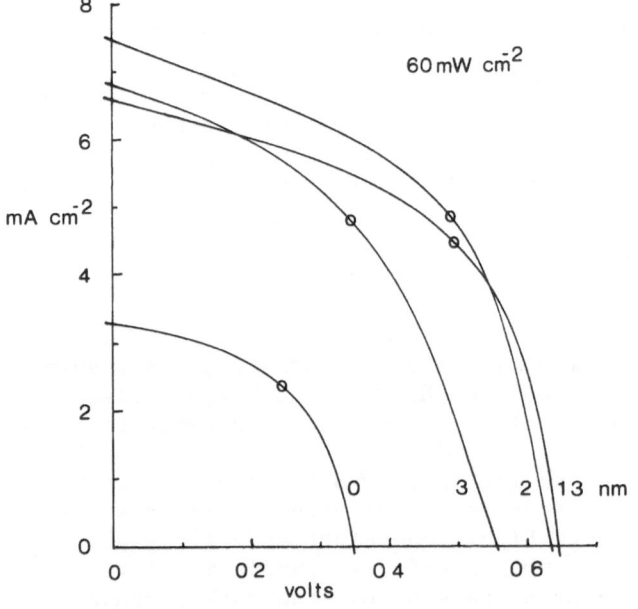

Fig. 13 The current-vol-tage characte-ristics of the MIS cell of Figure (12), showing the improvement of V_{oc} by suppressing the dark cur-rent. The circles mark the maximum power points (39).

whilst majority carriers are drifted away from the junction by the
built-in field and so do not have a great chance of tunneling
through the oxide, the minority carrier holes are driven against

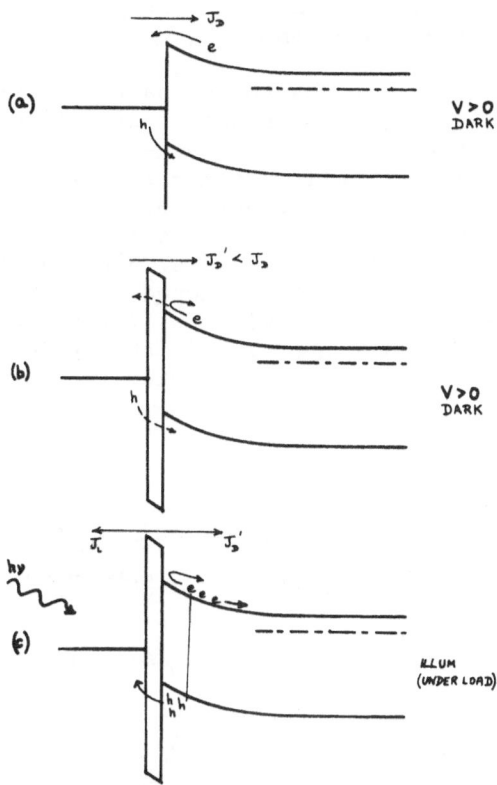

Fig. 14   The electron-blocking effect of a thin insulating layer
          on an n-type semiconductor Schottky cell. J_D is the dark
          current in forward bias.

the barrier and accumulate there until they tunnel through. (Card
has recently shown that in crystalline Si MOS junctions the tunneling barrier for holes is actually larger than for electrons,
because of charges within the oxide[40]. The actual voltage drop
across the thin insulator is usually negligible.

6. AMORPHOUS SILICON SOLAR CELLS

A crystalline Si p-n cell will have a short circuit current density of at least 30 mA cm^{-2}, an open circuit voltage of $\sim$ 560mV, and a fill factor of $\sim$ 0.75, in 100 mW cm^{-2} sunlight ("air mass one", AM1, irradiance). From the tables it is obvious that the voltages from these cells exceeds that generally obtained from crystalline Si cells, but that their poorer performance originates in the fill factor and photocurrent.

The poor photocurrent is due to the narrower collection depth of a -Si:H cells, caused by the absence of diffusion current from the field-free bulk. Since the space-charge region has a maximum width of $\sim$ 0.3 μm, and most cells are 1 - 1.5 μm thick, there is much room for improvement. Under reverse bias the collection region increases its width, and this looks like a shunt resistance effect on the illuminated characteristics. Because of the charges residing in the gap-states, even in undoped a-Si:H it is impossible to extend the zero bias depletion width as far as the back face in Schottky-type cells. An advantage of this structure over bipolar structures, as well as being easier to make, is that it has a

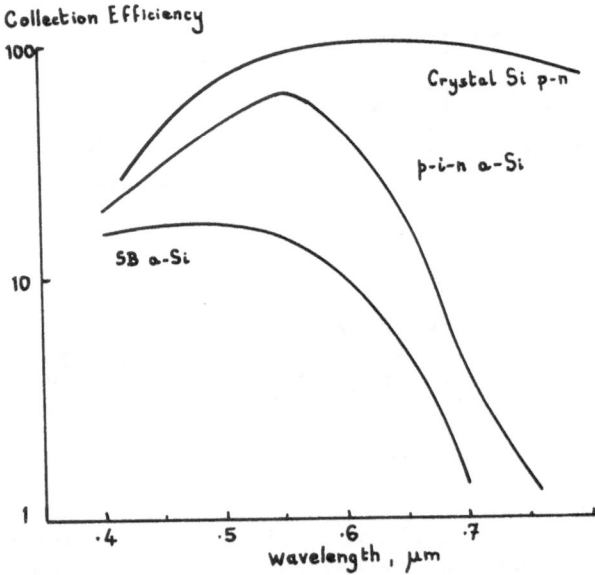

Fig. 15   The spectral responses of three solar cells, showing
          the relatively better response of Schottky cells to
          short-wavelengths (although an antireflection coating
          is necessary to increase the overall response).

better short-wavelength response since the built-in field lies
at the site of maximum optical absorption. In the p-i-n devices,
the p-layer must be as thin as possible, for even a 30 nm layer
at the surface has drastic effects on the short-wavelength
response owing to its high recombination velocity. A wide- gap
p-layer would avoid this limitation, but would require a high
hydrogen content without the usual accompanying poor conductivity.
A possible addition to the simple i-n-structure is graded doping,
to introduce a drift field to the normally undoped bulk of
Schottky-type cells, but due to the gap-states it is difficult
to predict the shape of any such built-in field.

The fill factor could be improved by reducing R_s. This is
composed of a resistance due to the undoped bulk of the cell and
the contact resistance at the n^+/metal junction. As stated earlier,
there is a limit to the conductivity attainable by gaseous doping
during film deposition, and a possible solution is to crystallise
the shallow n^+ region alone, either by thermal annealing or by
laser. If this is to be a laser process, then the cell would be
better constructed with the opaque ohmic n^+ contact on top, and
the thin barrier metal on glass substrate.

It has been proposed that fluorinated a-Si is a better material[41], for the Si-F bond is stronger than the Si-H bond and
so is stable at higher temperatures. There have not yet been any
reported solar cells on this material with efficiencies exceeding
those on a-Si:H, but doping is apparently more effective and the
space-charge region is wider[42,43]. There are possibilities for
alloying amorphous germanium, silicon and other semiconductors,
with hydrogen or fluorine passivation, to produce graded band-gap
cells with wider spectral responses, but a-Ge:H itself is less
successful a device material than a Si:H because of the weaker Ge-
H bond.

Apart from the ease with which the composition of these films
can be changed during growth, it is comparatively easy to produce
multi-layers, or monolithic arrays of series/parallel-connected
cells on a single substrate. These can increase the output voltage
from a module to help overcome contact resistances. The main problems with stacked junctions are ensuring that photocurrent passes
easily through the whole module, and ensuring that each section
generates the same photocurrent. In the module developed at Osaka
University[48], the top p-i-n-cell is much thinner than the bottom
cell, to allow for the attenuation of the illumination as it passes
through. Cell interconnections are the p-n junctions between adjacent p-i-n cells, at which large recombination currents flow via
gap-states. The open circuit voltage from these modules was proportional to the number of cells, and the short-circuit varied inversely, with an almost constant efficiency of 4% for up to five cells.

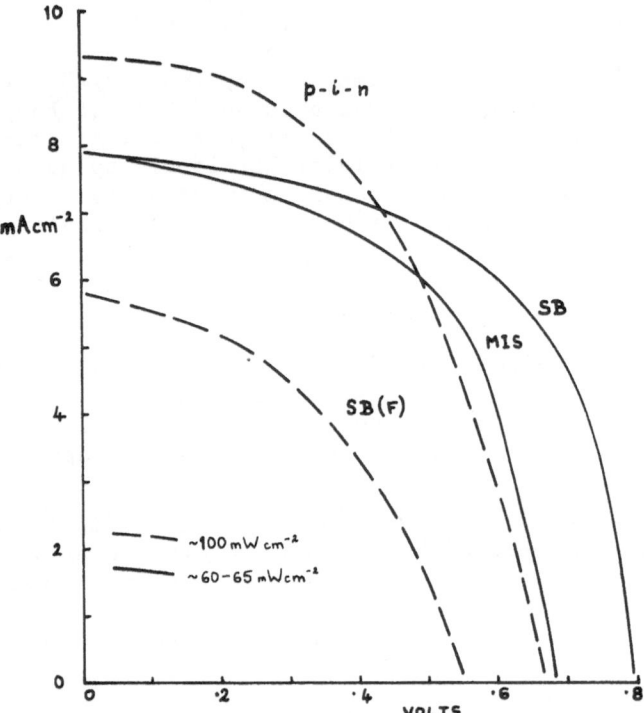

Fig. 16   The current-voltage characteristics of a Schottky
          cell[44], and MIS cell[45], and a p-i-n cell[46]
          on a-Si:H, and of a Schottky cell on a-Si:F[47].
          These are the best representative cells of their
          kind to date.

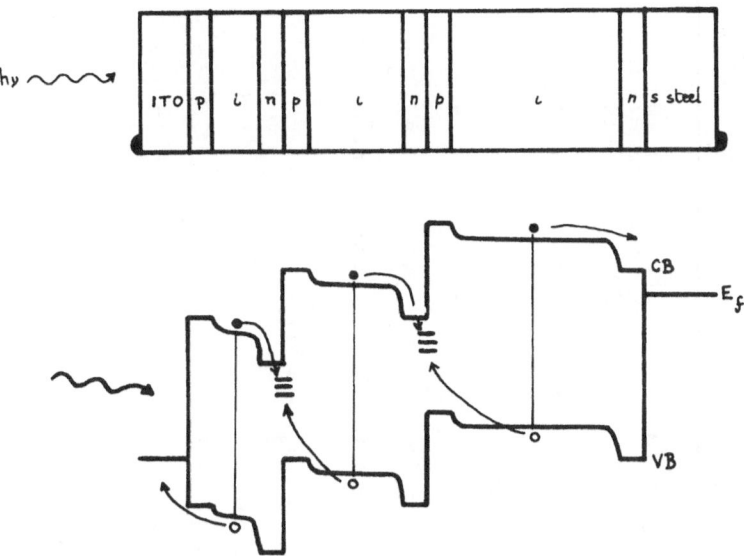

Fig. 17 The multilayer cell developed at Osaka University[48], with a band diagram showing the recombination current route at the cell interconnections. The series-added voltages are indicated by the shift in Fermi level.

Table 1. Glow Discharge a-Si:H Cells; White Light Performance

Laboratory,Year Reported	Cell Structure	V_{oc} (mV)	J_{sc} (mA cm^{-2})	FF	n(%)	Illumination (mW cm^{-2})
1. RCA,1976	glass/ITO/p$^+$in$^+$/Al	580	~10	0.40	2.4	100
2. RCA,1977	Pt/in$^+$/st.steel	800	12	0.58	5.5	100
3. Heriot-Watt Univ., 1978	Ni/TiO$_x$/in$^+$/st.steel	680	~8	0.51	4.8	60
4. Rutgers Univ.,1978	{Pd/I/i/st.steel (50 cm^2)	600	2	0.50	~0.6	100
5. Sanyo, 1979	{glass/ITO/p$^+$in$^+$/Al (16 cell module in 300 lux fluorescent→ 4 volts °/c; 14 µAS/c)	586	9.4	0.58	3.2	100
6. RCA, 1979	glass/ITO/Pt-SiO$_2$/p$^+$in$^+$/Mo (cement) (9 cell module)	6.49	5.8	0.55	2.6	89
7. Dundee Univ.,1979	Au/n$^+$ip$^+$/Cr	600-700	~10	0.51	3-4	100
8. Osaka Univ., 1979	{ITO/in/st.steel	380	11.5	0.55	2.7	90
	ITO/p$^+$in$^+$/st.steel	590	12	0.51	4.5	80
	(2 cell module:	1.35 volts	5	-	4.1	80

Table 2.   Sputtered a-Si:H; White Light Performance

Laboratory, Year Reported	Cell Structure	V_{oc} (mA)	J_{sc} (mA cm^{-2})	FF	n(%)	Illumination (mW cm^{-2})
1. Sheffield Univ.,	Pt/i/nIcr or Al	610	~ 7	~ 0.40	~ 2	100
2. Grenoble (CNRS), 1979	Pt/i/Sb-Cr/glass	700	4	0.42	1.3	100
3. Harvard Univ., 1979	Pt/in^{+}/Cr/glass	670	2	~ 0.50	< 1	100
4. Exxon, 1979	Pd/I/in^{+}/metal	> 800	~ 5	–	–	100

One final point which we have not discussed is that of
energy payback period. If solar cells are to be net producers of
energy they must deliver more energy during their lifetime than
was consumed during their manufacture. We have analysed the
energy required by the manufacturing plant, the process itself,
and the materials, for our MIS structure, assuming a semicon-
tinuous plasma deposition unit taking thirty-six 5 x 5 cm^{-2} substrates, and with an 80% yield[49]. The calculated figures for each
energy component are given in the table.

Table 3.   The Energy Requirement of MIS Amorphous Silicon Cells

	MJ(t) per cell
Machine Energy (15 year life)	1.95
Process Energy (vacuum pumps,rf power,heating, sputtering)	1.16
Materials Energy (Ar,SiH$_4$,PH$_3$,NiC$_r$,Ni,Al,TiO,glass)	0.35
Total	3.46

In contrast to crystalline silicon cells (which consume $\sim$ 26 MJ(t)
for materials and processing alone) the machine energy is the
larg st component. Since the best quality amorphous silicon films
are deposited at $\sim$ 40 nm min^{-1}, there is at present no way of
increasing the production rate to reduce this energy input per
cell. Nonetheless, if as is usual we consider the most
favourable location (with five hours per day of AM1 sunlight
throughout the year), then a 25 cm^2 cell of 5% efficiency would
give 0.82 MJ(e) per year. (Note that this is electrical energy
which is equivalent to four times the thermal energy used by
fossil fuel routes to electrical power). Thus even a 5% a-Si:H
cell is a net producer of energy, since it is expected to last
much longer than the $\sim$ 4.25 years required to repay its energy
(w ithout including the conversion from MJ(e) to MJ(t)).

7. LASER PROCESSING OF a-Si:H

a-Si:H grown at fast rates contains excessive amounts of
SiH$_2$, and it would be useful if this could be removed after deposition. As we have already seen, normal thermal diffusion of
hydrogen depletes SiH and SiH together, but we have found that
Nd:glass laser annealing at 1.06 μm decreases the absorption at

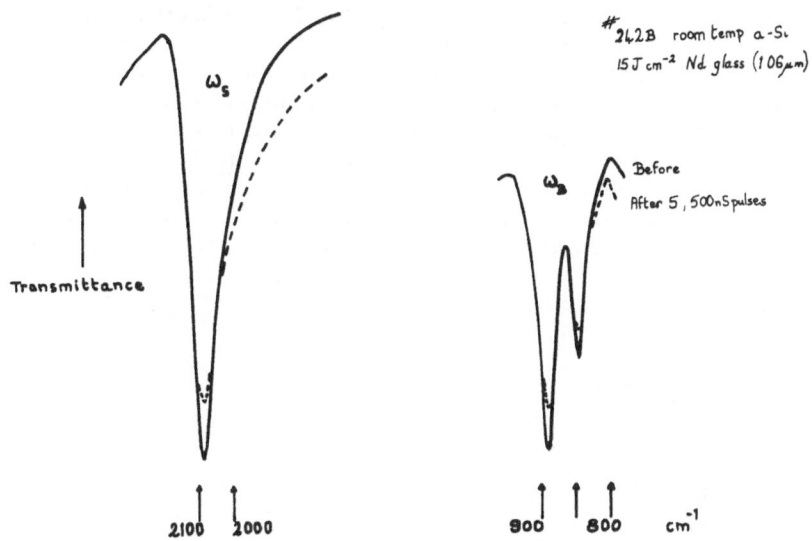

Fig. 18 The change in infrared absorption produced by five
successive pulses from a Nd:glass laser on an a-Si:H
low temperature film.

2100 cm^{-1} (SiH$_2$ stretch mode) with a complementary increase in
the absorption at 2000 cm^{-1} (SiH). (This 38% decrease and 77%
increase is not readily seen from the unconvoluted infrared data).
The laser pulses were 500 nS packets of $\sim$ 150 mJ, with a spot
diameter of $\sim$ 2 mm, and the absorption coefficient of a-Si:H at
this wavelength is less than 10^3 cm^{-1}, so absorption took place
through the whole film thickness. We know of only one other result
in which a Q-switched Nd:glass laser was used to recrystallise
amorphous silicon (produced by Si$^+$ ion bombardment) into which
H$_2^+$ ions were implanted[50]. The report showed that even with high
laser energies (> 5 J cm^{-2}) hydrogen was still retained in
these films and that it retarded complete crystallisation.
Further, only SiH groups were present before annealing, but SiH$_2$
groups were produced concurrently with a decrease in the SiH
content.

It is worth pointing out that most laser irradiation studies
on the various forms of amorphous silicon have had the aim of
recrystallising the amorphous layer (e.g. the study by Bertolotti
and Vitali using a ruby laser)[51], but for solar cells it might be

preferable to <u>retain</u> the optical properties of a-Si:H. Two other
studies of laser-beam annealing have also produced structural
changes in a-Si:H. The RCA study [52]used the 488 nm Ar laser line
to remove hydrogen from very thin a-Si:H deposited on to room temperature substrates. In contrast to our own experience wit a similar energy 514.5 nm Ar laser, all films <u>darkened</u> appreciably
upon exposure and remained amorphous. This was proven to be a
thermal effect, equivalent to heating to $\sim$ 200°C, and was connected with an increased oxygen content and a decreased hydrogen

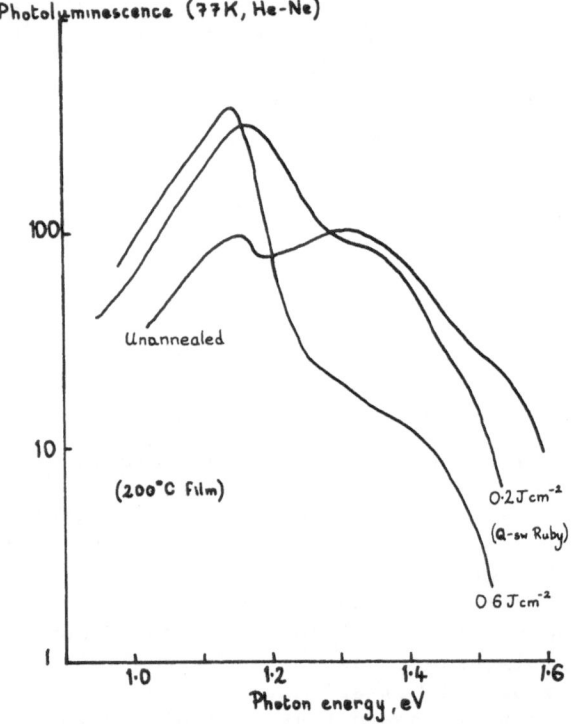

Fig. 19 The
He-Ne laser-
excited photo-
luminescence of
an a-Si:H film
before and
after recrystal
lisation with
a Q-switched
ruby laser (53)

content. In addition, a phosphorus-doped film was bleached
slightly when the laser energy was low. This was thought to be
due to removal of defects states. A group at Plessey believe the
opposite occurs under Q-switched ruby laser irradiation, from
their photoluminescence studies of recrystallised a-Si:H[53]. In
their case also SiH bonds were removed by the process, and this
accompanied a decrease in the 1.33 eV luminescent peak and an
increase in the 1.15 eV peak. The 1.33 eV transition is thought to

be between conduction and valence band-tails, which is quenched when the non-radiative recombination centres are reactivated by the loss of hydrogen. The 1.15 eV transition is thought to be between the conduction band-tail and the maximum in the gap state density near the valence band.

We have been annealing n^+ a-Si:H (0.5% PH_3 to SiH_4) with a 514.5 nm Ar laser, and obtain a 15% reduction in its sheet resistivity after $\sim$ 20 scans of a 1 mm diameter spot at $\sim$ 1 cm s^{-1}, with less than 300 mW. The absorption coefficient at this wave-

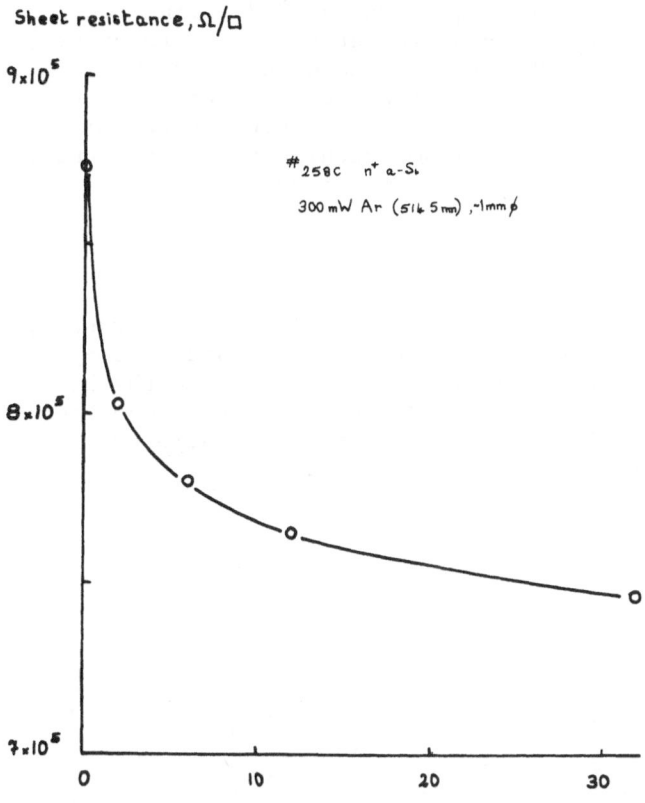

Fig. 20   The decrease in sheet resistance of a heavily-doped a Si:H film on 7059 glass, after Ar laser annealing.

lenght is $\sim 10^5 cm^{-1}$, which gives an absorption depth equal to that of the n^+ layer in our MIS cells. The simultaneous deposition and alloying of metal contacts by laser is an attractive process for these cells[54-56].

8.   CONCLUSIONS

As well as being a fascinating material in its own right, hydrogenated amorphous silicon offers a low energy route to the large-scale production of cheap thin film solar cells. In order to improve their performance beyond the 5-6% already reached after $\sim$ 5 years research, it is necessary to improve the collection of photocarriers by increasing minority carrier lifetimes and perhaps by adding drift fields.The requires a greater knowledge of the way in which hydrogen and dopant atoms are included in the structure during deposition, and hence of ways in which the gap state density might be modified even further. The electron-phonon interaction at laser-irradiated dangling bonds and Si-H bonds has not yet received much attention by theorists, but is essential requirement for understanding laser-induced structural rearrangements. Laser annealing and deposition are compatible with the technology being developed,which is becoming more acceptable to industry as plasma etching and laser anneal of ion implants are adopted for routine production.

9.   ACKNOWLEDGEMENTS

The work at Heriot-Watt University is supported by UK Science Research Council grants, and SRC Studentships in cooperation with Hughes Microelectronics Ltd., and the British Petroleum Co. Ltd. Professor S D Smith, Ian Boyd, John McGill, Fiona Riddoch, Alex Wallace and Professor D L Weaire have all contributed to the success of the project.

REFERENCES

1. D.E. Carlson, C.R.Wronski, Appl. Phys. Lett., 28, 671 (1976).
2. P.G.Le Comber, W.E. Spear, A.Ghaith, Electron. Lett., 15 ,
    179 (1979).

3. Y.Imamura, S.Ataka, Y. Takasaki, C.Kusano, T.Hirai, E.Maruya-
    ma, Appl. Phys. Lett., 35, 349 (1979).

4. H. Hayama, M. Matsumura, Appl. Phys.Lett. 36, 349 (1979).
5. R.A.Lemons- Appl. Phys.Lett., 36, 919 (1980).
6. R. Hill, "Thin Film Solar Cells" in "Active and Passive Thin
    Film Devices", ed.T.J. Coutts, Academic Press, London (1978).

7. J.A. Zoutendyk, Solar Energy, 20, 249 (1978).
8. A.R. Kirkpatrick, J.A.Minnucci, A.C.Greenwald, R.H. Josephs,
    pp. 706-710, Proc. 13th IEEE Photovoltaic Specialists Conf.,
    Washington D.C. (1978).

9. R.T. Young, C.W. White, J. Narayan, R.D. Westbrook, R.F.Wood,
    W.H. Christie, pp. 717-723, ref. (8).

10. R.G. Forney, pp. 81-91 Photovoltaic Solar Energy Conversion
    Conference (C21), UK-ISES, London (1979).

11. Proc. 8th Int. Conf. on Amorphous and Liquid Semiconductors,
    Cambridge, USA (1979), J. Non-Cryst.Solids, 35/36 (1980).

12. "Amorphous Semiconductors" ed.M.H. Brodsky, Springer-Verlag
    (1979), (volume 36 of " Topics in Applied Physics").

13. W.E. Spear, Adv. in Phys., 26, 811 (1977).
14. R.C. Chittick, J.H. Alexander, H.F. Sterling, J. Electrochem.
    Soc. 116, 77, (1969).

15. W.E. Spear, P.G. Le Comber, Solid State Comm., 17, 1193
    (1975).

16. T.D.Moustakas, D.A. Anderson, W. Paul, Solid State Comm., 23,
    155 (1977).

17. H. Wiesmann, A.K. Ghosh, T. McMahon, M. Strongin, J.Appl.Phys.
    50, 3752 (1979).

18. C.H. Seager, D.S. Ginley, Appl. Phys. Lett. 34, 337 (1979).
19. A.Madan, S.R. Ovshinsky, E. Benn, Phil. Mag. B., 40, 259
    (1979).

20. C.P. Christensen, K.M. Lakin, Appl. Phys.Lett., 32, 254
    (1978).

21. M. Hanabusa, A.Namiki, Appl. Phys.Lett., 35, 626 (1979).
22. V. Baranauskas, C.I.Z. Mammana, R.E. Klinger, J.E. Greene,
    Appl. Phys. Lett., 36, 930 (1980).

23. D. Weaire, N.Higgins, P. Moore, I. Marshall, Phil. Mag. B.,
     40, 243 (1979).

24. J.C. Knights, G.Lucovsky, R.J. Nemanich, J. Non-Cryst.
     Solids, 32, 393 (1979).

25. P. John, I.M. Odeh, M.J.K. Thomas, M.J. Tricker, F. Riddoch,
     J.I.B. Wilson, Phil. Mag., to be published.

26. H.Fritzsche, M. Tanielian, C.C. Tsai, P.J. Gaczi, J. Appl.
     Phys., 50, 3366 (1979).

27. M.H. Brodsky, M.A. Frisch, J.F. Ziegler, W.A. Lanford, Appl.
     Phys.Lett., 30, 561 (1977).

28. P. John, I.M. Odeh, M.J.K. Thomas, M.J. Tricker, J.I.B.
     Wilson, J.B.A. England, D. Newton, submitted to J.Phys. C.

29. C.W. Magee, D.E. Carlson, pp. 151-158, in "Semiconductor
     Characterisation Techniques", ed. Barnes and Rozgonyi,The
     Electrochemical Society, Vol. 78-3 (1978).

30. M.H. Brodsky, M. Cardona, J.J. Cuomo, Phys. Rev. B., 16,
     3556 (1977).

31. P.J. Zanucchi, C.R. Wronsky, D.E. Carlson, J. Appl. Phys.,
     48, 5227 (1977).

32. J.C. Knights, G. Lucovsky, R.J. Nemanich, Phil. Mag. B., 37,
     467 (1978).

33. G.Lucovsky, R.J. Nemanich, J.C. Knights, Phys. Rev.B., 19,
     2064 (1979).

34. G. Muller, F. Demond, S. Kalbitzer, H. Damjantschitsch,
     H. Mannsperger, W.E. Spear, P.G. Le Comber, R.A. Gibson,
     Phil. Mag., B., 41, 571 (1980).

35. W.E. Spear, P.G. Le Comber, Phil. Mag. B., 33, 935 (1976).
36. R. Williams, R.S. Crandall, RCA Review, 40, 371 (1979).
37. H.J. Hovel, "Solar Cells", Vol. 11, in "Semiconductors and
     Semimetals", ed. R.K. Willardson and A.C. Beer, Academic
     Press (1975).

38. J.I. B. Wilson, J.McGill, Solid-State and Electron Devices,
     2, S7 (1978).

39. J. McGill, J.I.B. Wilson, S. Kinmond, J. Appl. Phys., 50,
     548 (1979).

40. H.C. Card, K.K. Ng, Solid State Comm., 31, 877 (1979).
41. H. Matsumura, Y. Nakagome, S. Furukawa, Appl. Phys. Lett.,
     36, 439 (1980).

42. A. Madan, S.R. Ovshinsky, J. Non-Cryst. Solids, 35/36, 171
     (1980).

43. M. Shur, W. Czubatyi, A. Madan, J. Non-Cryst. Solids,
44. D.E. Carlson, IEEE Trans. Electron. Devices, ED-24,449
     (1977).

45. J.I.B. Wilson, J. McGill, S. Kimond, Nature, $\underline{272}$, 152
     (1978).

46. R.A. Gibson, W.E. Spear, P.G. LeComber, A.J.Snell, J. Non-
     Cryst. Solids, $\underline{35/36}$,725 (1980).

47. M. Konagai, K. Takahashi, Appl. Phys. Lett., $\underline{36}$, 599 (1980).
48. Y. Hamakawa, H. Okamoto, Y. Nitta, Appl. Phys.Lett. , $\underline{35}$,
     187 (1979).

49. F. Riddoch, J.I.B. Wilson, J. Solar Cells, to be published.
50. P.S. Peercy, H.J. Stein, pp. 331-336 in "Laser-Solid
     Interactions and Laser Processing 1978", American Institute
     of Physics, Conf. Proc. 50, New York (1979).

51. M.Bertolotti, G.Vitali, W.E. Spear, pp. 492-495, ref. 50.
52. D.L. Staebler, J. Appl. Phys., $\underline{50}$, 3648 (1979).
53. R.S. Sussmann, A.J. Harris, R. Ogden J. Non-Cryst. Solids,
     $\underline{35/36}$, 249 (1980).

54. T.F. Deutsch, D.J. Ehrlich, R.M. Osgood,Appl. Phys. Lett.,
     $\underline{35}$, 175 (1979).

55. T.F. Deutsch, D.J. Ehrlich, R.M. Osgood, Z.L. Liau, Appl.
     Phys. Lett. , $\underline{36}$, 847 (1980).

56. D.J. Ehrlich, R.M. Osgood, T.F. Deutsch, Appl. Phys. Lett.,
     $\underline{36}$, 917 (1980).

PULSED ELECTRON BEAM APPLICATIONS FOR SEMICONDUCTOR ANNEALING

Armando Luches

University of Lecce
Physics Department
I-73100 Lecce, Italy

ABSTRACT

Pulsed electron beams are widely used in many fields of research. The most recent use is for semiconductor wafer annealing. Like pulsed laser beams, pulsed electron beams are successfully used to remove lattice damage and to restore electrical properties of ion-implanted devices. There is also the possibility to produce metastable metallurgical phases in silicon single crystal wafers covered with thin layers of metals, because of the rapid heating and cooling possible with pulsed beams of submicrosecond duration. Published and unpublished data about semiconductor annealing and non-equilibrium compound formation with pulsed electron beams are presented together with electron source characteristics for large-area electron beams.

INTRODUCTION

Pulsed electron beams were initially developed for simulation of nuclear weapon effects on materials. At present, they are widely used for many different purposes, such as confinement and heating of plasma, pellet heating for fusion studies, excitation of high power gas lasers and production of gigawatt bursts of microwaves[1]. Now it is possible to control the voltage, current and pulsewidth of these intense bursts of energy over very broad ranges.

The most recent use of pulsed electron beams is for semiconduc-

tor wafer annealing. Thermal annealing in a high temperature furnace
is the technique normally used to remove lattice damage and to restore
electrical properties of ion implanted silicon devices. Because of
the long duration of the annealing process, the devices can suffer
changes or degradation of their parameters. Heating the entire wafer
can change its bulk properties, the doped impurities may diffuse and
the surface my be contaminated by uncontrolled impurities.

To overcome these drawbacks, new methods such as the use of
pulsed lasers, pulses of incoherent light, pulsed high power electron beams, mechanically scanned cw lasers, electrically or mechanically scanned electron beams have recently been considered. These
methods allow a greater annealing speed and a reduction in energy
consumption.

Laser beams are most widely used, but also electron beams are
attracting growing interest. Large-area pulsed electron beams have
been developed which can raise the surface of a silicon wafer to
high temperature in a single sub-microsecond pulse. Because of the
very short pulsewidth, a depth of the order of only a few microns
is heated, thereby keeping the overall device structure at low temperature. So thermal distortion and warping of the wafer and redistribution of implanted dopants can be avoided.

Very attractive is also the possibility of producing otherwise
unobtainable metallurgical phases, because of the rapid heating and
cooling possible with pulsed beams.

COMPARISON BETWEEN PULSED ELECTRON BEAMS AND LASERS

The use of pulsed electron beams is very promising because,
compared to the techniques based on the use of coherent or incoherent
light pulses, it presents the following advantages:
a) The absorption of energy is more predictable and less dependent
   on uncontrollable variables, such as doping density and surface
   reflectivity, than absorption of energy from light beams. Most
   of the energy of the electron beam is lost in scattering with
   atomic electrons and the energy lost per unit path length at the
   energies used in these cases may be written[2]

$$-dE/dx \;=\; kE^{-2/3}NZ \qquad\qquad \text{where N is the material density.}$$

The stopping power is much less selective than the laser light absorption dI/dx. The limited penetration depth of the laser light and the reflectivity of the melted layer do not allow reaction of relatively thick (of the order of 100 nm) films of certain metals with underlieing materials. As an example, to circumvent this problem it was decided to use Si-Pd (100 nm)-Si (50 nm) sandwiches in which laser reflection from the front surface of the metal can be very strongly reduced[3]. Also multiple layers of metal and silicon alternatively deposited onto substrates are used[4].

b) It is possible to define the depth of the process from simple range-energy relations, like[5]:

$$R = 0.412E^{1.265-0.0954\log E}$$

or, more accurately, from spatial distribution of the energy loss, obtained by using Monte Carlo methods[6]. Figure 1 shows an example of energy loss profiles in silicon, obtained in this way. ΔE is expressed in eV/layer per electron; each layer has a thickness of 50 nm. Electron backscattering is taken into account.

c) It is possible to process large areas (up to about 50 cm) in a single pulse with very good uniformity of the process[7](within ±7%).

d) It is possible to vary over relatively broad ranges the current, energy and pulsewidth of the electron beam and the energy fluence on the samples.

There also some drawbacks. Generally electron beam sources are more complex and perhaps more expensive than laser sources. Also the maintenance may be more troublesome and it is mandatory to process the materials in vacuum.

Although the interaction mechanisms with materials are very different, the results obtained up till now with pulsed electron beams and pulsed lasers are very similar. So it is apparent that it is the heating effect which is mainly responsible of the epitaxial regrowth of annealed layers.

It is possible to calculate the spatial distribution of temperature in a silicon wafer for various energies E and current densities I of the incident electron beam. In the remaining part of this section we report the results obtained by P.G.Merli[6]. The values of the energy densities D necessary to reach melting temperature of the sample as a function of current density are shown in Fig.2 for different electron energies. For high values of I (1000 A/cm^2) D varies

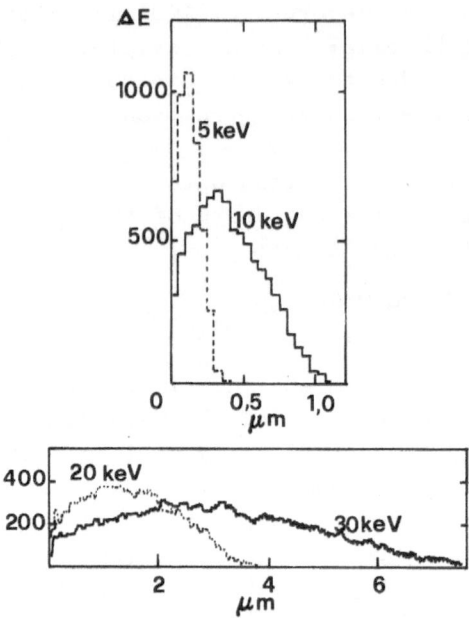

Fig. 1 Energy loss in eV/layer per electron in Si at different
       values of E (Courtesy of P.G. Merli).

from $0.4 \, J/cm^2$ at 5 keV to $1.8 \, J/cm^2$ at 30 keV and increases with E.
This is due to the "quasi-adiabatic" rise of temperature and to the
decreasing energy loss per unit path length of electrons of increasing energy. Melting temperature is reached in under-surface layers
before than at the surface itself.

    The energy densities of interest in wafer annealing may there-
fore be very different for different electron energies, and also
very different from those used in laser annealing (about $1 \, J/cm^2$).
Thus comparison between the two methods based on the energy densities absorbed by the sample is not significant. To specify D without
specifying two of the three quantities E, I and pulswidth T does
not supply useful information to reckon the temperature reached by
the sample.

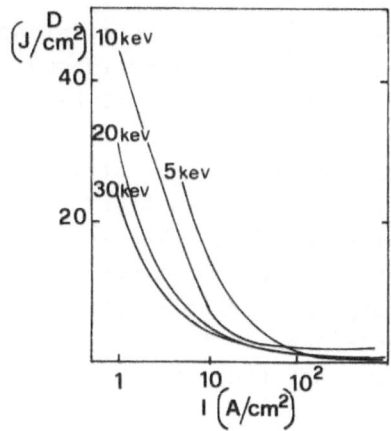

Fig. 2 Energy density necessary to reach melting temperature as a
function of current density for different electron energies
(Courtesy of P.G.Merli).

It must also be considered that in pulsed electron beam sources
the electron emission is generally due to field effect. Therefore it
is impossible to have electron emission at very low accelerating
voltages. Practically the lowest value of the voltage which may be
usefully applied is about 10 kV. Then, using pulsed electron beams
for epitaxial regrowth in liquid phase, a small enlargement of the
profile appears unavoidable, with redistribution of the dopant to a
depth of about 0.4 μm. This enlarging of the profile represents a
result not desired, particularly in solar cell technology, where the
introduction of pulsed techniques appears more promising for the increased speed of the processes and lower production costs. It is
therefore necessary to keep at the lowest level the energy of the
incident beam, because the thickness of the liquid layer depends not
only on the exposure time or current density, but essentially on
beam energy. A possibility of overcoming this drawback is to compel
the electrons to hit the wafers at a high incidence angle.

PULSED ELECTRON BEAM SYSTEMS AND RESULTS

Pulsed electron beams specially devoted to semiconductor wafer
annealing were first developed at Spire Co. by A.R. Kirkpatrick and

co-workers[8]. They claim that their patent[9] for the use of pulsed
electron beams to regrow the crystal lattice radiation damage in
silicon predates the use of lasers for similar applications[10]. More
recently they developed a large-diameter (75mm) pulsed electron
beam[11]. This is part of a high-throughput, high efficiency process
proposed for the automated production of silicon cells. The apparatus
consists of a coaxial capacitive energy store which is charged to a
high voltage (100-400 kV) with an electrostatic generator and dis -
charged into a field emission diode. The anode of this diode consists
of a highly transparent mesh, which allows the cathode emitted electrons to propagate through and drift to the target to be irradiated.
Strong self-electric-and-magnetic fields of the beam affect the
electron trajectories so that they impinge upon the sample at oblique
angles.

In a recent paper they give evidence of improved crystal structure from pulsed annealing compared to an optimized furnace annealing
schedule. Further evidence for improved crystal structure from TEM
shows a lower density of dislocations in pulse annealing. The reproducibility of processing was demonstrated in successful fabrication of
500 3-in diam silicon solar cells by ion implantation and pulse
annealing. Annealing of ion-implanted damage in GaAs and InP is also
reported. The electron beam used for these processings has a pulsewidth of 40-100 ns, average particle energy between 5 to 150 keV and
a peak current between 1 and 50 kA[12]. On a sample the beam has a
usable diameter up to 76 mm and a variable fluence from 0.04 to 20
J/cm^2. The uniformity of the processing is claimed to be better than
±7%, as measured by sheet resistivity with a four-point probe.

The electron sources of the Spire Co. are used by many authors
for annealing of their samples. As an example, J. Narajan[13] used this
source to remove displacement damage in ion implanted Si and GaAs,
to remove dislocations, loops and precipitates in silicon and to
study impurity segregation and solubility limits; D.E. Davis et al.[14]
observed heavily conducting layers in InP subjected to pulse electron
beam annealing.

Our high-voltage pulser[15] is a two stage Marx circuit (Fig.3).
To obtain a time compression and high power outputs our Marx circuit
is coupled to a Blumlein line. This line gives the pulses an almost
square shape and a length of about 50 ns (fwhm). The peak voltage of
the pulses can be varied from 10 to 50 kV. This short voltage pulse
is applied to a field emission diode. Depending on the cathode, the

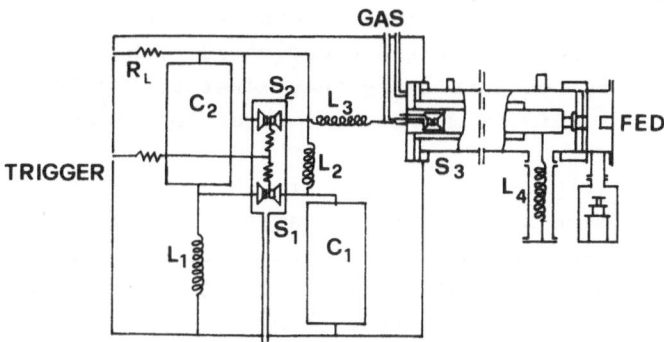

Fig. 3   Schematic diagram of our pulse generator. C_1, C_2: capacitors;
R_L charging resistor; L_1, L_2: charging inductors; L_3: coupling
inductor; S_1, S_2, S_3: spark-gaps.

beam diameter can vary from 2 to 120 mm. The beam current intensity
can vary between 1 and 20 kA, depending on voltage and diode impedance.
Energy densities varying from 0.1 to 100 J/cm^2 are obtained on the
anode. Most commonly we use cathodes 15-30-mm-diameter, made of finely
machined graphite or razor blade arrays to have good beam homogeneity,
and work at 20 keV electron energy and at 0.4 to 10 J/cm^2 energy
density. The anode is formed by the silicon wafer, stuck to a movable
graphite absorber, placed directly in front of the cathode. Sometimes
a transparent mesh is used.

We use our beam generator mainly to study metal-silicon interactions. The impetus to study metal-silicon interactions is due to the
requirements placed on the performance of integrated circuits where
these reactions are mainly used to produce either ohmic or rectifying
contacts[16], barriers to interdiffusion, reactive or stable overlayers,
interconnections and so on. The phenomena occurring at the semicon-
ductor-metal interface are still poorly understood because of the
lack of knowledge of some of the basic properties of the compounds.

Metal-silicide formations have been investigated by solid phase
reaction in conventional furnaces[17]. Recently considerable interest
has arisen in phenomena induced by irradiating the metal-silicon
structure with energy pulses[18]. The most interesting aspect of this

technique is the very short times necessary to reach high temperatures
and to cool. The cooling rates are estimated to be of the order of
10^{10}°K/s, so it is not surprising that metastable alloys never
observed in usual thin film interactions are obtained.

In our investigation we used near-noble and refractory metals
deposited on silicon single crystal wafers. The thickness of metal
layers range between 70 and 250 nm. The electron energy is maintained
generally at 20 keV, the pulsewidth is 50 ns, the beam dimater vary
between 10 and 30 mm and the energy density between 0.4 and 2.4 J/cm^2.
In the final part of this paper a short review of some results is
given.

PALLADIUM. After electron irradiation the Pd layer interacts
with Si. PdSi, Pd$_2$Si, Pd$_3$Si and Pd$_4$Si phases are observed[19], like
after laser irradiation[19,20]. Thermal post-annealing at 600°C for
30' of these samples produces the disappearance of Pd$_4$Si, the decrease of Pd$_3$Si and PdSi and the increase of Pd$_2$Si. After 2 hours
Pd$_2$Si is the only phase detected[21]. Also a (111)Si/Pd$_2$Si structure
was irradiated at 1.8 J/cm^2. The silicide layer, 110 nm thick,
reacted with Si to form a uniform layer richer in Si, in which only
PdSi is present.

PLATINUM. The (111)Si/Pt structure were irradiated at several
energy densities (0.4-1.6 J/cm^2). As in the previous case, many
phases were observed (PtSi, Pt$_2$Si, Pt$_3$Si ...). Similar results were
obtained also with laser irradiation[19,20]. After thermal annealing
(2 hours at 500°C) PtSi is the only phase detected. Also Si/Pt$_2$Si
and Si/PtSi structures were annealed at 1.8 J/cm^2. The two samples
behave substantially in the same way: the silicide layer interacts
with Si. In the sample originally with PtSi, the outermost layer has
a composition close to PtSi, while the innermost is close to Pt$_2$Si .
The same occurs in the second sample, but more silicon is contained
in the reacted layer . Post annealing in furnace produces a better
separation of the two layers.

TUNGSTEN and MOLYBDENUM. In (100)Si/W samples both a silicon-
rich phase WSi$_2$ and a metal-rich phase W$_5$Si$_3$ are present, likewise
in (100)Si/Mo samples, where MoSi$_2$ and Mo$_5$Si$_3$ have been detected.
By conventional thermal annealing only the silicon-rich phase MoSi$_2$
is formed.

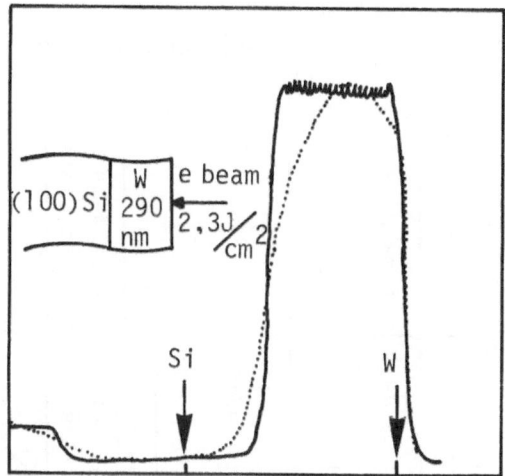

Fig. 4 ^{4}He^{+} backscattering spectra of a (100)Si/W sample (energy
vs. yield). —— as evaporated; ‾‾‾ after 2.3 J/cm^{2} e-beam
annealing.

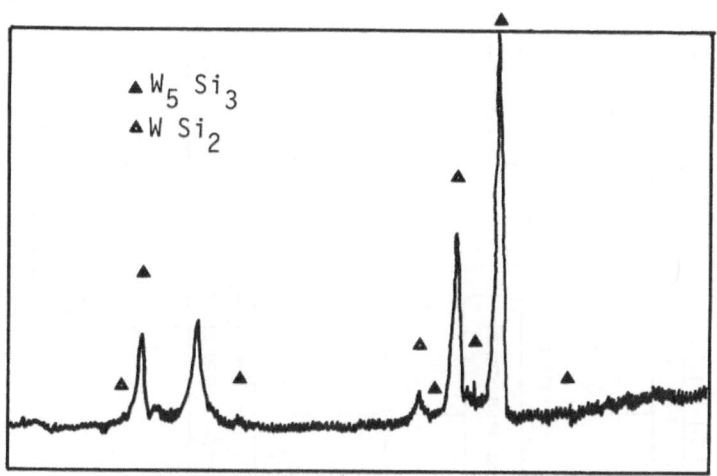

Fig. 5   x-ray diffraction spectrum of a (100)Si/W(290 nm) sample
after 2.3 J/cm^{2} e-beam annealing.

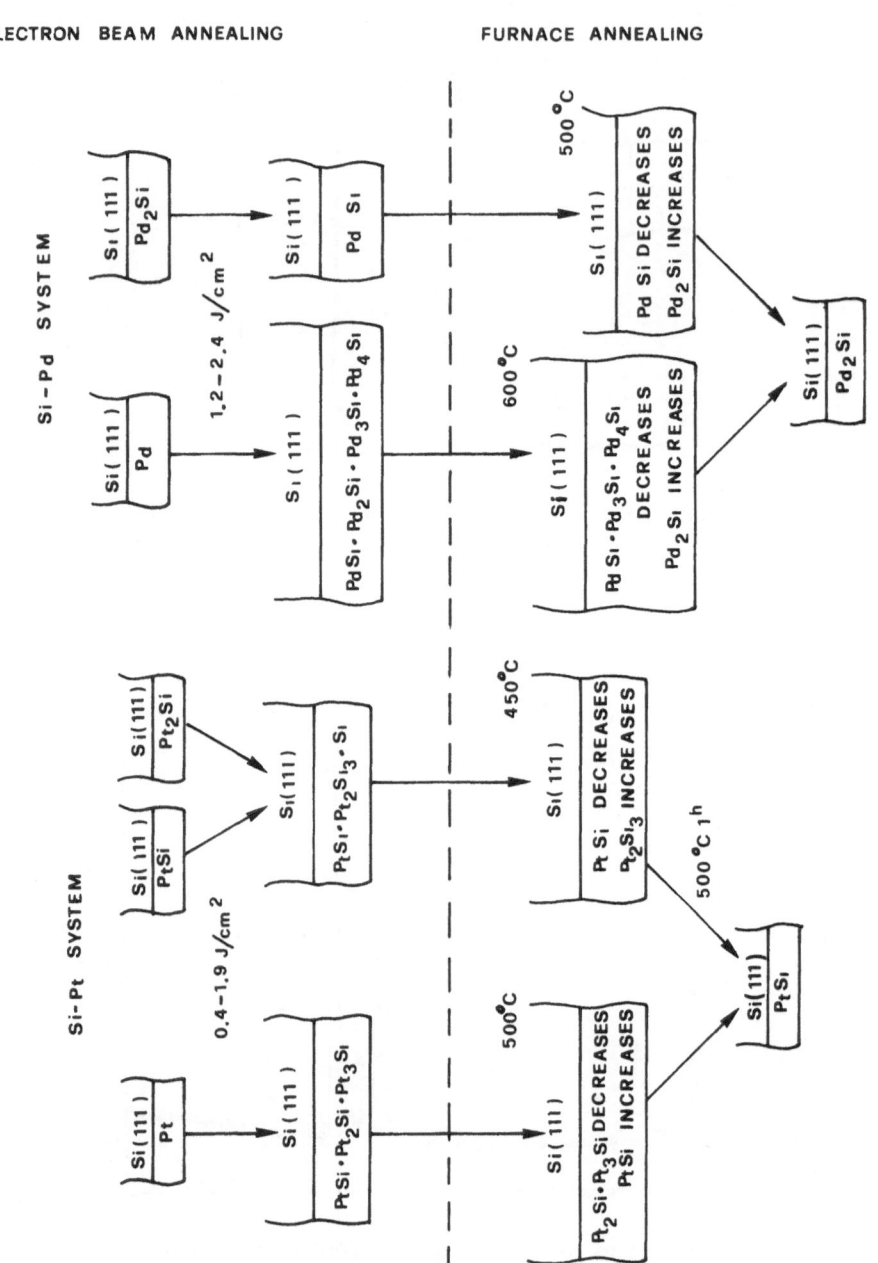

Fig. 6 Phase evolution in Si/Pt and Si/Pd systems.

Figure 4 shows an example of a Rutherford backscattering spectrum of a Si/W sample. The compounds are positively identified by x-ray diffraction method. An example is shown in Figure 5.

It has to be noted that we reacted successfully layers of refractory metals up to 250 nm thick. By using laser pulses[20] it was not possible to react Mo layers 170 nm thick, even at the highest possible power density (40 MW/cm^2, 180 ns). Then it is evident that the high energy delivered by the electron beam pulse rises the temperature in the film to values high enough for liquefaction. The high diffusivity in the liquid produces the intermixing between silicon and metal. The subsequent fast solidification is responsible for the phase formation.

The results obtained for the Si/Pd and Si/Pt systems[21] are schematically resumed in Figure 6. Almost all the phases predicted by the phase diagram are observed in the case of silicon/metal structures, while only the phases richer in silicon are formed in the case of silicon/silicide structures. A possible way to interpret these results is to assume a liquid having, at least in the case of silicon/metal structures, a non uniform composition in depth. The subsequent solidification which occurs starting from the silicon substrate enhances the inhomogeneity and produces almost all the compounds predicted by the phase diagram. The compounds richest in silicon seem to be located in the innermost side of the reacted layer and the ones richest in metal in the outermost side. The hypothesis of a non-homogeneous melting is supported also from the results obtained in conventional thin film interactions, where metal or silicon rich compounds can be formed only in excess of metal or silicon. A more homogeneous liquid is obtained in the case of Si/silicide structures. Consequently only few phases are formed. Post annealing of the irradiated samples shows that the phases formed after electron beam irradiation are unstable[21].

CONCLUSION

The first utilization of pulsed electron beams in material processing was for recrystallization of ion implanted semiconductor surfaces. They are now being investigated for a wide range of possible applications, not only as a convenient replacement for existing processes, but also for totally new capabilities.

ACKNOWLEDGMENTS

    I want to thank G.Ottaviani and P.G.Merli for helpful suggestions
and for the permission to use published and unpublished data.

REFERENCES

1. J.A. Nation, Particle Accelerators 10, 1 (1979).
2. K. Kanaya and S. Okayama, J. J. Phys. D 5, 43 (1972).
3. J. F. Gibbons, "Laser and Electron Beam Processing of Electronic
    Materials," C. L. Anderson, G. K. Celler and G. A. Rozgonyi
    Eds., The Electrochemical Society Inc. 1980, pg. 13.
4. M. von Allmen, S. S. Lau, M. Maenpaa and B. Y. Tsaur, Appl. Phys.
    Lett. 36, 207 (1980).
5. L. Katz and A. S. Penfold, Rev. Mod. Phys. 24, 28 (1952).
6. P. G. Merli, Optik, 56, 205 (1980).
7. A. C. Greenwald, A. R. Kirkpatrick, R. G. Little and
    J. A. Minnucci, J. Vac. Sci. Technol. 16, 1838 (1979).
8. A. R. Kirkpatrick, J. A. Minnucci and A. C. Greenwald, 1EEE Trans.
    Electron Devices ED-24, 439 (1977).
9. U. S. Patent 3950187 (1974).
10. G. A. Kachurin, N. B. Pridachin and L. S. Smirnov, Sov. Phys.
    Semicond. 9, 946 (1975).
11. A. C. Greenwald, A. R. Kirkpatrick, R. G. Little and
    J. A. Minnucci, J. Appl. Phys. 50, 783 (1979).
12. R. G. Little, A. C. Greenwald and J. A. Minnucci, IEEE Trans.
    Nucl. Sci. NS-26, 1683 (1979).
13. See ref. 3, pg. 294.
14. D. E. Davies, E. F. Kennedy, J. J. Comer and J. P. Lorenzo,
    Appl. Phys. Lett. 36, 922 (1980).
15. A. Luches, V. Nassisi, A. Perrone and M. R. Perrone, Physica C
    (to be published).
16. G. Ottaviani, J. Vac. Sci. Technol. 16, 1112 (1979).
17. K. N. Tu and J. W. Mayer, "Thin Films-Interdiffusion and Reactions,
    J. M. Poate, K. N. Tu and J. W. Mayer Eds., John Widely,
    New York, 1978, pg. 359.
18. See ref. 3, pg. 485-536.
19. F. Nava, G. Majni, A. Luches, V. Nassisi and E. Janniti,
    J. Physique 5, C4-97 (1980).
20. G. J. von Gurp, G. E. J. Eggermont, Y. Tamminga, W. T. Stacy
    and J. R. M. Gijsbers, Appl. Phys. Lett. 35, 273 (1979).
21. G. Majni, F. Nava, G. Ottaviani, E. D'Anna, G. Leggieri,
    A. Luches and G. Celotti, J. Appl. Phys. (to be published).

LASER CHEMICAL VAPOR DEPOSITION

Susan D. Allen

Center for Laser Studies
University of Southern California
University Park, Los Angeles, CA   90007

INTRODUCTION

Laser chemical vapor deposition (LCVD) is an adaptation of conventional CVD using a laser heat source. As shown in Fig. 1, the laser is focused through a transparent window and the transparent gaseous reactants onto an absorbing substrate, creating a localized hot spot at which the deposition reaction takes place.

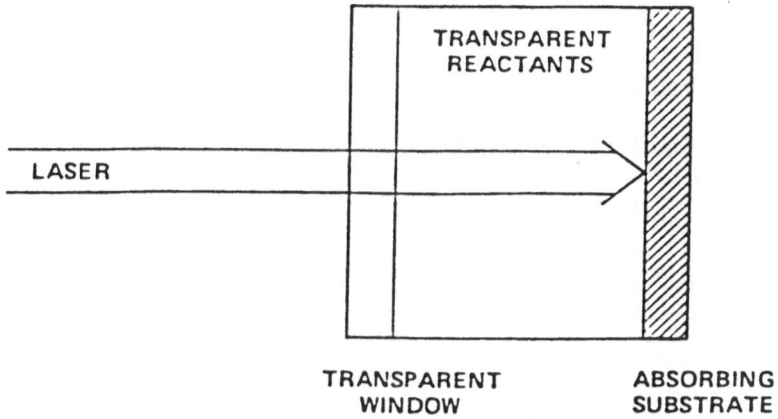

Fig. 1.   Idealized schematic of LCVD apparatus.

455

The optical absorptivities of the reactants and the substrate
determine the laser wavelength which is used. In this paper the
characteristics of several LCVD metal and dielectric films are
reported. Because LCVD is a new deposition technique, the film
characteristics are, as yet, not optimized, but do serve to
illustrate some of the important parameters which must be taken
into account in LCVD.

The use of a laser as a heat source for CVD offers several
distinct advantages, many of which are shared by other laser
processing techniques: a) spatial resolution and control;
b) limited distortion of the substrate; c) the possibility of
cleaner films due to the small area heated; d) availability of
rapid, i.e., nonequilibrium, heating and cooling rates; e) the
capability of heating the substrate to higher temperatures than
available conventionally; and f) the ability to interface easily
with laser annealing and diffusion of semiconductor devices and
laser processing of metals and alloys. It should be possible to
generate deposits of almost any material that can be deposited by
conventional CVD and probably some which cannot. Possible applications of this technique include: writing interconnects, formation of ohmic contacts and localized protective coatings in
semiconductor devices; repair of integrated circuits and integrated
circuit masks; localized coatings and dopants for waveguide optics;
surface hardening and alloying of machine surfaces, localized wear
and corrosion resistant coatings and the repair thereof, and welding
of ceramic materials.

EXPERIMENTAL

Initial experiments in LCVD on quartz substrates were
carried out using the experimental apparatus illustrated in Fig. 2.

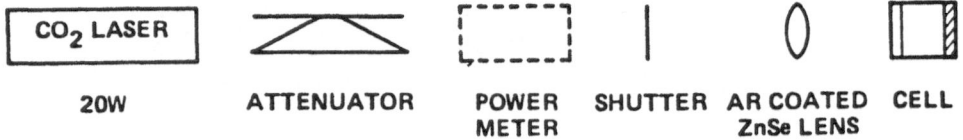

Fig. 2. Low power LCVD apparatus.

The laser was a cw CO_2 rated nominally at 20 W, but usually run
at about 10 W. Attenuation was provided by a ZnSe Brewster angle
polarizer using two Brewster angle plates at opposing angles to
minimize beam walk-off. Although the beam profile without the
attenuator is approximately TEM_{00}, multiple reflection in the
thin ZnSe plates introduce some distortion in the attenuated beam
profile. The removable power meter and/or fluorescent viewing

plates were used to check power stability and beam quality. Irradiation time was controlled by a mechanical shutter with speeds of 2.9 msec to 1.0 sec. After the shutter, the beam was focused onto the absorbing substrate with a 10" focal length AR-coated ZnSe lens through a NaCl window. The LCVD cell was mounted on a motorized translation stage so that both spots and line geomettries could be deposited.

The reactions studied to date include:

$$M(CO)_x \longrightarrow \underline{M} + CO \text{ on quartz}$$
$$M - Ni, W$$

$$M(CH_3) + H_2 \longrightarrow \underline{M} + CH_4 \text{ on quartz}$$
$$M = Al, Sn$$

$$TiCl_4 + H_2 + CO_2 \longrightarrow \underline{TiO_2} + HCl + CO \text{ on quartz}$$

$$TiCl_4 + \text{hydrocarbon} \longrightarrow \underline{TiC} + HCl \text{ on quartz and stainless steel}$$

Because of the different optical properties of the LCVD film and substrate, these ractions can be divided into three types: deposition of a reflecting film on an absorbing substrate, metal/SiO$_2$; an absorber on an absorbing substrate, TiO$_2$ and TiC/SiO$_2$; and an absorber on a reflecting substrate, TiC/stainless steel. The deposition characteristics of these three cases will be shown to be different.

LCVD Ni

The deposition of Ni from Ni(CO$_4$) is a well known process and forms the basis for a purification for Ni, the Mond process.[1] The equilibrium kinetics of the CVD reaction have been studied by Carlton and Oxley,[2] who concluded that both heterogeneous and homogeneous reaction mechanisms contribute to the deposition. The absorption spectra of Ni(CO)$_4$ in the 9-11 μm range is dominated by a combination band at 920 cm^{-1}[3] but at the pressures used (40 torr) the absorption of a 1 cm pathlength at 10.6 μm (943 cm^{-1}) is less than 1%.

Two parameters of importance in evaluating LCVD are the resolution attainable, i.e., the minimum spot size and the thickness of the deposit as a function of the irradiation conditions. For the LCVD of Ni on SiO$_2$ the spot diameter is proportional to the square root of the irradiation time,[4] t, as shown in Fig. 3 and is less than the diameter of the depositing laser beam, FWHH = 0.35 mm, over most of the range of deposition conditions.

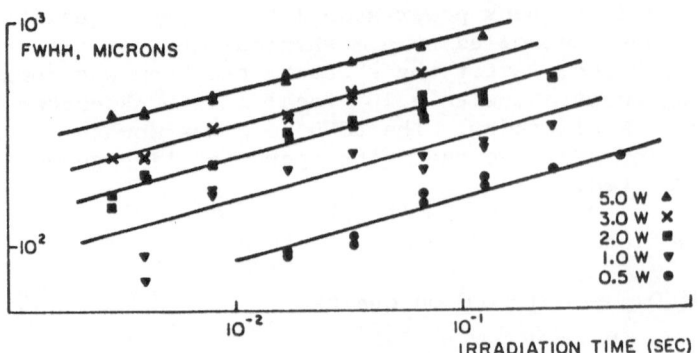

Fig. 3.   Spot diameter (full width at half height, FWHH)
of Ni films deposited on quartz as a function
of irradiation time.

This resolution enhancement can be understood by referring to
Fig. 4. The surface temperature distribution is a Gaussian and
the reaction occurs over that part of the Gaussian that is above
the threshold temperature, effectively selecting the center of
the Gaussian beam.

Figure 5 is a profilimeter trace[5] of such a deposited spot illustrating the truncated Gaussian shape.

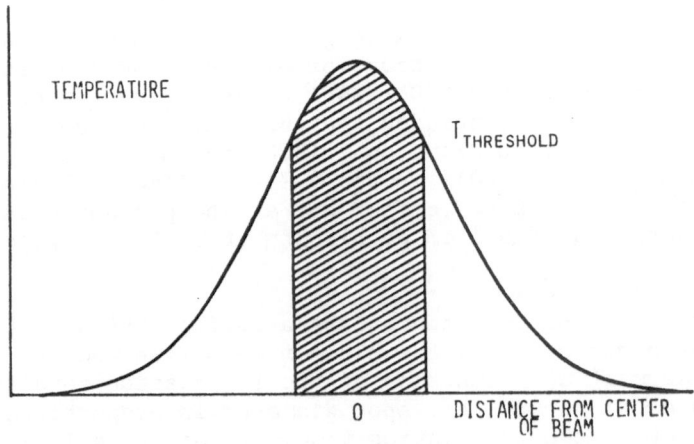

Fig. 4.   Temperature distribution on a surface irradiated
by a Gaussian laser beam.

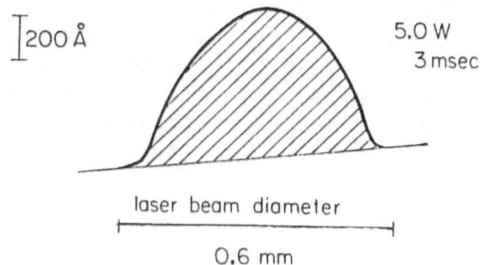

Fig. 5. Profilimeter trace of LCVD Ni film.

The thickness of the deposited Ni films, as a function of
time, t, for various laser powers is given in Fig. 6. The functional dependence of the film thickness on irradiation parameters
is more complex than in the case of the deposited spot diameter.
The initial deposition rate is linear in t and varies from 1000
μm/min for incident power, P_O = 5 W, to 60 μm/min for P_O = 0.5 W.
The deposition rate at all incident laser powers decreases at a
point corresponding to a change in the cross sectional shape of
the deposited film.

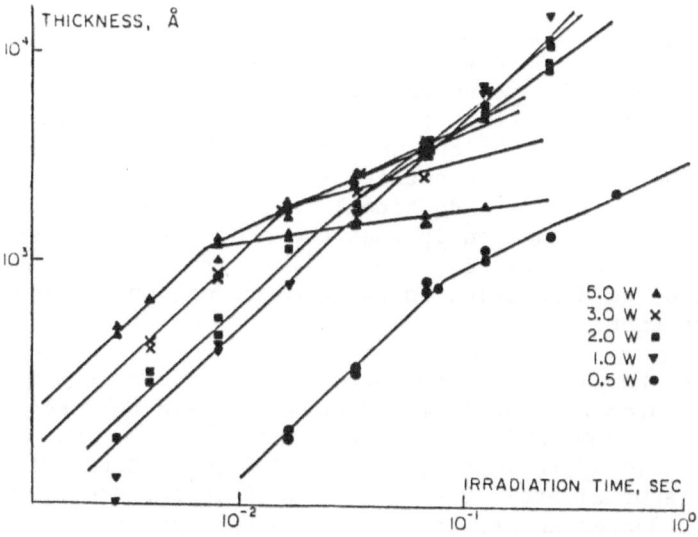

Fig. 6. Thickness of LCVD Ni films as a function of irradiation
time. For 2.0, 3.0, and 5.0 W data, the upper curve is the maximum thickness and the lower curve is the thickness at the center
of the spot.

The truncated Gaussian shape observed for relatively short irradiation times changes smoothly to the double-humped "volcano" shape
shown in Fig. 7 for longer irradiation times. The thickness plot
for these data therefore splits into two curves when the profile
shape changes; the upper curve corresponding to the maximum thickness and the lower to the thickness at the center of the deposited
spot.

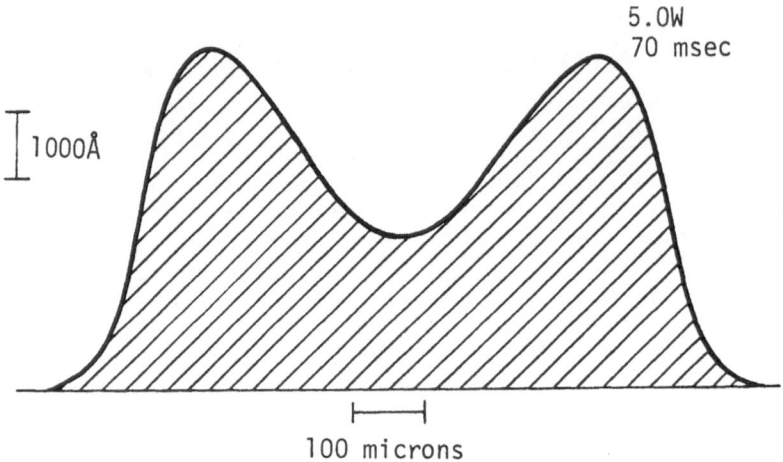

Fig. 7. Profilimeter trace of LCVD Ni films.

This change in shape is probably due to the onset of convection processes and is observed for most LCVD films. Stoichiometry-
driven convection (the production of more molecules of product
than of reactant) would tend to suppress deposition at the center
of the spot by creating a flow away from the heated area and thereby lowering the reactant concentration and enhance deposition
around the perimeter where additional reactant would be entrained.
Another possible explanation is that the center of the spot becomes too hot and the sticking coefficient for the reactants
decreases favoring deposition in the cooler regions outside the
center of the spot.

From the knowledge of the optical constants of quartz and
Ni, one would predict that film thicknesses would be "self-limited"
to approximately 200Å using an optical heat source. The absorptivity   of SiO_2 is approximately 0.9 at 10.6 μm[6] and the absorptivity of Ni is approximately 0.05.[7] One would therefore expect that
once several hundred angstroms of metal were deposited on the substrate, the absorptivity of the surface would decrease by a
factor of 18, the laser power absorbed and the surface temperature
would decrease correspondingly, and metal deposition would cease.
This is obviously not the case as films of Ni on SiO_2 up to 1 μm
in thickness have been deposited. Some evidence of thickness

self-limiting is seen in the Ni deposition at low incident laser powers, i.e., $P_0 = 0.5$ W, for long irradiation times. It is also observed in cases of multiple irradiation of a single site. The film thickness produced for n irradiations is less than n X (the thickness for a single irradiation).

Preliminary results with other systems show that Ni is probably an unusual case.[8] LCVD films of Fe and W do show self-limiting but thick films can be deposited under the proper choice of irradiation conditons. The deposition rate is thus a complicated function of the changing optical and thermal properties of the film/substrate system.

The hardness of LCVD Ni films was demonstrated by their resistance to scratching by a hardened steel pin. The exception were those coatings deposited considerably above threshold which were softer. The LCVD Ni films passed the scotch tape adherence test with the same exception as in the hardness tests. Coatings produced by irradiation at high intensities for long times partially peeled when scrubbed with an alcohol-soaked cotton swab. The thickness of the LCVD Ni films ranged from 100Å to 1 μm. The electrical conductivity was measured as less than 4×10^{-5} ohm-cm for a 550Å thick, large area film. Grain size, as measured by SEM, was less than 100Å for most films. Nodules 2000Å in diameter appeared on the inside slopes of the double humped films deposited at long irradiation times.

LCVD W, Al, Sn

W films were deposited from $W(CO)_6$ but the room temperature vapor pressure of $W(CO)_6$ is less than 1 T[9] and the deposition rates were very slow. The films produced were shiny, metallic in appearance but thin and several millimeters in diameter due to thermal spreading during the long irradiation times necessary. A heated cell under construction should enable us to raise the vapor pressure and therefore the deposition rate significantly.

Two metal alkyls, $Al(CH_3)_3$ and $Sn(CH_3)_4$ were used as LCVD reactants. The metal alkyl was vacuum distilled into a side arm of the deposition cell. The Al films produced from $Al(CH_3)_3$ at room temperature (vapor pressure = 10 T)[10] had initially a shiny metallic appearance but oxidized with exposure to air. It is expected that the as deposited films were also heavily oxidized as the sample handling system was only mechanically pumped and the metal alkyls are known to be oxygen sensitive. The Sn films were deposited at room temperature and $0°C$ (vapor pressures = 300 and 100 T respctively[10] and were granular in structure and easily removed from the substrate. Additional experiments with these materials will be necessary to produce films with good physical properties.

LCVD TiO$_2$

The reaction for the TiO$_2$ deposition in the initial experiments belongs to a class known as "water gas" reactions. In these reactions, H$_2$ and CO$_2$ combine at high temperatures to generate H$_2$O which reacts locally with an easily hydrolyzable compound, in this case, TiCl$_4$, to yield the product oxide. TiCl$_4$ was vacuum distilled into a sidearm tube of the deposition cell and 205 torr each of CO$_2$ and H$_2$ added. The partial pressure of TiCl$_4$ was 12.4 torr, the room temperature vapor pressure. TiCl$_4$ is transparent at the CO$_2$ depositing laser wavelength.[11]

The range of irradiation conditions over which TiO$_2$ can be deposited on quartz is narrower than that achieved with Ni/SiO$_2$, because the higher reaction temperature is close to the SiO$_2$ melting point. No evidence of self-limiting is observed or would be expected on the basis of the bulk optical constants. TiO$_2$ has a greater absorptiviy than SiO$_2$ at 10.6 μm and multiple, n, irradiations of the same site produced thicker films than would be predicted from n X (a single irradiation). Measured deposition rates range from 2 - 20 μm/min.

The resolution enhancement observed in the spot diameter of the Ni/SiO$_2$ system is also seen in the TiO$_2$/SiO$_2$ but no "volcano" thickness profiles were observed. The spatial profile could be tailored depending on irradiation conditions as shown in Fig. 8 from flat-topped to sharply peaked. These thickness profiles were measured using the interference colors observed in reflection microscopy and were checked by stylus profilimetry.[5] For a single irradiation under conditions considerably above threshold, the sharply peaked profile of Fig. 8a was observed. Multiple irradiation just barely above threshold produced the flat, uniform pro-

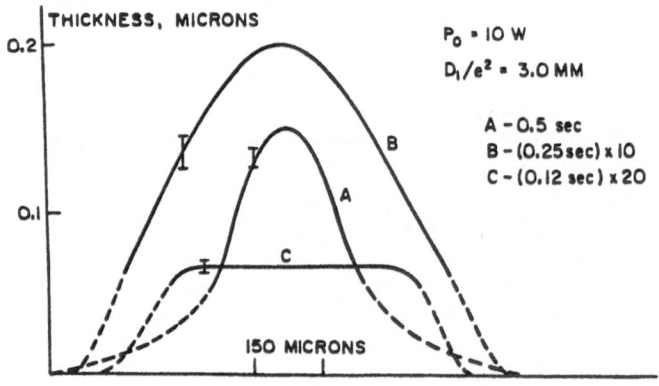

Fig. 8. Thickness profiles of LCVD TiO$_2$ films.

file shown in Fig. 8c. For deposition conditions lying between the two extremes, the broadly peaked profile of Fig. 8b resulted. Various film thickness profiles are thus possible with a Gaussian laser intensity profile. This result is probably due to a combination of both diffusion and heat flow kinetics.

The TiO_2 films were clear and adherent. None were removed with vigorous scrubbing with an alcohol-soaked cotton swab. Preliminary results[12] on the composition of the deposited films using scanning Auger microscopy (SAM) indicate that the composition TiO_x, changes somewhat over the diameter of the larger deposited spots.

LCVD TiC

TiC has been deposited by LCVD on quartz, stainless steel and carbon steel substrates. For LCVD on quartz substrates the range of irradiation conditions is limited as in the TiO_2 LCVD due to the proximity of the deposition temperature to the melting point of quartz. Three different hydrocarbons have been used: CH_4, C_2H_2 and C_4H_{10}. Satisfactory films were produced from both CH_4 and C_2H_2 with most of the data obtained using CH_4. C_4H_{10} exhibited gas phase decomposition for some irradiation conditions due to an absorption at the depositing laser wavelength, 10.6 μm. The films were shiny and black in appearance over a large range of thicknesses.

For deposition on metallic substrates, a 1500 W CO_2 laser was used. The increase in power is necessary due to the high reflectivity and therefore decreased absorptivity and increased thermal diffusivity of the metallic substrates, and the larger laser spot size used. Film thicknesses have been determined by both cross sectional polishing and fracture cross sections with EDAX analysis of the Ti containing layer. Deposit thickness to date ranges from 0.25 - 1.7 μm. The films have been identified as TiC by x-ray diffraction, TEM and electron microprobe. The hardness of the film itself is difficult to measure but a composite hardness of the film plus substrate can range from KHN_{500} = 1000 - greater than 2000. The upper limit of these measurements is not very reproducible probably due to slight variations in thickness over the sample. In any case, hardness measurements greater than KHN_{500} = 850 - 950 indicate the presence of a film harder than the base material.

No evidence of self-limiting of the film thickness was observed or would be expected for TiC deposition on steel. In fact, multiple irradiations of the same site sometimes produced melting on the second or subsequent irradiation due to the increased coupling efficiency of the laser energy into the substrate once a TiC layer had been deposited.

Pulsed LCVD

    In order to measure the time scale of the LCVD process a
short pulse CO_2 TEA laser was used to heat the substrate. With
a pulsed laser heat source, it is possible to heat the substrate
to the deposition temperature before significant deposition takes
place and then monitor the growth of the film after the laser
pulse. Quartz substrates were chosen for the deposition because
of their transparency in the visible, allowing optical monitoring
of the film, and because quartz has a very low thermal diffusivity.
The time constant for the decay of the surface temperature of a
laser heated quartz substrate is on the order of milliseconds.
Thus the short pulse laser heating will create a step function in
the surface temperature which will remain approximately constant
over a period of hundreds of microseconds. This is the ideal
situation for studying the dynamics of LCVD.

    The apparatus used in these preliminary experiments is shown
in Fig. 9. It consists of a CO_2 TEA laser operated as a stable
resonator and generating 0.3J.

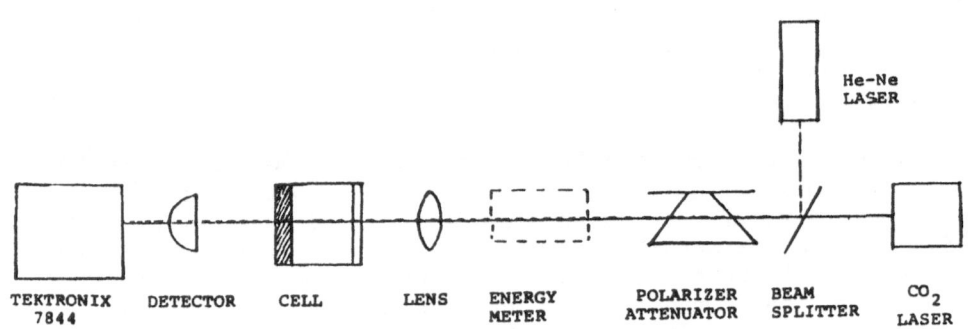

Fig. 9 Schematic diagram of the pulsed LCVD apparatus.

As measured by burn patterns, the laser spatial distribution was
approximately TEM_{00} but the temporal waveform (Fig. 10) had some
structure in the envelope of the nitrogen-pumped tail of the
pulse. This indicated that some higher order modes were present.
The total length of the laser pulse was about 1 μsec. Different
values of the incident energy were obtained by setting a Ge
Brewster angle polarizer/attenuator with Cu beam turning mirrors.
The beam splitter used to fold in the He-Ne monitoring beam was
a Brewster angle ZnSe plate. The transmission of the depositing
film and therefore the optical thickness at the He-Ne wavelength
was measured with a Si photodiode on a Tektronix 7844 oscillo-
scope.

Initial experiments with this system were performed using the Ni from $Ni(CO)_4$ deposition system. The depositions were slightly about 1 mm in diameter. The focal length of the AR coated ZnSe lens is 7% shorter for He-Ne than for CO_2 and this experimental arrangement therefore yielded a large CO_2 spot and a small He-Ne monitoring beam at approximately the center of the CO_2 beam.

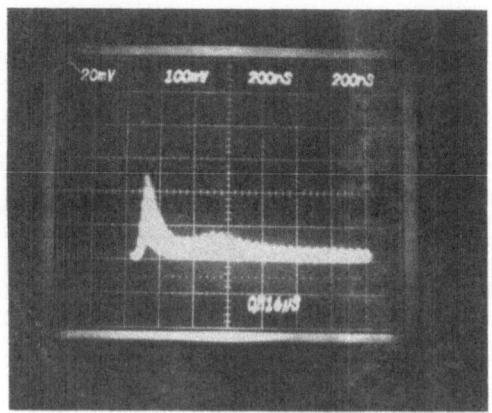

Fig. 10.   Temporal waveform of the CO_2 TEA laser.

The transmission vs time for several different incident energies is shown in Fig. 11. Instead of the expected single exponential decay curve, two decay curves are observed and neither is a simple exponential. They are both characterized by an initial induction period before changes in transmission begin.

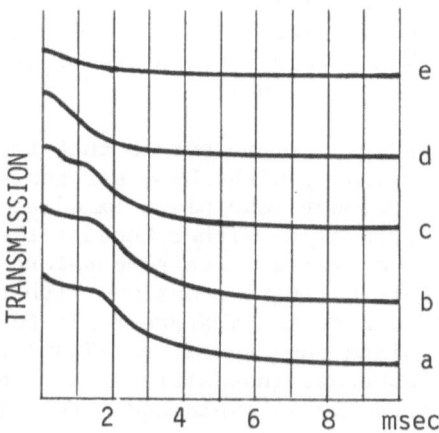

Fig. 11.   Transmission vs. time curves for LCVD Ni deposition on quartz for different incident energies: a) 0.082 J, b) 0.076J, c) 0.063 J, d) 0.051 J, and e) 0.038 J.   (Curves are offset verti-cally.)

Using a simple diffusion model, if every collision resulted in deposition of a Ni atom, the time required to deposit a 10Å thick film would be from 0.1 to 1 µsec. For the msec deposition times observed, the assumption that the laser pulse heats the surface before any deposition occurs is certainly valid but the further assumption that the surface temperature remains essentially constant for long enough to measure the deposition rate is not. The surface temperature as a function of time must now be considered explicitly.

Even if the deposition rate is changing as the surface temperature changes, the initial slope of the exponential decay curve will give a measure of the initial deposition rate. Table 1 lists the initial slopes of the transmission curves, dT/dt; the final transmission, the optical thickness of the initially deposited film where applicable, the optical thickness of the total film and the physical thickness of the total film as measured by stylus profilimetry as functions of the incident energy (see the diagram accompanying Table 1). The optical thickness of the films were determined from the thin film optical constants[7] and the full trnasmission equation.[7]

It is apparent from the data of Table 1 that the total thickness of the deposited film is a function of the incident energy but that the deposition rate characterized by dT_2/dt is not. The rate of the initial deposition, dT_1, is difficult to measure and no specific trend is seen in the values given in Table 1. There are two obvious questions posed by the data: 1) why are there two distinct exponential decay curves observed in some cases? and 2) if we are measuring a reaction rate or adsorption rate-limited deposition, why isn't dT_2/dt a strong function of the surface temperature? (For pulse lengths much less than 0.75 msec, the surface temperature of quartz is proportional to the incident energy density.)

The working hypothesis at present is that the reactants already adsorbed react rapdily with the laser heated surface. Subsequent deposition does not take place immediately because the surface temperature is too high, either because the reaction rate at the surface exhibits a maximum with temperature or, more probably, the adsorption rate exhibits a maximum with temperature. After a time, t_1, the surface temperature falls to T' and the deposition proceeds. Since the deposition always starts at the same temperature, the apparent insensitivity of the deposition rate to incident laser energy is explained. The time, t_1, is given in Table 2 as a function of incident energy (temperature) and does increase with increasing energy (temperature) as would be expected.

Table 1. The incident energy of the laser pulse, E; the slopes of the two exponential curves, dT_1/dt and dT_2/dt; the transmission of the total film, $T_f = I_f/I_0$; the optical thickness of the total film, d_f; and the physical thickness, h.

E(J)	$-dT_1/dt \times 10^{-3}$ sec^{-1}	$-dT_2/dt \times 10^{-3}$ sec^{-1}	T_f	Optical Thickness d_1 (Å)	d_f (Å)	Tencor Thickness h (Å)
0.082*	0.11	0.198	0.455	5	65	
0.076	0.09	0.219	0.46	5.5	65.5	80
0.063	0.16	0.183	0.56	6.5	45.4	60
0.051		0.196	0.70		27	
0.038			0.946		3	

*One data point only, others melted.

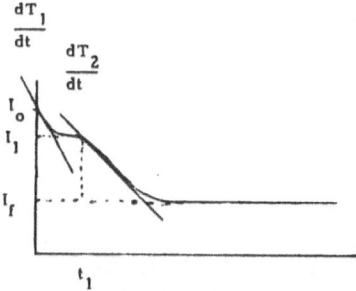

Table 2. The incident energy of the laser pulse, E and the time, t', before the beginning of the second decay curve, dT_2/dt.

E(J)	t_1 (ms)
0.082	1.6
0.076	1.0
0.063	0.45
0.051	
0.038	

In conclusion, laser chemical vapor deposition has been demonstrated for a wide variety of films and substrate materials. Area resolution of the deposited films can be better than the incident laser beam spot size. Thicknesses of up to several microns have been obtained to date for most of the deposition systems. Plans for future work include analysis of the effect of the various parameters on the LCVD films and investigation of other laser/substrate systems.

The author wishes to acknowledge the assistance furnished by Prof. A. Burg in the handling of $Ni(CO)_4$. This work was supported in part by a grant from AFOSR under the technical cognizance of H. Schlossberg.

REFERENCES

1. C. F. Powell, I.E. Campbell, and B. W. Gonser, Vapor Plating , Wiley, New York (1955).
2. H. E. Carlton and J. H. Oxley, A.E. Ch. E. Journal 12:86 (1966).
3. B. L. Crawford, and P. C. Cross, J. Chem. Phys. 6:535 (1938).
4. M. Sparks, J. Appl. Phys. 47:837 (1976), M. Sparks and E. Loh, J. Opt. Soc. Am. 69:847 (1979), M. Lax, Appl. Phys. Ltr. 33:786 (1978).
5. Talystep Profilimeter measurements performed courtesy of J. Bennett, NWC, China Lake, CA. Tencor profilimeter measurements courtesy of Griot Assoc., Los Angeles, CA
6. L. E. Howarth, and W. G. Spitzer, J. Amer. Ceram. Soc. 44:26 (1961).
7. D. E. Gray, American Institute of Physics Handbook, McGraw-Hill, New York, (1972).
8. S. D. Allen and A. B. Trigubo, to be published.
9. I. Wender and P. Pino, Organic Syntheses via Metal Carbonyls, Vol. I, Wiley, New York
10. Alfa Division, Ventron Corporation, P.O. Box 299, Davers, MA 01923.
11. W. B. Person and W. B. Maier II, J. Chem. Phys. 69:297 (1978).
12. SAM measurements courtesy of J. E. Green, Univ. of Illinois, Urgana, Illinois

CUTTING WITH CO_2-LASERS IN THE AUTOMOBILE INDUSTRY

G. Fritzsche

Daimler-Benz Aktiengesellschaft

Stuttgart

GENERAL PROBLEMS

Let us take a short, incomplete look at some very simple problems concerning cutting with CO_2-Lasers from the viewpoint of an engineer.

The problems are very simple, as opposed to the solution of some theoretical equations. However, in this field you have to pay attention not only to physical but also to technical and economical aspects.

a) there are no, or at least very few, complete Laser machining systems (s.Fig.1) for realistic applications

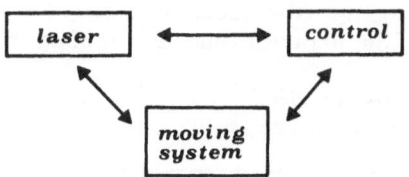

Fig. 1 Complete laser machining system.

b) there is a gap between theory and practice
c) laser machining technology is developing very quickly. The decision to buy today is difficult

MATERIALS AND PARTS WHERE LASERS CAN BE BEST USED

Materials and parts are listed in Tables I and II

Table I

materials	
nonmetals	metals
plywood	mild steel
other wooden materials	stainless steel
leather	coated steel
cotton	– Al
plastics	– Zn
	– Cr
thickness: 3 – 15 mm	max 8 mm
	(1 – 2,5 mm 70%
	pass. cars)

Table II

parts
- for parts to be produced in small quantities (many different types)
- shape still changing (development)
- at start of series production if cutting tools are not yet available
- failure of cutting tools
- production of spare parts

SPECIAL PROBLEMS

Fig.2 shows a problem sketch of CO_2-Laser cutting. The most important parameters are focal position (a), material thickness (d) material properties, coatings (Zn,Al,Cr) and absorption of the radiation, and flatness of the workpiece.

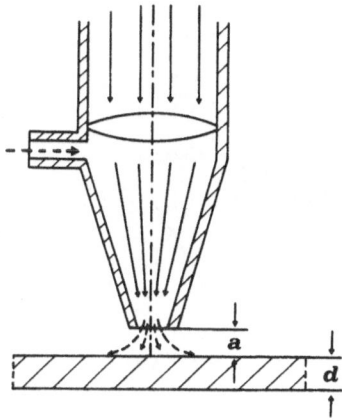

Fig. 2   Problem Sketch CO_2-Laser cutting ($\sim$ 500 w).

When cutting in a plane without riding optics, the workpiece has to be maintained flat to within a tolerance of $\pm$ 0.5 mm (focal length 5").

When cutting in 3 dimensions there are two problems.

a) as the beam normally goes in the Z-axis, with a greater angle of inclination, the thickness of the cut increases (see fig. 3)

b) defocusing
   when using a riding optic with two small wheels
   defocusing occurs when climbing up an incline (see fig. 4)

when using wheels of 3 mm radius the critical defocusing point is reached at an angle of 31°. In the region of about 30° incline bad quality is already evident.

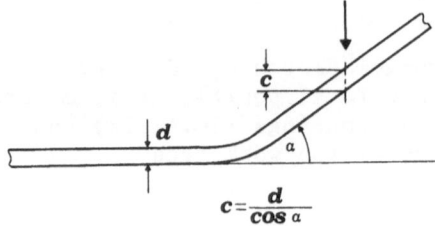

$$c = \frac{d}{\cos \alpha}$$

Fig. 3   cutting problem

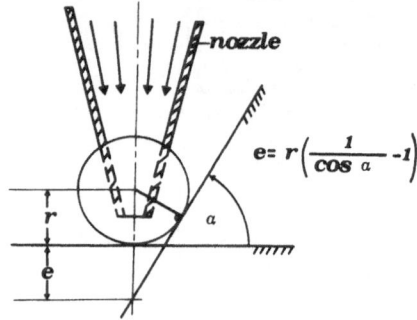

$$e = r\left(\frac{1}{\cos \alpha} - 1\right)$$

Fig. 4   defocusing problem

EXAMPLES

a) brakebands for automatic gears, 3 mm stell (s.fig. 5). Prototype production. Several changes in shape (especially the radius).

b) stencils, 3 mm steel (s.fig. 6) a few parts were produced with a line follower control.

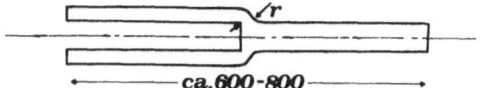

Fig. 5 brakebands

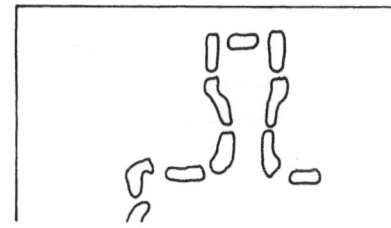

Fig. 6 stencils

c) supports (bus production)
   steel 2-3 mm, coated steel (Zn,Al) 1 - 3 mm a few parts for
   tests (s.fig.7).

d) pressed parts (3 dimensions) 2 mm steel (600x200x100 mm)

   prototype production, shape still changing, cutting tool not
   yet available, a few hundred produced, maximum incline $\sim$ 30°.

EQUIPMENT USED

     The cutting jobs were done on 5 different laser systems. There
were great differences in the quality of lasers, controls and
moving systems (s. Table III).

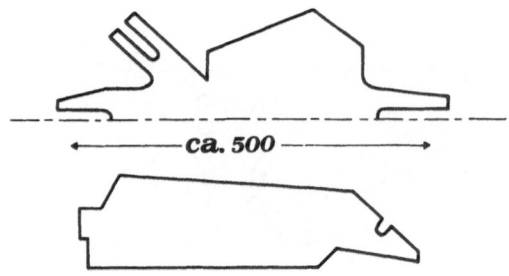

Fig. 7 supports

Table 3

	workpiece	moving optics	laser	control
A		(Z)	x,y	line follower
B	x,y	(Z)		CNC
C	x	y (Z)		CNC
D		x,y,(Z)		CNC
E		x,y,(Z)		line follower

Recently four companies started to offer combined laser/punching
presses. With such a machine you can always chose the best
technique to suit your problem. Thus holes of the same size, which
are required very often, are punched, whereas shapes and large
outcuts are done by the laser.

INDUCED EFFECTS OF A LASER PULSE ON DEFECTS AND DISORDER
CREATION IN SOME ORDERED ALLOYS

E.L. Mathé, J.P. Eymery

Laboratoire de Métallurgie Physique, Faculté des Sciences
L.A.131 du C.N.R.S.
40, avenue du Recteur Pineau - 86022 POITIERS

In this paper, an irradiation by a laser pulse is considered as the origin of :
- a rapid elevation of the temperature of the sample followed by a rapid cooling ($\approx 10^{9}°$/sec)
- a stress wave due to the rapid expansion of the vaporized zone on the surface [1,2,3].

The purpose of this paper is to show the effects of these facts on the structural state (point and linear defects concentration, degree of order) of some ordered alloys : Fe-Al 40 at.% Ni_2Cr, Fe-Co 50-50, Ni_3Fe.

I- EXPERIMENTAL

I.1 - Methods

Three methods were used :

- transmission electron microscopy of thin disks of alloy electropolished on the lower face after irradiation of the upper face.
- resistivity measurements by potentiometric method on samples with parallelepipedic shape ($20 \times 0,5 \times 0,5$ mm^3). All the measures were done at liquid N_2 temperature.
- Mossbaüer spectrometry of thin plates ($a = 0,2$ cm^2, thickness ≈ 25 µm) with a ^{57}Co source. The spectra were taken with constant acceleration at room temperature.

I.2 - <u>Laser</u>

The laser is a Yag : Nd one working in TEM$_{00}$ mode with 20 ns and 0.150 J pulses or with 150µs and 0.85J pulses. The beam is focused by a lens (f = 11 cm). Irradiations were done by successive juxtaposed pulses without overlapping on the surface of the sample.

II - RESULTS

II.1 - <u>Transmission electron microscopy</u>

After laser irradiation, all the alloys contain a large dislocation density. They are arranged as a network ("cell" structure) as after strong traction or shock deformation.

Moreover, the orientation of the microfractures appearing on the surface of a Fe-Al sample after a 20 ns pulse can be explained by a mechanism of rupture by cleavage which is unusual at high speed and high temperature[4]. In Ni_2Cr and Ni_3Fe twins of classical orientation were observed. Some of them, in Ni_3Fe are parallel to the walls of the dislocations cells[5]. T.E.M. gives also information on the atomic order in the samples. If disorder occurs, superstructural spots disappear from the diffraction patterns and the dislocations are no more associated. Table 1 summarizes the obtained results.

The destruction of the order depends on the initial state and on the pulse duration. A too short pulse does not disorder the sample may be because the diffusion time is not long enough.

II.2 - <u>Resistivity measurements</u>

II.2.1 - <u>Fe-Al 40 at.%</u>

These results are given in [6].
After a 20 ns pulse laser irradiation, the resistivity of the sample was measured after each step of an isochronal annealing (t = 15 mn, T = 20 K), between ambient temperature and 500°C. The resistivity decreases mostly in two stages centred on 330°C and 420°C.

The comparison of the temperature of the restauration stages after quenching, deformation by traction and neutron irradiation [7,8] is made in table 2.

The persistance of stage III after laser irradiation shows that despite the high dislocation concentration the irradiation creates an amount of vacancies higher than what is required to sature dislocations.

Table 1

Alloy	Fe-Al	Ni_2Cr	Ni_2Cr	Fe-Co	Ni_3Fe
Preliminary heat treatment	1000°C annealing 1 h 10^{-6}torr	800°C annealing 1 h 10^{-6}torr	quenched from 800°C + 500°C reco- vering 5000 h	850°C annealing 1 h 10^{-6}torr	quenched from 600°C one month recovering at 500°C
So : initial order parameter	So $\neq$ 0.8	Short range ordered	Long range ordered	S $\approx$0.98	S $\neq$ 1
S after 20ns pulse	S $\neq$ So			ordered	
S after 150ns pulse	S $\neq$ So	Short range order destroyed	Long range order destroyed	disordered	disordered

Table 2

Treatment	Stage I	Stage II	Stage III
Quenching from 900°C			688 K
Traction ε = 0.6 %	500 K	615 K	680 K
Traction ε = 5.3 %	500 K	615 K	
Neutron irradiation	510 K		653 K (weak)
Laser irradiation		603 K	693 K
possible origine of the stage	elimination of interstitials on vacancies and/or dislocations	elimination of vacancies on dislocations	free vacancies elimination (on surface or in loops)

II.2.2 - Ni₂Cr.

    In this alloy two types of order can be observed [9,10] depen-
ding upon the heat treatment as mentionned in table I. The restauration of the short range order (S.R.O.) is correlated with an increase
of resistivity of the sample [9].

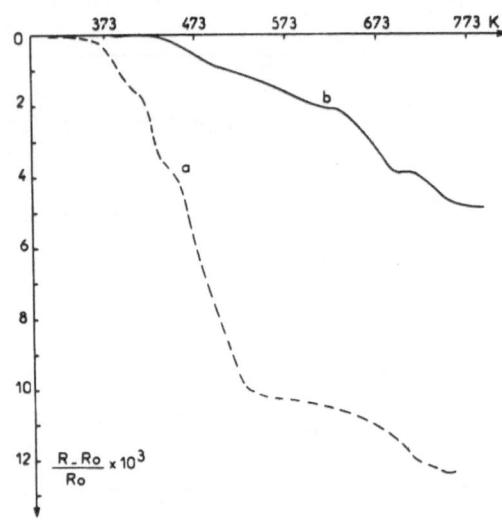

Fig.1 : Resistivity
variation during
isochronal annealing
of L.R.O. Ni₂Cr after:

a) 15 % deformation
b) laser irradiation

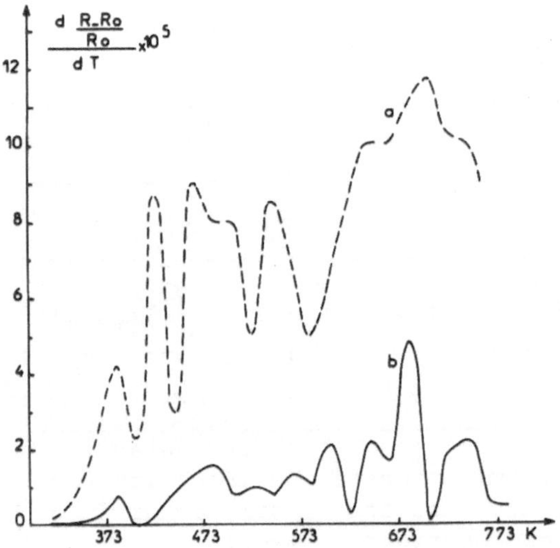

Fig. 2 : Annealing
stages of defects
in a L.R.O. Ni₂Cr
after 15 % deforma-
tion (a) or laser
irradiation (b)

A quench from 1000°C destroyed both kinds of order and during a further isochronal annealing from room temperature, ρ increases up to a maximum then decreases from about 500°C during the instauration of the long range order (L.R.O.).

- L.R.O. samples.

On fig. 1 the values of $\dfrac{R-R_o}{R}$ are plotted versus T during isochronal annealing after

- irradiation by laser pluses (150 ρ s, 2.10^{-6} W/cm^2)

- deformation by traction ($\varepsilon = \dfrac{\Delta l}{l} = 15$ %, velocity : 0.2 cm/mn)

R_o is the electrical resistance after irradiation
R is the electrical resistance after each step of the annealing ($\Delta t = 15$ mn, $\Delta t = 20°C$).

The first observation is that ρ decreases, it means that only the L.R.O. is destroyed by the laser pulse as already shown by T.E.M.

The second result concerns the origin of the restauration stages. From comparison between derivation curves of fig.2 and results obtained on samples quenched or irradiated by neutron beam[10] we can identify the migrating defects at each stage.

$350 \longrightarrow 610$ K interstitials going towards vacancies or dislocations

$610 \longrightarrow 700$ K ⎫
$\qquad\qquad\qquad$ vacancy migration ⎬ ↗towards dislocations
$700 \longrightarrow 773$ K ⎭ ↘random elimination

As for Fe-Al the division of the vacancy stage into two parts show that the laser irradiation is responsible for the presence of a high vacancy density.

- S.R.O. samples.

As after L.R.O. we compare samples after

- laser irradiation

- deformation ($\dfrac{\Delta l}{l} = 0.64$ % and 15 %)

The isochronal restauration of ρ is shown in fig. 3 and 4.
In this case, laser irradiation destroyed S.R.O. as did deformation and neutron irradiation (10).

The resistivity increase occurs during the S.R.O. restauration.

The persistance of a stage between 700 and 773 K i.e. the same temperature range as the found on a L.R.O. sample, confirms the formation of an important vacancy concentration.

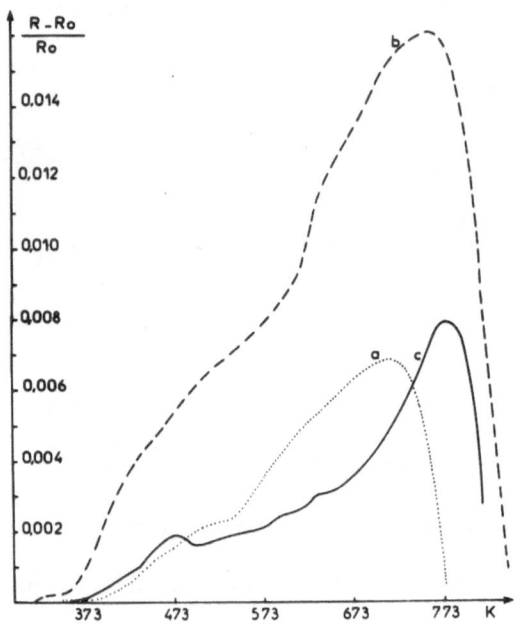

Fig. 3 : Resistivity variations during isochronal annealing of S.R.O. Ni_2Cr after deformation (a), 15% deformation (b),0.64% laser irradiation (c).

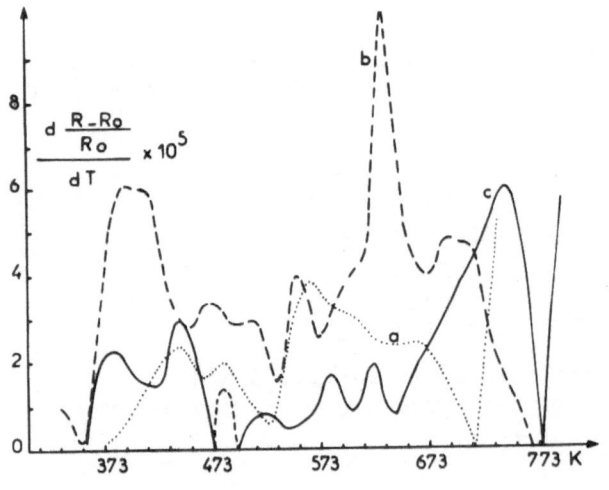

Fig. 4 :

Annealing stages in a S.R.O. Ni_2Cr.

II.2.3 - Conclusion

On both alloys the structural state created by the pulse differs from a quenched or a strained state since there are more vacancies than after deformation but less than after quench (the amplitude of the stage is smaller). The effect on the atomic order
is the same for the laser beam and the neutron beam.

III - MOSSBAUER SPECTROMETRY

The Mössbauer spectra of the Fe-Co (50-50) and Ni_3Fe alloys
exhibit 6 lines from which can be calculated the average hyperfine
field H and the isomeric shift IS. Disorder is responsible for an
asymetric breadening of the lines and the hyperfine field is a
decreasing function of the order parameter S [11].

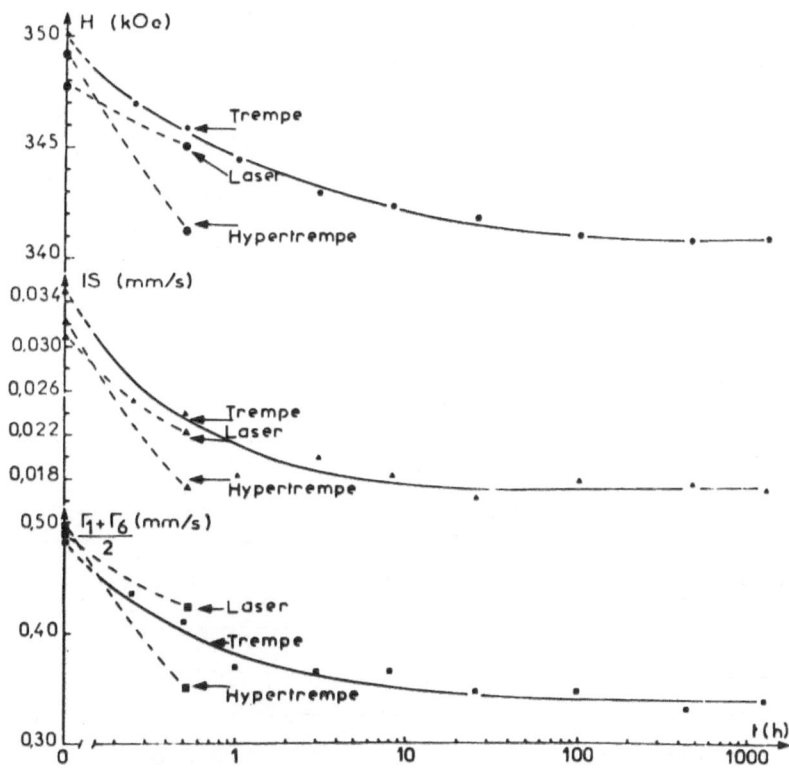

Fig. 5 : Comparison of order kinetics after quench (trempe),
         splat cooling (hypertrempe) and laser irradiation,
         detected by Mössbauer spectrometry.

III.1 - <u>Fe-Co 50-50</u>.

The results are exposed in[12] and can be summarized as follows:

- a 20 ns pulse never disordered the initially ordered alloy; after a 150 μs pulse irradiation with 11.2 pulses/cm^2 the value of H is slightly smaller than after quenching and splat cooling but higher than after annealing : so it can be said that the alloy is notably disordered.

- the velocity of the restauration of order during isothermal annealing at 480°C is nearly the same than after quenching but smaller than after splat cooling (Fig. 5). So we can deduce that in the case of laser irradiation a smaller amount vacancies contribute to the order restauration. Two reasons can be proposed : either the time during which the sample was near the melting point was not long enough to obtain the equilibrium vacancy concentration or the vacancies are captured by the dislocations during the pulse itself. The recovering rate may be slowered also if the order was not completely destroyed by the laser pulse.

III. 2 - Ni_3Fe.

An ordered sample was irradiated by 150 μs pulses with a 2 MW/cm^2 fluence. The values of the hyperfine field after different treatments are listed in table 3.

Table 3

Treatment	$(H \pm 0,3)$ k O$_e$
Annealed for 1 month at 500°C S ≠ 1 (domain size ~ 280 Å)	278.2
Quenched from 800°C	288.9
Laser irradiated	290.1
Splat cooling	290.8
90 % cold rolling at 300 K	292.5

This table shows that laser irradiation disordered Ni_3Fe nearly as well as did splat cooling but less than a strong deformation.

CONCLUSION

All the preceding results show that the laser irradiation has a considerable influence on the location of the atoms, creating dislocations, point defects, disorder.

It gives the possibility to obtain a new structural state differing from all that is observed after quench, deformation, neutron irradiation.

REFERENCES

1. R.M. Measures Acoustical propagation of thermally induceed stresses, Acustica 15, 133 (1965).
2. S.A. Medtz, F.A. Smidt, Appl. Phys. Lett. 19, 207 (1971).
3. B. Steverding, H.P. DUDEL, J. Appl. Phys. 47, 1940 (1976).
4. P. Bonneau, Thèse, Poitiers, 1978.
5. J.P. Eymery, M.F. Richard, H. Garem. To appear in "Phil Mag A.
6. P. Bonneau, E.L. Mathé, P. Moine. Script. Met.13, 267 (1979).
7. J.P. Rivière, J. Grilhé. Acta Met. 20, 1275 (1972).
8. J.P. Rivière, H. Zonon, J. Grilhé. Acta Met. 22, 929 (1974).
9. R.J. Taunt, B. Ralph. Phys. Stat. Sol. (a), 29, 431 (1975).
10. J.P. Rivière, J.P. Eymery. Rad. Eff. 37, 155 (1978).
11. J.P. Eymery, P. Moine. J. Phys. Lett. 39, L. 23 (1978).
12. J.P. Eymery, E.L. Mathé, P. Bonneau. Script. Met. 12, 969 (1978).

SUPERSATURATED SOLID SOLUTION IN ION IMPLANTED SEMICONDUCTORS

Salvatore Ugo Campisano

Istituto di Struttura della Materia dell'Università

Corso Italia 57 I95129 Catania, Italy

INTRODUCTION

Irradiation of ion implanted semiconductors with high power laser pulse has been used to produce heavily supersaturated solid solutions[1,2]. Impurity concentration in excess of 10^3 times the maximumin equilibrium solubility have been obtained in Si implanted with Te atoms[2]. Supersaturated solid solutions can be also obtained during thermal annealing of the amorphous surface layer produced by the implantation[3]. Both processes are characterized by a moving interface separating an ordered phase from a disordered one (liquid during laser and amorphous during thermal annealing). We will show that the basic mechanism leading to the formation of the supersaturated solid solution is the same in both cases, i.e. solute trapping[4-6].

Laser annealing

The effect of the irradiation of an ion implanted semiconductor with an high power laser pulse is illustrated in Fig.1. Silicon single crystals, <111> oriented were implanted with 5×10^{15} at/cm^2 As ions at 400 keV. The crystals were irradiated with ruby laser pulses with energy density up to 2J/cm^2. The channeling analysis, reported in Fig.1, reveals that for irradiation up to 1.5 J/cm^2 no change is detected in both the Si surface peak and As profile. T.E.M. observations[7] have shown that the amorphous surface layer has been transformed into a polycrystalline one.

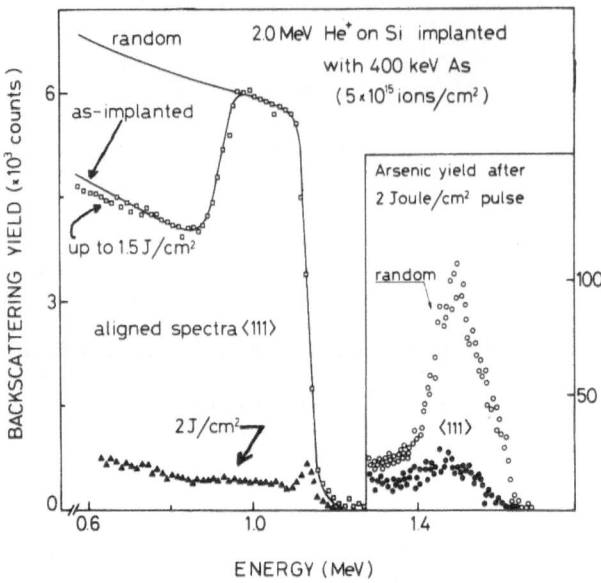

Fig. 1 Channeling analysis of a Si crystal implanted with 5×10^{15}
        As ions/cm^2 at 400 keV. The insert shows the As yield.

For irradiation above 2.0 J/cm^2 the Si surface minimum yield drop
to about 5%; the As profile shows a little broadening and the
atoms are substitutionally located, as shown by the decrease of
the aligned yield.

Assuming that all the absorbed photon energy is instantaneously converted into heat, we can solve[8] numerically the that equation including the source term. In Fig.2 is reported the result
concerning the formation of a surface liquid layer and the kinetics of the liquid-solid interface. For a very thick amorphous
layer and 50ns duration of the ruby laser pulse the surface starts
to melt at about 0.8 J/cm^2. At larger energy densities the melt
front penetrates inside the sample. When no more energy is available from the external pulse, the surface cools and the liquid-
solid interface moves back to the sample surface. If the maximum
thickness of the molten layer exceed the value of the amorphous
one, the liquid wets the underlying single crystal and liquid pha-

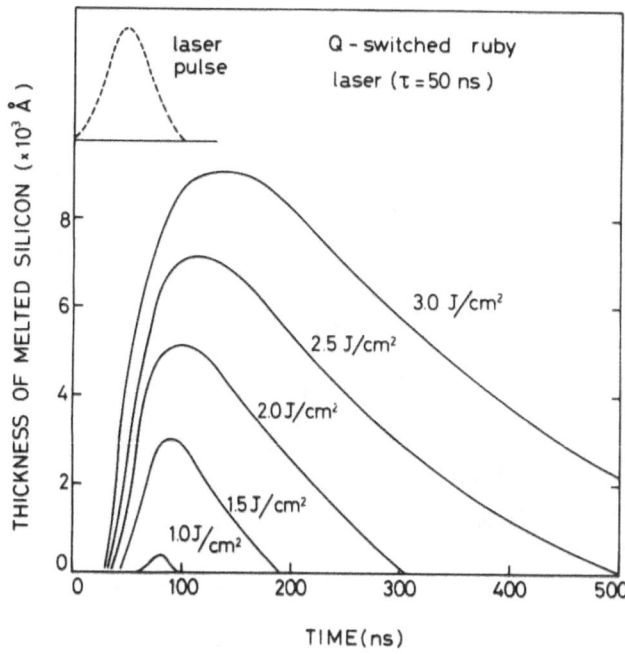

Fig. 2   Kinetics of the liquid-solid interface due to a 50ns ruby
laser pulse of various energy densities on amorphous
silicon.

se epitaxy can occur.

Impurities, if present, can diffuse in the liquid layer and
can be redistributed by the movement of the liquid-solid interface. An example is shown in Fig. 3 where the experimental depth
profiles of 400 keV As ions implanted in silicon are shown. Up
to 1.5 J/cm^2 irradiation no large broadening is detected and the
surface layer is polycrystalline (see Fig.1). At 3.0 J/cm^2 the
As profile broadens and reach the sample surface. The full lines in Fig.3 represent results of numerical solutions of the diffusion equation[9] assuming a diffusion coefficient of 10^{-4}cm^2/s

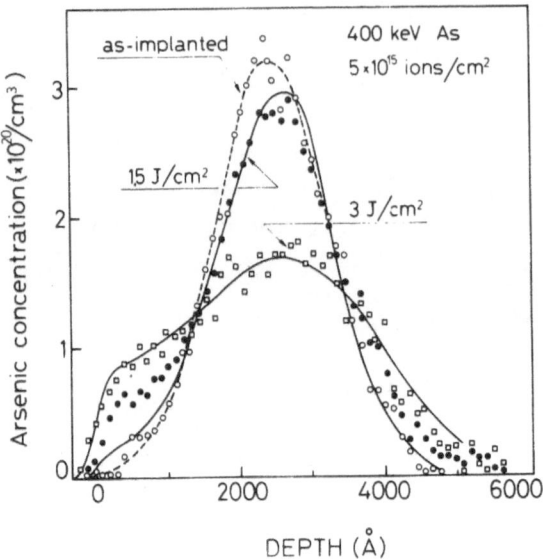

Fig. 3 Depth profiles of 400 keV As implanted in Si after laser
        annealing. The full lines represent the result of a
        numerical fit.

for As in liquid silicon and a negligible value for diffusion in
solid material. The liquid solid boundary moves with a kinetic
shown in Fig.2 and no segregation effects were included in the
calculations.

    More striking is the result reported in Fig.4 concerning las-
er annealing[2] of 400 keV Te implantation in Si. After irradiation with 2.5 J/cm^2,50ns ruby laser pulse the silicon surface
reorder in a perfect crystal layer. The Te atoms show little
broadening; moreover they are substitutionally located as demonstrated by the yield decrease for analyses along the <111> and
<110> directions of the substrate. In spite of the very small

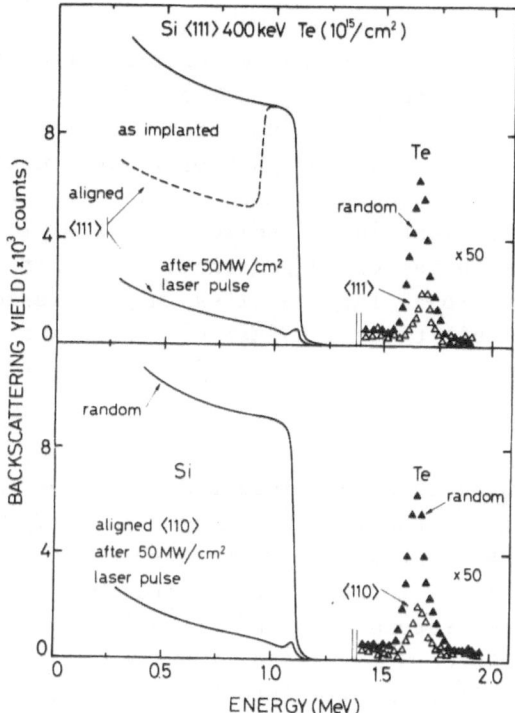

Fig. 4   Channeling analysis of a Si crystal implanted with 400 keV
         Te ion and annealing by a 2.5 J/cm^2 ruby pulse.

value of the equilibrium segregation coefficient (K_o=4×10^{-6}) no
surface accumulation is observed. The Te profile can be fitted
using a diffusion coefficient in liquid silicon of 10^{-4}cm^2/s and
a segregation coefficient of 0.55 at the interface[2] .

The large increase of the segregation coefficient in laser
irradiated ion implanted silicon crystals has been found for several impurities such as Bi, In, Ga and etc. Other impurities such as
Cu,Fe etc show instead large surface accumulation. The mechanism
leading to such high value of the interfacial segregation coefficient is the kinetic trapping of impurities at the interface[4-6] .
The time required to grow a layer of thickness δx comparable with
that of the interface is: $\delta t = \delta x / v$ where v is the liquid-solid interface velocity. Its value is of the order of 10^{-10}s being v
of the order of a few m/s. Impurities reside in the interface
region for a time $\tau = \delta x^2 / D_i$ being D_i their diffusion coefficient

at the interface. If τ is large compared to δt they are trapped
by the moving interface and no large segregation effects are observed. For τ small compared to δt, they can diffuse in the remaining liquid layer and will be accumulated on surface, as it
occurs for Cu,Fe etc. If we assume that D_i is determined by the
diffusivities in the two adjacent phases[6] , i.e. $D_i = \sqrt{D_s D_\ell}$ being
D_s the diffusion coefficient in the solid, this kinetic description account also for the specie dependence. Slow diffusers can
be trapped in substitutional lattice sites in great excees of
their equilibrium solubility whilst fast diffusers are segregated
at the surface.

Thermal annealing

The crystalline properties of implanted amorphous layers can
be restored by a suitable thermal treatment. For temperature
above 500⁰C solid phase epitaxy occurs through an amorphous-crystal
interface movement. The velocity of this interface depends expo-
nentially from the temperature with an activation energy of 2.35
eV and its value at 550°C is 1.5 Å/sec, i.e. ten order of magnitude smaller than the liquid-solid interface velocity during
laser annealing. Impurities, if present inside the amorphous can
give rise to two results[6] : slow diffusers are incorporated in
substitutional lattice sites, also in great excess of the solubility limit, fast diffusers inhibit the regrowth. As example in
Fig.5 is shown the channeling analysis of a Si crystal implanted
with 40 keV Te ions to a fluence of 7×10^{14} at/cm^2. After annealing at 550°C all the amorphous layer has regrown and Te atoms are
substitutionally located, as in the case of laser annealing. An
example of fast diffuser is reported in Fig.6 where thermal annealing and laser annealing results are shown in the upper and lower
part respectively for the case of Ag implantation in silicon.
Furnace annealing produce the transition of the surface layer from
amorphous to polycrystal; laser annealing produce transition to
single crystal, but the impurity is completely rejected at the
sample surface.

Applying the same kinetic trapping concept, as in the case
of laser annealing, the time δt to grow the interface layer results of the order of 1s. The impurity residence time τ can range
from 10^{-2} to 10^8s for fast and slow diffusers respectively. Trapping conditions are then fulfilled for slow diffusers, as in the
case of laser annealing. Heterogeneous nucleation is responsible

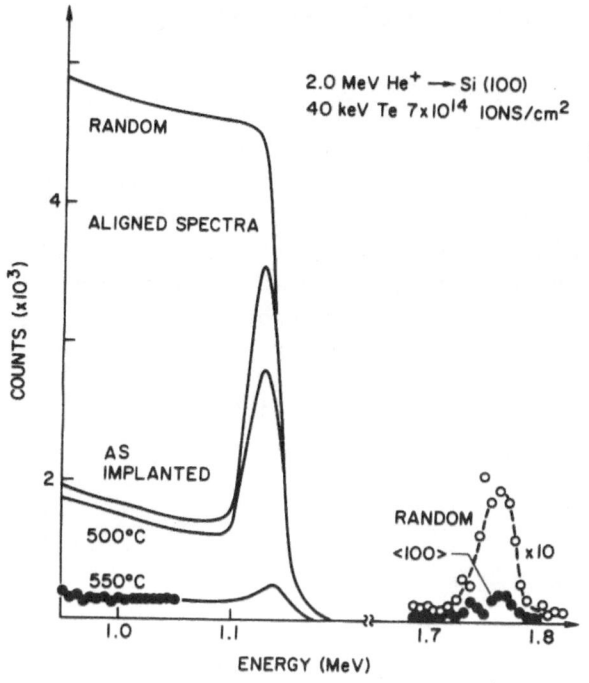

Fig. 5 Channeling analysis of a Si crystal implanted with 40 keV Te ions thermally annealed.

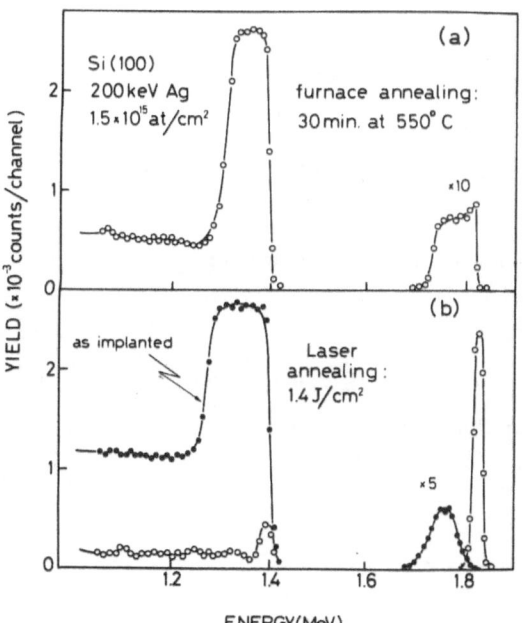

Fig. 6 Channeling analyses of a Si crystal implanted with 200 keV Ag ions:
a) thermal annealing;
b) laser annealing

for the transition to polycrystal in Si implanted with fast diffusers.

CONCLUSIONS

   The basic mechanism leading to the formation of supersaturated
solid solutions in both laser and thermal annealed implanted silicon is the same: kinetic trapping at the moving interface. This
process is controlled mainly by the diffusion coefficient of the
dopant in the solid phase and by the interface velocity itself.

   Thanks are due to Mrs.G.Giuffrida for her typing aid.

REFERENCES

1. C.W.White,S.R.Wilson,B.R.Appleton,F.W.Young;
    J.Appl.Phys. 51,738(1980)

2. S.U.Campisano,P.Baeri,M.G.Grimaldi,G.Foti,E.Rimini;
    J.Appl.Phys. 51,3968(1980)

3. S.U.Campisano,E.Rimini,P.Baeri,G.Foti;
    Appl.Phys.Lett.(July 1980)

4. D.Turnbull; J.de Physique 14 suppl.5 C4,209(1980)

5. D.Turnbull; Proceeding of this school

6. S.U.Campisano,G.Foti,P.Baeri,M.G.Grimaldi,E.Rimini;
    Appl.Phys.Lett. (Oct.1980)

7. W.F.Tseng,J.W.Mayer,S.U.Campisano,G.Foti,E.Rimini;
    Appl.Phys.Lett.32,824(1978)

8. P.Baeri,S.U.Campisano,G.Foti,E.Rimini;
    J.Appl.Phys. 50,788(1979)

9. P.Baeri,S.U.Campisano,G.Foti,E.Rimini;
    Appl. Phys.Lett.33,137(1978).

THIN FILM REACTION INDUCED BY CW SCANNED

LASER AND E-BEAM

J.L. Regolini,[+] T. Shibata,[++] T.W. Sigmon,[++] and J.F. Gibbons[++]

[+]Centro Atomico Bariloche, Argentina
[++]Stanford Elec. Labs., U.S.A.

INTRODUCTION

The reaction of thin metal films with both single and poly-silicon to form silicide compounds has received wide attention in the past few-years. The properties of thin-film metal silicides formed at temperatures below the melting points of the components is of interest from both technological and fundamental aspects.

The bulk of the research to date has been carried out in thermal equilibrium furnace reactions and an excellent review is given by reference [1].

Within the last years the use of energetic beams to react metal silicon films has been investigated by a number of research groups [2,3].

The main features offered by the laser or electron-beam processing of those films include: shorted reaction times, local-ised heat treatment and rapid heating and cooling rates.

Similar to the case of energetic beam annealing of semicon-ductors, a natural division of the physical mechanisms and results can be made for pulsed vs scanned cw beams.

We will describe results using both cw laser and e-beam to react metal/Si systems. In contrast to the results found for the pulsed case, uniform, essentially single phase silicide films are obtained with the scanned cw beams.
However, under certain conditions, metastable mixed phase systems are also obtained.

Both MeV Rutherford Backscattering and glancing angle X-ray diffraction analysis have been utilized for characterizing the structure of the layers. Four point probes resistivity measurements have been utilized to obtain bulk resistivity using the film thickness data obtained from RBS.
Optical microscopy and SEM were also used.

EXPERIMENTAL

Samples were prepared by e-beam evaporation in an oil-free system. Thin amorphous Si layers were deposited on the metal layer to be reacted with the laser to provide an antireflection coating, if not, higher laser power is required to compensate the reflected power. However the smaller reflection of the silicide film results in an increased power absorption in the film during the reaction, leading to difficulty of control of the reaction. The following table shows laser and e-beam parameters used during the experiments:

	Laser	e-beam
Type	Argon,Spectra Physics - 12 w	Hamiltonian Standard Welder EBW 7.5
Lens	130 mm	---
X,γ	by mirrors	Electronically
Spot ϕ	$\sim$ 50 μm	$\sim$ 100 μm
Scan speed	$\sim$ 12 cm/sec	$\sim$ 12 cm/sec
Pressure	Atm.	$\sim$ 10^{-4} Torr
Substrate Temp.	R.T.	$\sim$ 50°C
V, I	---	30 kV, $\sim$ 0,2 mA

RESULTS AND DISCUSSION

Pd/Si

In Fig.1 we show an optical micrograph of a sample consisting of Si (200 Å)/Pd (1300 Å)/Si <100 > laser beam scanned as a

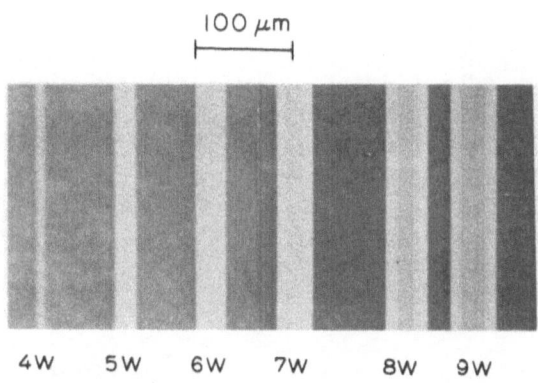

Fig. 1   Optical micrograph of a Si/Pd/Si sample after single
          laser scans at various laser power level.

function of laser power levels. It can be seen that the reaction
is initiated at about 4 watts of laser power. This reaction is
easily detected by the unaided eye through a surface color change,
from blue (unreacted film) to yellow (reacted one, low power phase).
Above a threshold power of about 7 watts a high power phase
appears in the center of the scan line, going to a grey color
characteristics of that phase. For the remainder of this paper we
will normalize the output power level of the laser (P) to the
power at which the surface of a virgin silicon wafer just begins
the melt (P_0).

Backscattering spectra for the Pd/Si samples which were
laser annealed at p = 0.71 (low power phase) and p = 1.1 (high
power phase) are shown in Fig.2.

The peak in the solid line near 1.2 MeV on the energy scale
is the signal from the 200 A of Si overcoating. The peak at
higher energy is the signal from the Pd and the step at about
1.1 MeV represents the signal from the underlying silicon substrate. From the areas under the spectra for the different atomic
species, one can determine the average composition of the reacted
materials. This was determined to be Pd_2Si for the low power
phase and PdSi for the high power phase.

Fig.3 shows X-ray diffraction pattern obtained with a Read
camera. We can see that the compound is essentially single phase
PdSi containing a trace amount of Pd_2Si. Similar results were

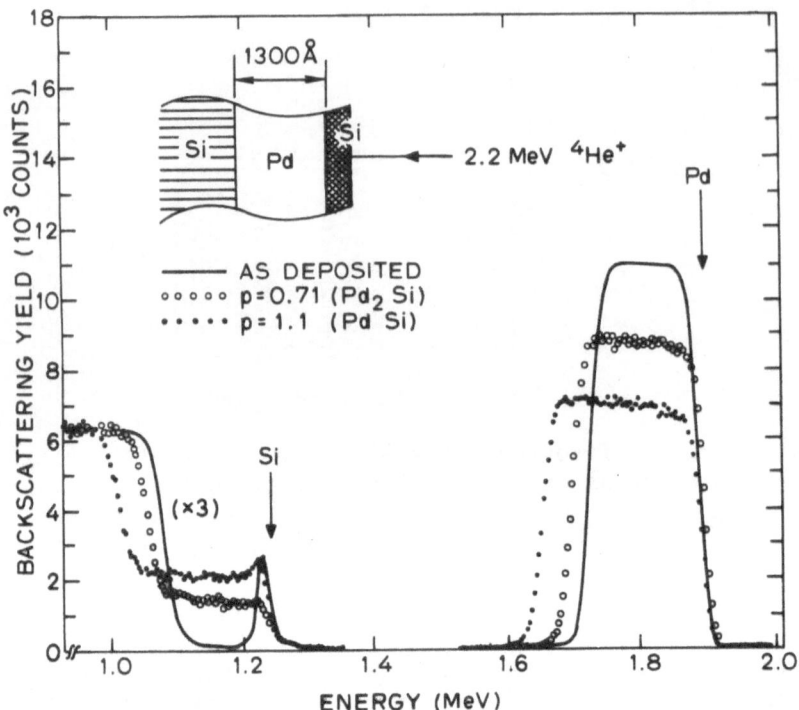

Fig. 2   RBS spectra for Si/Pd/Si laser annealed samples.
Solid line represents unreacted sample. Open circles
low power reaction and closed circles high power
reaction.

found for the sample reacted at low power were the compound was
mainly Pd_2Si. Fig.4 shows e-beam reacted Pd/Si. Again single
phases were observed as a function of power. In this case no antireflection coating is needed.

For a constant-temperature furnace anneal of amorphous Si,the
regrown-layer thickness can be expressed as[6] :

$$z = R_0 t \, \exp \, (-E_a/kT) \hspace{3cm} (1)$$

For diffusion-controlled reactions, such as the formation of
metal silicides the z in Eq.1 is replaced by z^2, T by "effective
annealing T", T_{eff} and t by "effective annealing time", t_{eff}[7].
Using a published data for the activation energy E_a in the Pd_2Si

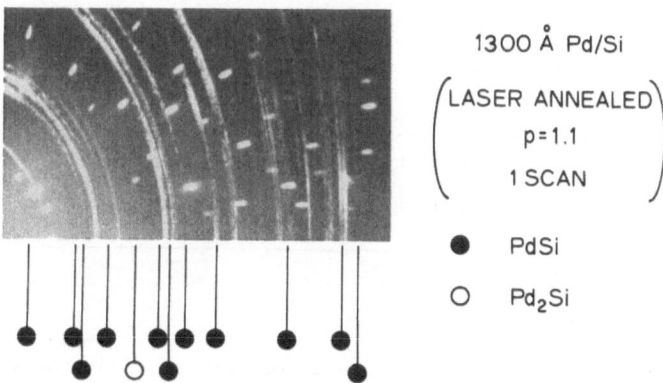

Fig. 3   X-ray diffraction pattern from a Read camera on PdSi.

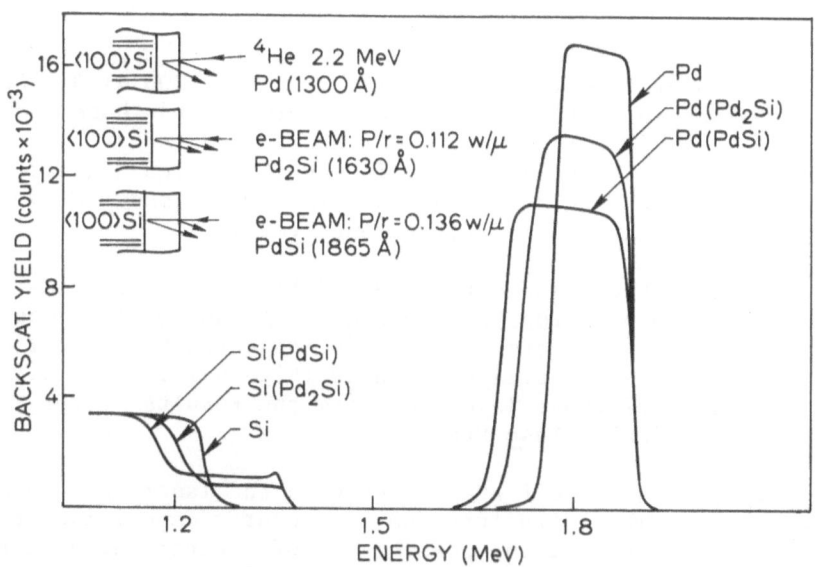

Fig. 4   RBS spectra for Pd/Si e-beam annealed at different power levels.

compound [8] we obtain the following results :

Pd_2Si	measured	calculated
Laser	$\sim$ 1900 Å	$\sim$ 1600 Å
e-beam	$\sim$ 1600 Å	$\sim$ 1700 Å

These calculations are in good agreement with the experimentally observed results as measured by RBS. We can conclude that the formation of this phase proceeds via a solid-state interdiffusion mechanisms.

The PdSi exhibit a laminar like surface morphology indicative of a spontaneous nucleation process then it doesn't follow the same mathematical formulation.

Pt/Si

Scanned laser reaction of the Si (200 Å)/Pt(1000 Å)/Si<100> samples resulted in surface color changes similar to those found for the Pd/Si samples. In Figure 5 we show the RBS spectra for the Pt/Si system. From this measurement it was calculated a reacted layer close to Pt_2Si but X-ray analysis (Figure 6) shows other nonstable phases like Pt_3Si, $Pt_{12}Si_5$ and $PtSi_2$.

On the other hand, RBS data obtained for the Pt/Si films reacted by scanned electron beam show a final and stable phase PtSi. That this is a uniform reaction is indicated by the uniformity of the platinum and silicon signals (Figure 7). X-ray diffraction analysis of those layers confirm the results of the RBS that we do have single phase PtSi.

The basic physical differences between the laser and e-beam experiments are sample structure and beam energy deposition profile. We cannot ruled out the possibility of interfacial effects (such as bond breaking) created by the e-beam at the interface.

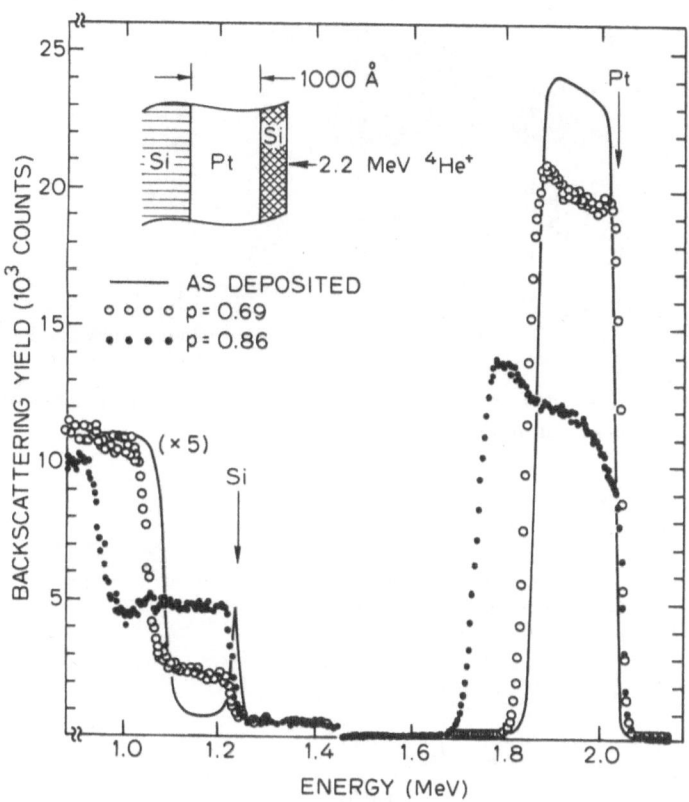

Fig. 5 RBS spectra for Si/Pt/Si as deposited and laser annealed at p = 0.69 and p = 0.89.

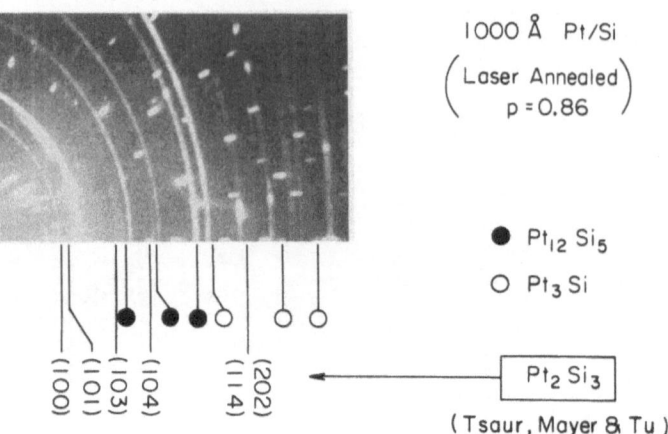

Fig. 6   X-ray diffraction pattern for Si/Pt/Si structure
         laser annealed.

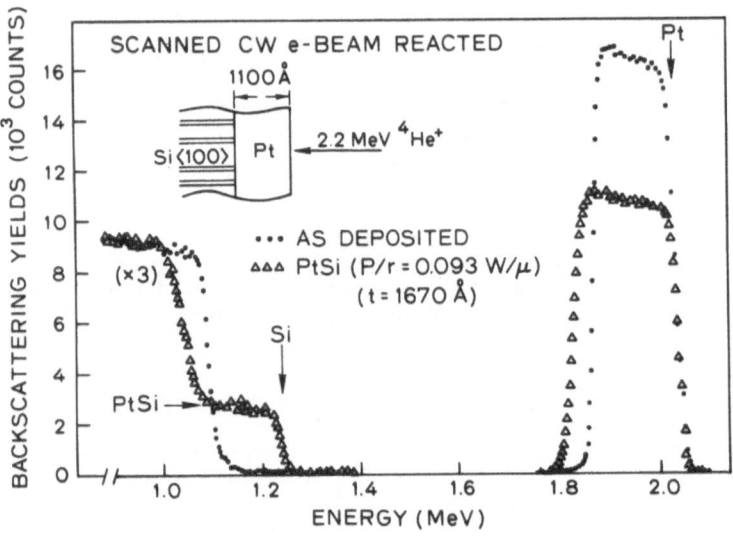

Fig. 7   RBS spectra shown for Pt/Si reacted by scanning e-beam.
         Both curves correspond to as deposited film and fully
         reacted PtSi.

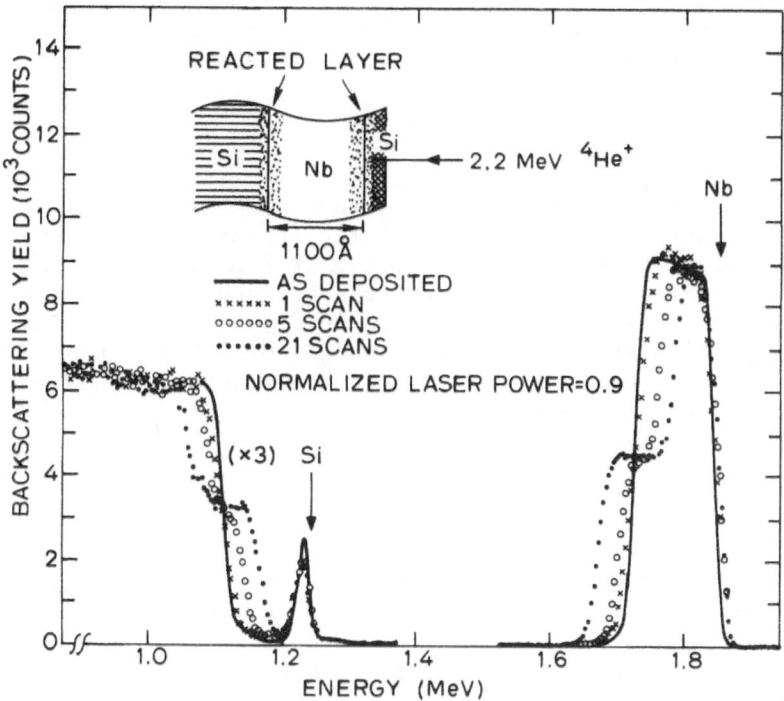

Fig. 8 RBS spectra for Si/Nb/Si with multiple laser scans.
The thickness of the silicide ($NbSi_2$) increases with
the number of scans.

Nb/Si

Figure 8 shows RBS spectra for the Nb/Si samples. There we
see the advancement of the interface occuring for constant laser
power as a function of the number of laser scans. The average composition of this reacted layer is found to be $NbSi_2$. A plot of
film thickness vs number of scans (proportional to an effective
annealing time) shows a parabolic growth behavior for $NbSi_2$. This
is in agreement with the general behavior of refractory metals
which form disilicides by furnace annealing[1] .

Mo,W/Si

We have successfully used the cw laser or e-beam to react Mo
and W with Si. The average composition of resulting layers were
found to be Metal/Si$_2$ after RBS analysis. X-ray diffraction patterns
also verifie the reacted layers are single phase.

CONCLUSION

Silicide forming reactions by scanned cw laser and e-beam
have been described. The characteristic features of the scanned
cw-beam process found differentiating it from those using pulsed
beam systems are: scanned cw beams can be used to form large area,
uniform, essentially single-phase silicide; different phases
within a particular system (such as Pd$_2$Si or PdSi) can be formed
by adjusting the beam power level; and the reactions proceeds
basically by a solid phase mechanism.

REFERENCES

1.   "Thin Films Interdiffusion and Reactions". Ed.by J.M.Poate,
     K.N.Tu and J.W.Mayer (N.Y.Wiley Interscience,1978).

2.   J.M.Poate,J.J.Leamy, T.T.Sheng and G.K.Celler. Appl.Phys.
     Lett.33; 918 (1978).

3.   Z.L.Liau, B.Y.Tsaur and J.W.Mayer, Appl.Phys.Lett. 34, 221
     (1979).

4.   T.Shib ata,J.F.Gibbons and T.W.Sigmon, Appl.Phys.Lett. 36,
     569 (1980).

5.   T.W.Sigmon,J.L.Regolini, J.F.Gibbons, S.S.Lau and J.W. Mayer,
     Proc. ECS Laser Symposium, Los Angeles (1979).

6.   L.Csepregi, E.F.Kennedy, J.W.Mayer and T.W.Sigmon, J.of Appl.
     Phys. 49, 3906 (1978).

7.   R.B.Gold and J.F.Gibbons, J.of Appl.Phys. 51, 1256 (1980).

8.   R.W.Bower, D.Sigurd and R.E.Scott, Solid State Electr. 16,
     1461 (1973).

PHYSICO-CHEMICAL BASIS OF LASER LITHOGRAPHY

S.M. Metev

Faculty of Physics, Sofia University

Sofia 1126, Bulgaria

The modern microelectronics needs new methods for pattern generation, which may ensure higher quality and resolution than the
photolithographic method. At the present the X-ray and the electron beam lithography methods are under fast development. The specific properties of the laser radiation make it very effective for
pattern generation and the results obtained are comparable with
these of the X-ray and the electron beam lithographies.

The laser radiation action on the material may activate some
physico-chemical processes, which may be used for pattern generation. The optical image of the pattern may be formed by contact,
projection, contour-projection or holographic methods. The projection and the contour-projection optical schemes are the most suitable for laser beam image formation. In the projection method
(Fig.1) the image of a mask is formed by a projection objective

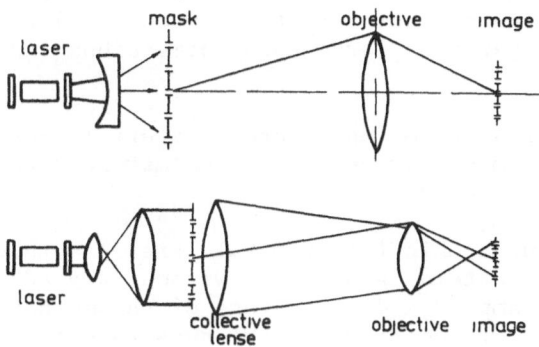

Fig. 1 Laser projection systems.

503

on the sample surface at suitable diminution. The projection scheme
with a collective lense, placed after the mask alows up to 100
times diminution of the mask image.

One characteristic peculiarity of the laser beam image formation is the constant value of the optical contrast function for
all space frequencies. This fact means that the different in size
pattern features will be transmitted with the same contrast through
the optical system, when coherent laser radiation is used. This
fact is of great importance for the lithographic process.

By the cotour-projection method (Fig.2) the optical image of

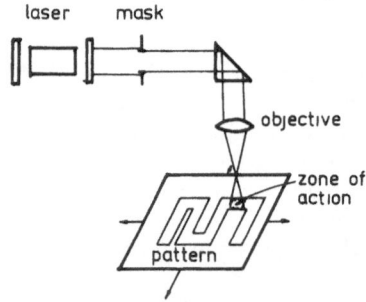

Fig. 2   Contour-projection scheme

the pattern is synthesized by laser beam scanning along the
contour of the pattern. In this case the zone of laser radiation
action has the form of a simple pattern feature (square,rectangle),
formed by a projection scheme. By the contour-projection method
one can obtain patterns of large total dimensions and high
resolution.

Laser radiation action on absorbing aterials may activate
several physico-chemical processes, which may be used for recording of the optical image.

1) Laser-induced thin film removal (Fig.3) – this is a
direct method for pattern generation without any additional
treatment of the sample. A disadvantage of the method is the
inevitable heating of the film and the substrate surface to very
high temperatures. It is necessary to take special measures in
this case to avoid the substrate surface damage or to reduce the
thermal distortion of the recorded image.

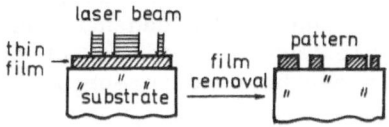

Fig. 3   Laser-induced thin film removal

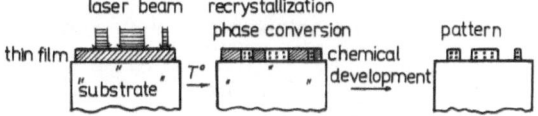

Fig. 4   Thermal action

2) Thermal treatment (Fig.4) - the thermal laser radiation action causes structural changes of the recording medium like recrystallization or phase conversions in solid state. As a result a latent image is recorded in the medium, which may be then developed chemically.

3) Diffusive-thermal treatment (Fig.5) - the laser radiation

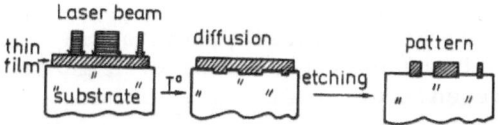

Fig. 5   Diffusive-thermal treatment

activates in this case diffusion of the film material in the
substrate, changing thus the physical properties of the substrate
surface layers. Sharp p-n boundaries of diffusive-alloyed character
may be obtained by this manner. The enhanced adhesion of the film
in the irradiated zones causes selective etching of the unirradiated zones. This fact may be used for obtaining of microrelief
on the substrate surface.

4) Thermochemical treatment (Fig.6) - the thermal action of

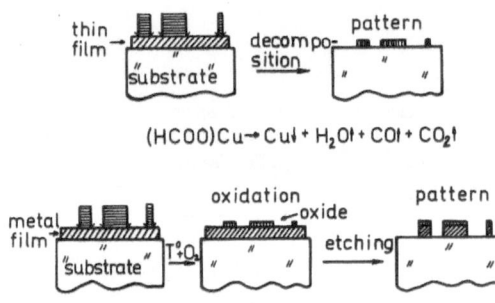

Fig. 6   Thermochemical action

the laser radiation may cause in this case activation of different
chemical reactions on the surface or in the bulk of the material.
Chemical reactions like decomposition, oxidation-reduction, poly-
merization or destruction and others may be used for image
recording. A typical example of thermal decomposition in solid
phase is the decomposition of $(ECOO)_2Cu$ in pure Cu and volatile
components

$$(HCOO)_2 \ Cu \xrightarrow{T^\circ} Cu \downarrow \ + \ H_2O \uparrow \ + \ CO \uparrow \ + \ CO_2 \uparrow$$

This process may be used for obtaining of metallic patterns on the
surface oq a dielectric substrate.

The thermal action of the laser radiation may activate an
oxidation reaction (Fig.6) on the surface of a metal thin film,
irradiated in the presence of oxidising environment (air, oxygen,
etc.). In this case a thin oxide layer builds up in the zone of
laser action on the metal surface. This protective layer enhances

the chemical stability of the irradiated film area. If the sample
is placed in a suitable etchant, the unirradiated zones of the
metal film dissolve, while the irradiated ones remain on the substrate.

  5) Photochemical treatment (Fig.7) - The chemical reactions,

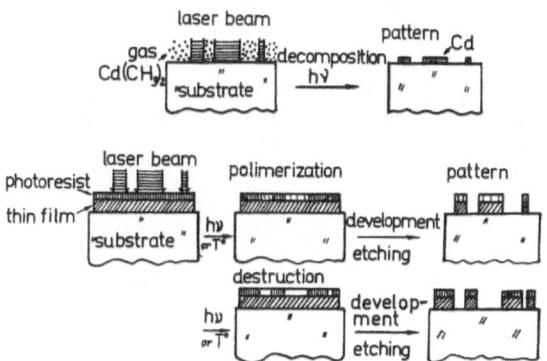

Fig. 7 Photochemical action

which may be used for image recording are activated in this case
by the resonant action of the laser radiation. Typical reactions
of this kind are photodecomposition in solid or in gas phases, the
photopolymerization and destruction of some organic substances
(photoresists) and others.

   The results in the field of laser lithography, obtained in
the Quantum Electronics Department of the Sofia University in collaboration with the Leningrad Institute of Fine Mechanics and
Optics are presented in the paper. Recently two processes have
been investigated theoretically and experimentally - the laser-
induced removal of thin absorbing films and the laser-activated
oxidation of thin metal films. This two processes may be used
very effectively for image recording.

   The laser-induced material removal is a basic method for laser machining of thin absorbing films, deposited on a dielectric
substrate. It is very important to know in detail the removal mechanism, when for the generation of high quality patters with microscopic features the removal method is used. Recently several
contradictory concepts on the laser-induced film removal mechanism

have been published[1-3].   Until now all theoretical removal models have been built on the basis of evaporation[4,5], which has
has been identified with sublimation in a certain sense as far as
the evaporating surface is considered nondeformable. This view on
the laser radiation effect on thin absorbing films has not taken
into consideration the circumstance that the vapour pressure of
the evaporating material influences considerably the removal
mechanism.

    The results of our investigations make possible the creation
of new theoretical model[6] of the laser-induced film removal.,
The model is called "wo-phase" as we suppose that the removed
material appears in two phases - vapour and liquid. It is based
on the following qualitative assumptions (Fig.8): the film is

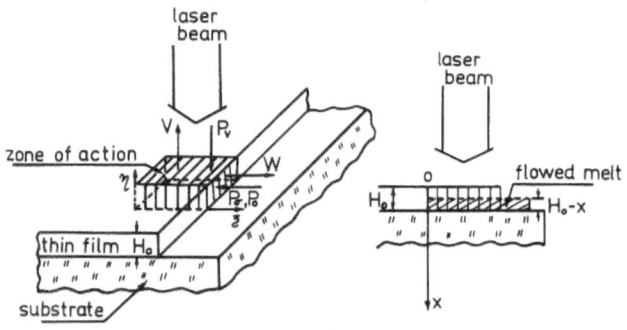

Fig. 8

successively heated to its melting point T_m under action of the
laser radiation; then it absorbs latent heat of melting L_m, melts
and in the process of further heating evaporates intensively,
especially after the film material boiling point T_b in air under
normal pressure is reached. The vapours of the evaporating material exert reactive pressure P_v on the melt surface, which
results in melt flow out of the irradiated zone with velocity W.
The melt motion is hindered by the surface tension forces P_σ,
adhesion, viscous friction etc. As a result of the joint action
of all forces the thin film partially evaporates (vapour-phase)
and partially flows out of the zone of action (liquid-phase).

    The model we have developed takes into account all stages of
the laser radiation effect on thin absorbing films: heating,mel-

ting, material removal and cooling. Because of the considerable
non-linearity of the process, the differential equations, describing the state of the film-substrate system have been solved
numerically.

The results of the calculations are illustrated in Fig. 9,

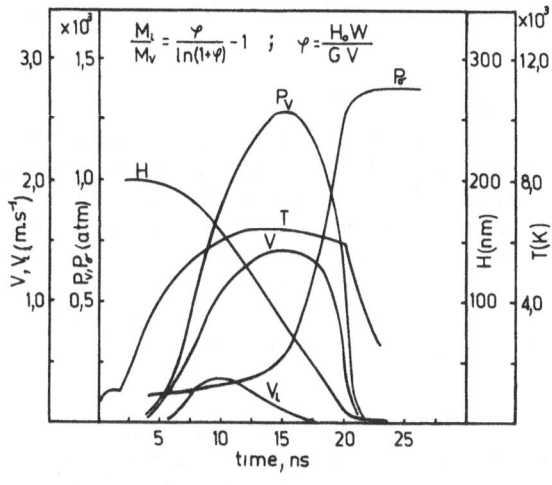

Fig. 9

which represents the basic version of computation of the process
kinetics for a silver film of initial thickness H_0 = 200 nm at
light flux density Q = 10^8W.cm^{-2} and pulse duration τ = 20 ns.
The graphs show, that in 0,3 ns the film surface temperature
reaches the melting point T_m = 1233°K and towards the end of the
first nanosecond the whole film turns out to be in the molten
state. Further increase of the temperature contributes to the
increase of the evaporation velocity , V, and the vapour pressure,
P_v. At the beginning there is no flow of the melt, which is
hindered by the great surface tension of the film. After 5 ns P
reaches the value of the surface tension pressure P_σ and the
melt starts flowing out of the zone of irradiation (of evaporation). After 15 ns the film temperature begins to decrease due to
the decrease in the thickness of the film, followed by decrease
in V and P_v. A result of the film thickness dropping and P_σ

increasing, the melt flow stops after 18 ns, when the rapidly
increasing P_σ reaches the decreasing P_v. At the 20th ns the
laser pulse ends and two nanoseconds later evaporation stops too.

It is possible to obtain from the calculations the ratio of
the mass M_L of the flowed melt and the mass M_v of the evaporated
melt in a single pulse, using a simplified phenomenological two-
phase removal model [6] :

$$\frac{M_L}{M_v} = \frac{\psi}{\ln(1+\psi)} - 1 \qquad \text{where } \psi = \frac{H_o \; W}{G \; V}$$

This equation shows that the parameter ψ characterises the relative contribution of the melt flowing and the evaporation to the
film removal process. At $\psi \gg 1$ (great initial film thickness H_o,
small dimensions of the irradiated zone G, small values of the
light flux density) the main contribution to the removal process
is due to the melt flowing under action of the reactive vapour
pressure. When $\psi \ll 1$ the effect of the liquid phase is brought to
a minimum and evaporation plays the principle role in the film removal mechanism.

The basic conclusions of the two-phase removal model have
been confirmed by qualitative and quantitative experiments [7].
The theoretical and the experimental investigations of the removal
process show that this process depends strongly on the irradiation
conditions and on the thermophysical properties of the film and
the substrate. It is very important previously to find the optimum
conditions for laser-induced thin film removal when necessary to
obtain by this method patterns of high quality and resolution.

The laser-activated oxidation is the second process, which is
studied in detail theoretically and experimentally. This process
may be used very effectively for image recording. The basic relations, describing the laser-induced oxidation of thin metal films
by pulsed laser radiation are obtained numerically. The oxidation
kinetics is described on the basis of Cabrer-Mott oxidation model,
which is valid for thin oxide layers. It is taken into account in
the calculations the dependence of the film reflectivity on the
oxide layer thickness, which is built up on the metal surface. In
Fig. 10 the theoretically obtained destribution of the temperature
and the oxide layer, which is built up during one laser pulse of
duration 50 ns, is presented. Our calculations show laser radiation pulse of wavelength 1,06 μm and duration τ = 50 ns causes
build-up of a 5 ÷ 7 A -thick oxide layer on the surface of a
2000 A° thick Cr film, irradiated in the presence of air. As the
experiments show this layer is thick enough to permit selective

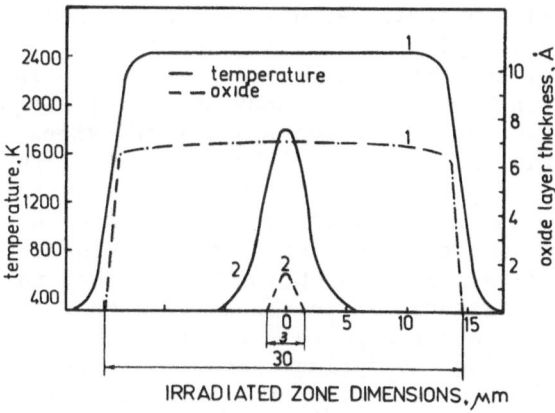

Fig. 10

etching of the unirradiated zone of the film. The thickness of the
protective oxide layer increases if multiple laser pulses are
used. The theoretical dependence of the oxide layer thickness on
the number of the laser pulses for different experimental conditions is presented in Fig. 11.

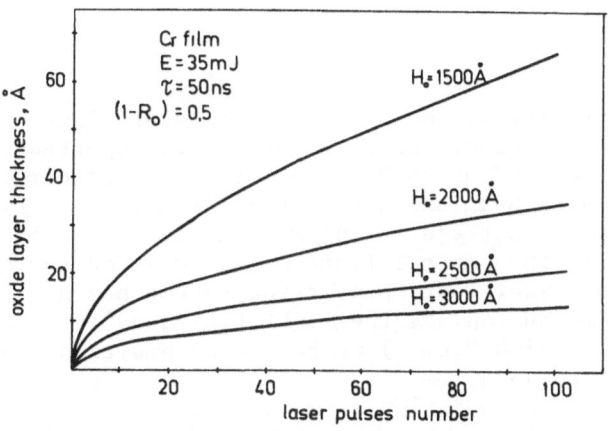

Fig. 11

Four different physical methods have been used by the expe-
imental investigation of the process. The thickness of the oxide
layer on the metal surface is measured directly, using the developed by us vibrating substrate method[8] and the resistance method. By the lateral method the oxide layer thickness is measured measuring the change of the film resistance. The oxide layer
thickness is measured also measuring the change of the film reflectivity during one laser pulse. The resistance and the reflectivity
methods have been used to study the oxidation kinetics during one
laser pulse. The X-ray spectroscopy method is used to identify the
oxide layer and to measure its thickness. The good correlation
between experimental and theoretical results show that the latter
may be used for finding the optimum conditions for image recording,
using the laser-induced oxidation of thin metal films.

In our laboratory experiments were carried out for using the
removal and the oxidation methods for image recording and for qualitative pattern generation. Different optical schemes have been
used to obtain qualitative image formation. The best results are
obtained in projection and in contour-projection optical systems.
The optical projection scheme used is shown in Fig.12. In the

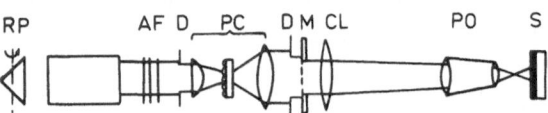

Fig. 12   Laser projection scheme

experiments a rotating-prism Q-switched Nd: glass laser with pulse
duration τ = 50 ns is used. The pulse energy is adjusted by means
of absorbing filters AF up to 3 J. An optical phase corrector PC
[9] widens the laser beam and causes the intensity distribution
to be uniform on the surface of the mask M. The projection
objective PO forms the optical image of the mask on the sample
surface S. In the experiments projection scheme has been used, allowing 3 ÷ 10X diminution. As the light flux density on the mask
surface is small ($\sim$100 W.cm^{-2}) we have used emultion photomasks
of dimentions up to 45x45 mm. In the contour-projection scheme a
Nd: YAG laser of pulse duration from 10 ns to 1 μs and pulse rep-
itition rate up to 50 kHz have been used. The zone of laser

action have had the form of a square of dimensions 10x10 μm
(Fig.2). The laser radiation and the motion of the sample in
X-Y direction were driven by a computer.

In the next figures some of the patterns obtained using the
removal or the oxidations methods are presented. Fig. 13 demon-

Fig.   13

strates these two methods on the example of thin Cr film, deposited on a glass substrate. The one half of the pattern is obtained
by direct removal of the film by its irradiation in projection
optical scheme. The other half of the pattern is obtained by laser
induced oxidation of the film and selective etching of the unirradiated zones (the white lines are metallic). In the first case
the positive and in the second case the negative image of the
mask is recorded on the film. The only difference in the experimental conditions in these two cases is the different value of
the light flux density.

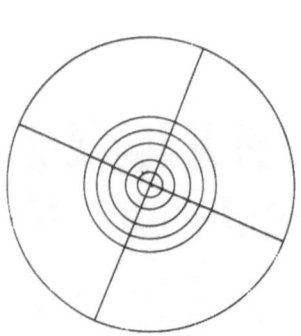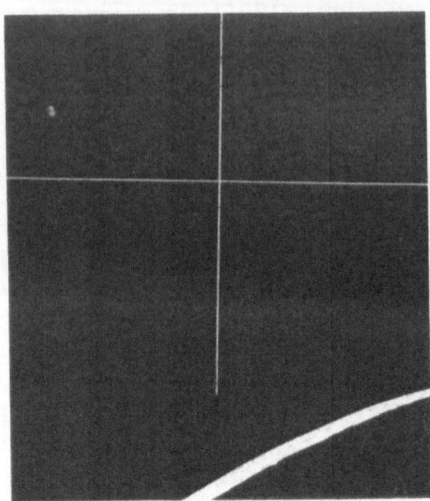

Fig. 14

In Fig.14 parts from two optical scales obtained point by
point, using the contour-projection scheme are shown. The line-
width in these two cases is 20 and 8 μm respectively.

In Fig. 15 a pattern is shown, which is obtained in projection scheme, using the oxidation method. This pattern may be multi-
plicated by mechanical motion of the sample in X-Y direction. The
overall pattern, obtained by this method is shown in Fig. 16. Thus
one can obtain patterns of large total dimensions and high resolution.

A photolithographic mask of an integral microcircuit, obtained
in five laser pulses by laser-induced oxidation in air of a 200 nm
Cr film, deposited on glass substrate 10 is shown in Fig. 17.This
photograph demonstrates the capabilities of the method. The total
dimensions of the metal pattern are 9x7 mm; the smallest feature
is 3 μm. A part from the same pattern is shown in Fig. 18 under

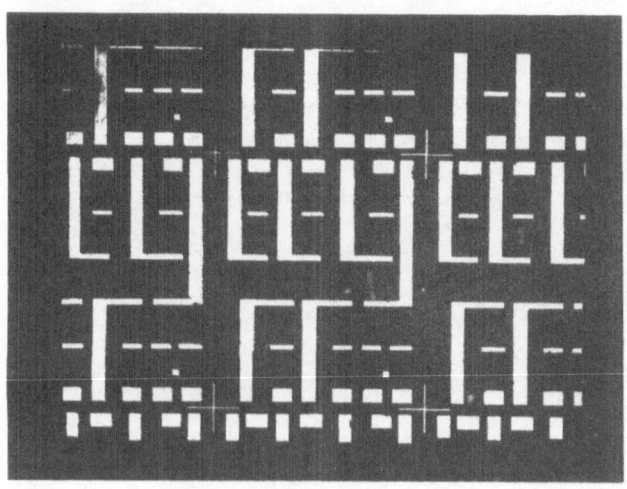

Fig. 15

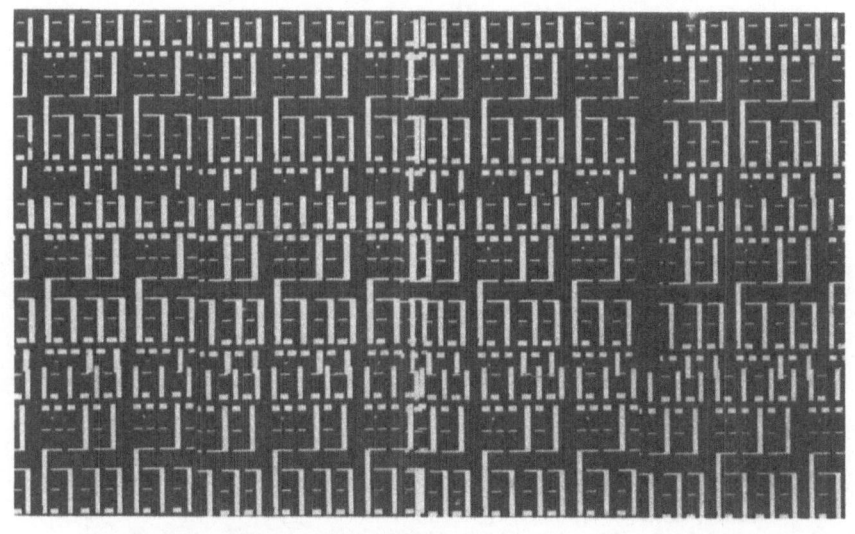

Fig. 16

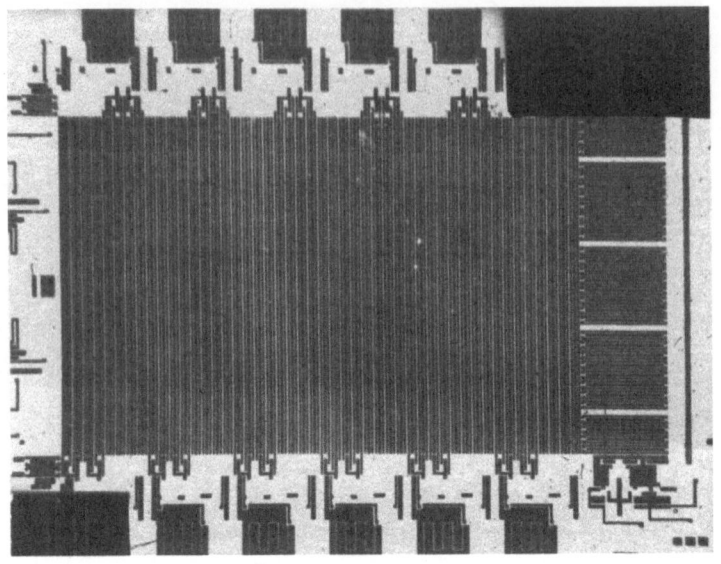

Fig.   17

Fig. 18

greater magnification. The edges of the metal lines are well
formed and their roughness is smaller than 0,3μ m at total line
width of 3 μm.

The laser lithography may be used in the microelectronics
for direct pattern generation without the intermediate photolithographic steps, required in conventional microelectronic fabrication. SHF integral circuits have been obtained, using the removal
method in combination with electrochemical etching of gold film,
deposited on suitable dielectric substrate. This method is very
effective at the design stage of such microelectronic devices,
economizing time and resources. The same method may be used for
engraving the electrodes of some piezoquartz devices like mono-
litic quartz filters, resonators and surface acoustic wave devices.

The laser-induced physico-chemical processes may be used for
photolithographic masks fabrication and the results obtained are
comparable with these of the X-ray and the electron-beam lithographies.

A very effective application of the laser lithography is the
direct correction of defective intermediate photolithographic
masks (Fig.19).Typical defects in this masks are the so called

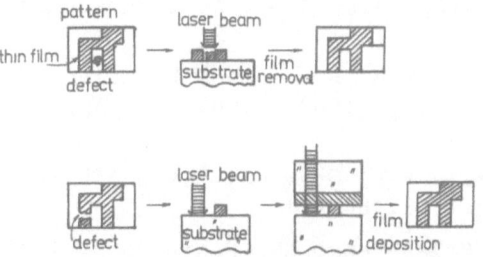

Fig. 19 Laser photomask correction

"black points" (unetched zones of the film) and the white points
(absence of masking layer in some zones). The first defect may be
corrected by laser-induced removal of the "black points." The
second defect may be corrected filling the hole in the film by
laser-induced deposition of masking layer. In this case the laser
beam must be coincided first with the defect in the film and

after that a transparant substrate with deposited thin film must
be placed on the mask with the back side to the laser beam. The
evaporated by the laser radiation film material is deposited on
the mask, correcting thus the defect.

A very effective application of the laser lithography is the
optical scales preparation. In this case the projection or the
contour-projection optical schemes may be used. Practically all
laser-induced physico-chemical processes are suitable for scales
preparation.

The laser lithography methods are very diverse and there is
no doubt that they will find great application in the modern microelectronic fabrication.

REFERENCES

1.   Sard R and Maydan D, 1971, J.Appl.Phys., 42, 5084-94

2.   Aboelfotoh M G and Gutfeld R J, 1972, J.Appl.Phys., 43,
       3789-94

3.   Murr L E and Payne R T, 1973, J.Appl.Phys., 44, 1722-6

4.   Libenson M N, 1968, Fiz.Khim.Obrab.Mater., (USSR), no.6,
       pp.67-72

5.   Paek U C and Kestenbaum A, 1973, J.Appl.Phys., 44, 2260-8

6.   Veiko V P, Metev S M, Kaidanov A I, Libenson M N and
       Jakovlev E B 1980, J.Phys.D : Appl.Phys., 13, to be publi-
       shed

7.   Veiko V P, Metev S M, Stamenov K V, Kalev H A, Jurkevitch
       B M and Karpman I M, 1980, J.Phys.D : Appl.Phys., 13, to be
       published

8.   Metev S M, Veiko V P, Stamenov K V and Kalev H A, 1977, Sov.
       J.Quant.Electr., 7, 863-71

9.   Gaponov S V, Salashtenko NN, Hanin Ja A, 1972, Kvant.Electr.,
       no.7, pp. 48-53

10.   Metev S M, Savtchenko S K, Stamenov K V, J.Phys.D : Appl.
       Phys., 13, L85-7

INDEX